6.95
W9-CKB-944

Fundamentals of ASTRODYNAMICS

ROGER R. BATE
Professor and Head

DONALD D. MUELLER
Assistant Professor of Astronautics

JERRY E. WHITE
Associate Professor of Astronautics

OF THE

DEPARTMENT OF ASTRONAUTICS

AND COMPUTER SCIENCE

UNITED STATES AIR FORCE ACADEMY

DOVER PUBLICATIONS, INC.

NEW YORK

Copyright © 1971 by Dover Publications, Inc.
All rights reserved under Pan American and International Copyright Conventions.

Published in Canada by General Publishing Company, Ltd., 30 Lesmill Road, Don Mills, Toronto, Ontario.
Published in the United Kingdom by Constable and Company, Ltd., 10 Orange Street, London WC 2.

Fundamentals of Astrodynamics is a new work, first published in 1971 by Dover Publications, Inc.

International Standard Book Number: 0-486-60061-0
Library of Congress Catalog Card Number: 73-157430

Manufactured in the United States of America
Dover Publications, Inc.
180 Varick Street
New York, N. Y. 10014

This textbook is dedicated to the members of the United States Air Force who have died in combat, are missing in action, or are prisoners of war.

This textbook is dedicated to the members of the United States Air Force who have died in combat, are missing in action, or are prisoners of war.

PREFACE

The study of astrodynamics is becoming a common prerequisite for any engineer or scientist who expects to be involved in the aerospace sciences and their many applications. While manned travel in Earth-Moon space is becoming more common, we are also concentrating on sophisticated applications of Earth and interplanetary satellites in the fields of communications, navigation, and basic research.

Even the use of satellites simply as transport vehicles or platforms for experiments requires a fundamental understanding of astrodynamics. We also recognize that for some time to come the ballistic missile, whose mechanics of flight has its foundation in astrodynamics, will be a front-line weapon in the arsenals of many countries.

Beginning with the first graduating class of 1959 the United States Air Force Academy has been in the forefront of astrodynamics education. Much has been learned from this experience concerning both the structure of the courses and what theoretical approaches are best for teaching the subject. Consequently, this text is particularly structured for teaching, rather than as an exhaustive treatment of astrodynamics.

The astrodynamic theory is presented with brief historical digressions. Although many classical methods are discussed, the central emphasis is on the use of the universal variable formulation. The theoretical development is rigorous but yet readable and usable. Several unpublished original derivations are included in the text. Example problems are used frequently to show the student how the theory can be applied. Exercises at the end of each chapter include derivations and quantitative and qualitative problems. Their difficulty ranges from straightforward to difficult. Difficult problems are marked with an asterisk (*). In addition to exercises at the end of each chapter, Appendix D includes several projects which are appropriate for a second course.

Chapter 1 develops the foundation for the rest of the book in the development of the basic two-body and n-body equations of motion. Chapter 2 treats orbit determination from various types of observations. It also introduces the classical orbital elements, coordinate transformations, the non-spherical earth, and differential correction. Chapter 3 then develops orbital transfer maneuvers such as the Hohmann transfer. Chapters 4 and 5 introduce time-of-flight with emphasis on the universal variable solution. These chapters treat the classical Kepler and Gauss problems in detail. Chapter 6 discusses the application of two-body mechanics to the ballistic missile problem, including launch error analysis and targeting on a rotating earth. Chapters 7 and 8 are further specialized applications to lunar and interplanetary flight. Chapter 9 is a brief introduction to perturbation analysis emphasizing special perturbations. It also includes a discussion of integration schemes and errors and analytic formulations of several common perturbations.

If the text is used for a first course in astrodynamics, the following sequence is suggested: Chapter 1, Chapter 2 (sections 2.1 through 2.7, 2.13 through 2.15), Chapter 3, Chapter 4 (sections 4.1 through 4.5), Chapter 6, and Chapter 7 or 8. A first course could include Project SITE/TRACK or Project PREDICT in Appendix D. A second course could be structured as follows: Chapter 2 (sections 2.8 through 2.12), review the Kepler problem of Chapter 4 and do Project KEPLER, Chapter 5 including projects GAUSS and INTERCEPT, and Chapter 9.

Contributions to the ideas and methods of this text have been made by many present and former members of the Department of Astronautics and Computer Science. Special mention should be made of the computer applications developed by Colonel Richard G. Rumney and Lieutenant Colonel Delbert Jacobs. The authors wish to acknowledge significant contributions by fellow faculty members: Majors Roger C. Brandt, Gilbert F. Kelley, and John C. Swonson, Jr. and Captains Gordon D. Bredvik, Harvey T. Brock, Thomas J. Eller, Kenneth D. Kopke, and Leonard R. Kruczynski for composing and checking example problems and student exercises and for proofreading portions of the manuscript; and Majors Edward J. Bauman and Walter J. Rabe for proofreading portions of the manuscript. Special acknowledgment is due Lieutenant Colonel John P. Wittry for his encouragement and suggestions while serving as Deputy Head for Astronautics of the Department of Astronautics and Computer Science.

Special thanks are also due several members of the Graphics Division of Instructional Technology at the USAF Academy for their excellent work in the composition of the text and the mathematical equations: Aline Rankin, Jan Irvine, Dorothy Fryar, Ronald Chaney (artist), and Edward Colosimo (Graphics Chief). The typing of the draft manuscript by Virginia Lambrecht and Marilyn Reed is also gratefully acknowledged.

United States Air Force Academy | R.R.B.
Colorado | D.D.M.
January 1971 | J.E.W.

CONTENTS

CHAPTER 1

TWO-BODY ORBITAL MECHANICS

> On Christmas Day, 1642, the year Galileo died, there was born in the Manor House of Woolsthorpe-by-Colsterworth a male infant so tiny that, as his mother told him in later years, he might have been put into a quart mug, and so frail that he had to wear a bolster around his neck to support his head. This unfortunate creature was entered in the parish register as 'Isaac sonne of Isaac and Hanna Newton.' There is no record that the wise men honored the occasion, yet this child was to alter the thought and habit of the world.
>
> – James R. Newman[1]

1.1 HISTORICAL BACKGROUND AND BASIC LAWS

If Christmas Day 1642 ushered in the age of reason it was only because two men, Tycho Brahe and Johann Kepler, who chanced to meet only 18 months before the former's death, laid the groundwork for Newton's greatest discoveries some 50 years later.

It would be difficult to imagine a greater contrast between two men working in the same field of science than existed between Tycho Brahe and Kepler.

Tycho, the noble and aristocratic Dane, was exceptional in mechanical ingenuity and meticulous in the collection and recording of accurate data on the positions of the planets. He was utterly devoid of the gift of theoretical speculation and mathematical power.

Kepler, the poor and sickly mathematician, unfitted by nature for accurate observations, was gifted with the patience and innate mathematical perception needed to unlock the secrets hidden in Tycho's data.[2]

1.1.1 Kepler's Laws. Since the time of Aristotle, who taught that circular motion was the only perfect and natural motion and that the heavenly bodies, therefore, necessarily moved in circles, the planets were assumed to revolve in circular paths or combinations of smaller circles moving on larger ones. But now that Kepler had the accurate observations of Tycho to refer to he found immense difficulty in reconciling any such theory with the observed facts. From 1601 until 1606 he tried fitting various geometrical curves to Tycho's data on the positions of Mars. Finally, after struggling for almost a year to remove a discrepancy of only 8 minutes of arc (which a less honest man might have chosen to ignore!), Kepler hit upon the ellipse as a possible solution. It fit. The orbit was found and in 1609 Kepler published his first two laws of planetary motion. The third law followed in 1619.[3]

These laws which mark an epoch in the history of mathematical science are as follows:

KEPLER'S LAWS

First Law–The orbit of each planet is an ellipse, with the sun at a focus.

Second Law–The line joining the planet to the sun sweeps out equal areas in equal times.

Third Law–The square of the period of a planet is proportional to the cube of its mean distance from the sun.

Still, Kepler's laws were only a description, not an explanation of planetary motion. It remained for the genius of Isaac Newton to unravel the mystery of "why?".

In 1665 Newton was a student at the University of Cambridge when an outbreak of the plague forced the university to close down for 2 years. Those 2 years were to be the most creative period in Newton's life. The 23-year-old genius conceived the law of gravitation, the laws of motion and developed the fundamental concepts of the differential calculus during the long vacation of 1666, but owing to some small discrepancies in his explanation of the moon's motion he tossed his papers aside. The world was not to learn of his momentous discoveries until some 20 years later!

To Edmund Halley, discoverer of Halley's comet, is due the credit for bringing Newton's discoveries before the world. One day in 1685

Halley and two of his contemporaries, Christopher Wren and Robert Hooke, were discussing the theory of Descartes which explained the motion of the planets by means of whirlpools and eddies which swept the planets around the sun. Dissatisfied with this explanation, they speculated whether a force "similar to magnetism" and falling off inversely with the square of distance might not require the planets to move in precisely elliptical paths. Hooke thought that this should be easy to prove whereupon Wren offered Hooke 40 shillings if he could produce the proof within 2 weeks. The 2 weeks passed and nothing more was heard from Hooke. Several months later Halley was visiting Newton at Cambridge and, without mentioning the bet, casually posed the question, "If the sun pulled the planets with a force inversely proportional to the square of their distances, in what paths ought they to go?" To Halley's utter and complete astonishment Newton replied without hesitation, "Why, in ellipses, of course. I have already calculated it and have the proof among my papers somewhere. Give me a few days and I shall find it for you." Newton was referring to the work he had done some 20 years earlier and only in this casual way was his greatest discovery made known to the world!

Halley, when he recovered from his shock, advised his reticent friend to develop completely and to publish his explanation of planetary motion. The result took 2 years in preparation and appeared in 1687 as *The Mathematical Principles of Natural Philosophy*, or, more simply, the *Principia*, undoubtedly one of the supreme achievements of the human mind.[4]

1.1.2 Newton's Laws of Motion. In Book I of the *Principia* Newton introduces his three laws of motion:

NEWTON'S LAWS

> **First Law**—Every body continues in its state of rest or of uniform motion in a straight line unless it is compelled to change that state by forces impressed upon it.
>
> **Second Law**—The rate of change of momentum is proportional to the force impressed and is in the same direction as that force.
>
> **Third Law**—To every action there is always opposed an equal reaction.

The second law can be expressed mathematically as follows:

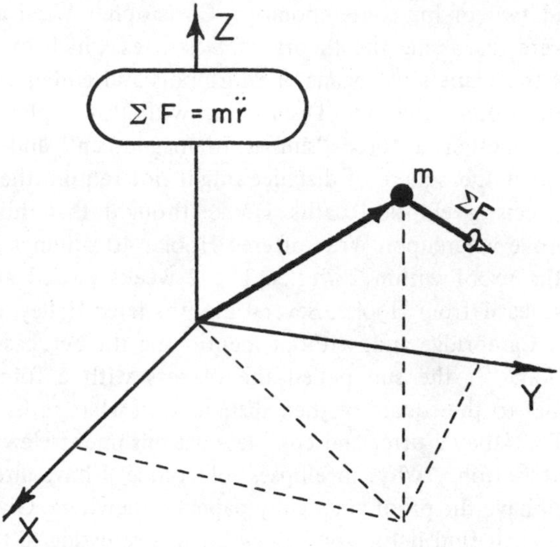

Figure 1.1-1 Newton's Law of Motion

$$\Sigma \mathbf{F} = m\,\ddot{\mathbf{r}} \tag{1.1-1}$$

where $\Sigma \mathbf{F}$ is the vector sum of all the forces acting on the mass m and $\ddot{\mathbf{r}}$ is the vector acceleration of the mass measured relative to an inertial reference frame shown as XYZ in Figure 1.1-1. Note that equation (1.1-1) applies only for a fixed mass system.

1.1.3 Newton's Law of Universal Gravitation. Besides enunciating his three laws of motion in the *Principia*, Newton formulated the law of gravity by stating that any two bodies attract one another with a force proportional to the product of their masses and inversely proportional to the square of the distance between them. We can express this law mathematically in vector notation as:

$$\mathbf{F}_g = -\frac{GMm}{r^2}\frac{\mathbf{r}}{r} \tag{1.1-2}$$

where $\mathbf{F}_g$ is the force on mass m due to mass M and $\mathbf{r}$ is the vector from M to m. The universal gravitational constant, G, has the value 6.670×10^{-8} dyne cm^2/gm^2.

In the next sections we will apply equation (1.1-2) to equation (1.1-1) and develop the equation of motion for planets and satellites.

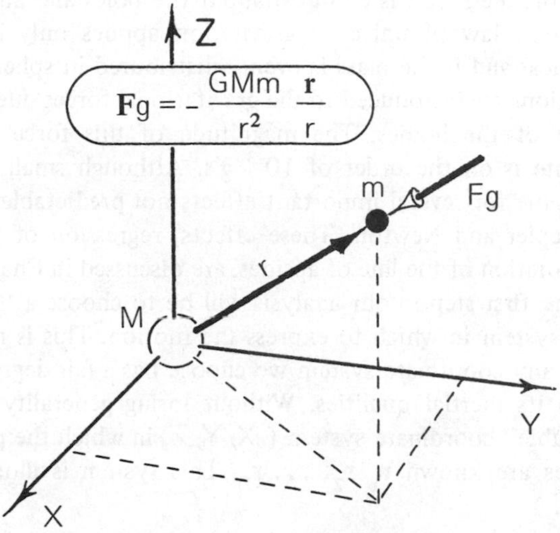

Figure 1.1-2 Newton's Law of Gravity

We will begin with the general N-body problem and then specialize to the problem of two bodies.

1.2 THE N-BODY PROBLEM

In this section we shall examine in some detail the motion of a body (i.e., an earth satellite, a lunar or interplanetary probe, or a planet). At any given time in its journey, the body is being acted upon by several gravitational masses and may even be experiencing other forces such as drag, thrust and solar radiation pressure.

For this examination we shall assume a "system" of n-bodies ($m_1, m_2, m_3 \ldots m_n$) one of which is the body whose motion we wish to study—call it the ith body, m_i. The vector sum of all gravitational forces and other external forces acting on m_i will be used to determine the equation of motion. To determine the gravitational forces we shall apply Newton's law of universal gravitation. In addition, the ith body may be a rocket expelling mass (i.e., propellants) to produce thrust; the motion may be in an atmosphere where drag effects are present; solar radiation may impart some pressure on the body; etc. All of these effects must be considered in the general equation of motion. An important force, not yet mentioned is due to the nonspherical shape of the planets. The earth is flattened at the poles and bulged at the equator; the moon is elliptical about the poles and about the equator. Newton's law of universal gravitation applies only if the bodies are spherical and if the mass is evenly distributed in spherical shells. Thus, variations are introduced to the gravitational forces due primarily to the shape of the bodies. The magnitude of this force for a near-earth satellite is on the order of 10^{-3} g's. Although small, this force is responsible for several important effects not predictable from the studies of Kepler and Newton. These effects, regression of the line-of-nodes and rotation of the line of apsides, are discussed in Chapter 3.

The first step in our analysis will be to choose a "suitable" coordinate system in which to express the motion. This is not a simple task since any coordinate system we choose has a fair degree of uncertainty as to its inertial qualities. Without losing generality let us assume a "suitable" coordinate system (X, Y, Z) in which the positions of the n masses are known $\mathbf{r}_1$ $\mathbf{r}_2$, ... $\mathbf{r}_n$. This system is illustrated in Figure 1.2-1.

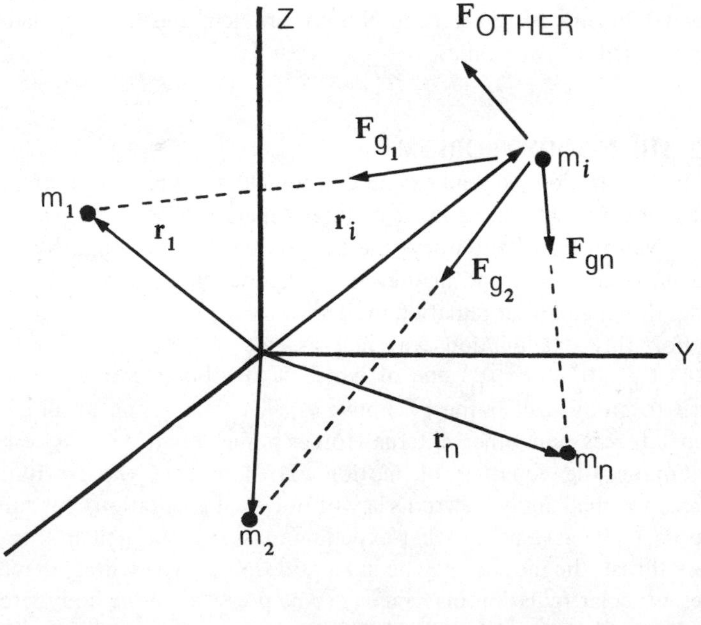

Figure 1.2-1 The N-Body Problem

Applying Newton's law of universal gravitation, the force $\mathbf{F}_{gn}$ exerted on m_i by m_n is

$$\mathbf{F}_{gn} = -\frac{Gm_i m_n}{r_{ni}^3}(\mathbf{r}_{ni}) \tag{1.2-1}$$

where

$$\mathbf{r}_{ni} = \mathbf{r}_i - \mathbf{r}_n \, . \tag{1.2-2}$$

The vector sum, $\mathbf{F}_g$, of all such gravitational forces acting on the ith body may be written

$$\mathbf{F}_g = -\frac{Gm_i m_1}{r_{1i}^3}(\mathbf{r}_{1i}) - \frac{Gm_i m_2}{r_{2i}^3}(\mathbf{r}_{2i}) - \cdots\cdots - \frac{Gm_i m_n}{r_{ni}^3}(\mathbf{r}_{ni}) \, . \tag{1.2-3}$$

Obviously, equation (1.2-3) does not contain the term

$$-\frac{Gm_i m_i}{r_{ii}^3}(\mathbf{r}_{ii}) \tag{1.2-4}$$

since the body cannot exert a force on itself. We may simplify this equation by using the summation notation so that

$$\boxed{\mathbf{F}_g = -Gm_i \sum_{\substack{j=1 \\ j \neq i}}^{n} \frac{m_j}{r_{ji}^3}(\mathbf{r}_{ji}) \, .} \tag{1.2-5}$$

The other external force, $\mathbf{F}_{OTHER}$, illustrated in Figure 1.2-1 is composed of drag, thrust, solar radiation pressure, perturbations due to nonspherical shapes, etc. The combined force acting on the ith body we will call $\mathbf{F}_{TOTAL}$, where

$$\mathbf{F}_{TOTAL} = \mathbf{F}_g + \mathbf{F}_{OTHER} \, . \tag{1.2-6}$$

We are now ready to apply Newton's second law of motion. Thus,

$$\frac{d}{dt}(m_i \mathbf{v}_i) = \mathbf{F}_{TOTAL} . \tag{1.2-7}$$

The time derivative may be expanded to

$$m_i \frac{d\mathbf{v}_i}{dt} + \mathbf{v}_i \frac{dm_i}{dt} = \mathbf{F}_{TOTAL} \, . \tag{1.2-8}$$

It was previously mentioned that the body may be expelling some mass to produce thrust in which case the second term of equation 1.2-8 would not be zero. Certain relativistic effects would also give rise to changes in the mass m_i as a function of time. In other words, it is not always true–especially in space dynamics–that $\mathbf{F} = m\mathbf{a}$. Dividing through by the mass m_i gives the most general equation of motion for the ith body

$$\boxed{\ddot{\mathbf{r}}_i = \frac{\mathbf{F}_{TOTAL}}{m_i} - \dot{\mathbf{r}}_i \frac{\dot{m}_i}{m_i}} \tag{1.2-9}$$

where

$\ddot{\mathbf{r}}_i$ is the vector acceleration of the ith body relative to the x, y, z coordinate system.

m_i is the mass of the ith body.

$\mathbf{F}_{TOTAL}$ is the vector sum of all gravitational forces

$$\mathbf{F}_g = -Gm_i \sum_{\substack{j=1 \\ j \neq i}}^{n} \frac{m_j}{r_{ji}^3} (\mathbf{r}_{ji})$$

and all other external forces

$$\mathbf{F}_{\text{OTHER}} = \mathbf{F}_{\text{DRAG}} + \mathbf{F}_{\text{THRUST}} + \mathbf{F}_{\text{SOLAR PRESSURE}} + \mathbf{F}_{\text{PERTURB}} + \text{etc.}$$

$\dot{\mathbf{r}}_i$ is the velocity vector of the ith body relative to the X, Y, Z coordinate system.

$\dot{m}_i$ is the time rate of change of mass of the ith body (due to expelling mass or relativistic effects.

Equation (1.2-9) is a second order, nonlinear, vector, differential equation of motion which has defied solution in its present form. It is here therefore that we depart from the realities of nature to make some simplifying assumptions.

Assume that the mass of the ith body remains constant (i.e., unpowered flight; $\dot{m}_i = 0$). Also assume that drag and other external forces are not present. The only remaining forces then are gravitational. Equation (1.2-9) reduces to

$$\ddot{\mathbf{r}}_i = -G \sum_{\substack{j=1 \\ j \neq i}}^{n} \frac{m_j}{r_{ji}^3} (\mathbf{r}_{ji}) . \tag{1.2-10}$$

Let us assume also that m_2 is an earth satellite and that m_1 is the earth. The remaining masses $m_3, m_4 \ldots m_n$ may be the moon, sun and planets. Then writing equation (1.2-10) for $i = 1$, we get

$$\ddot{\mathbf{r}}_1 = -G \sum_{j=2}^{n} \frac{m_j}{r_{j1}^3} (\mathbf{r}_{j1}) . \tag{1.2-11}$$

And for $i = 2$, equation (1.2-10) becomes

$$\ddot{\mathbf{r}}_2 = -G \sum_{\substack{j=1 \\ j \neq 2}}^{n} \frac{m_j}{r_{j2}^3} (\mathbf{r}_{j2}) . \tag{1.2-12}$$

From equation (1.2-2) we see that

$$\mathbf{r}_{12} = \mathbf{r}_2 - \mathbf{r}_1 \tag{1.2-13}$$

so that

$$\ddot{\mathbf{r}}_{12} = \ddot{\mathbf{r}}_2 - \ddot{\mathbf{r}}_1 \tag{1.2-14}$$

Substituting equations (1.2-11) and (1.2-12) into equation (1.2-14) gives,

$$\ddot{\mathbf{r}}_{12} = -G \sum_{\substack{j=1 \\ j \neq 2}}^{n} \frac{m_j}{r_{j2}^{\;3}} (\mathbf{r}_{j2}) + G \sum_{j=2}^{n} \frac{m_j}{r_{j1}^{\;3}} (\mathbf{r}_{j1}) \tag{1.2-15}$$

or expanding

$$\ddot{\mathbf{r}}_{12} = -\left[\frac{Gm_1}{r_{12}^{\;3}} (\mathbf{r}_{12}) + G \sum_{j=3}^{n} \frac{m_j}{r_{j2}^{\;3}} (\mathbf{r}_{j2}) \right]$$

$$-\left[-\frac{Gm_2}{r_{21}^{\;3}} (\mathbf{r}_{21}) - G \sum_{j=3}^{n} \frac{m_j}{r_{j1}^{\;3}} (\mathbf{r}_{j1}) \right] \tag{1.2-16}$$

Since $\mathbf{r}_{12} = -\mathbf{r}_{21}$ we may combine the first terms in each bracket. Hence,

$$\ddot{\mathbf{r}}_{12} = -\frac{G(m_1 + m_2)}{r_{12}^{\;3}} (\mathbf{r}_{12}) - \sum_{j=3}^{n} Gm_j \left(\frac{\mathbf{r}_{j2}}{r_{j2}^{\;3}} - \frac{\mathbf{r}_{j1}}{r_{j1}^{\;3}} \right) \tag{1.2-17}$$

The reason for writing the equation in this form will become clear when we recall that we are studying the motion of a near earth satellite where m_2 is the mass of the satellite and m_1 is the mass of the earth. Then $\ddot{\mathbf{r}}_{12}$

is the acceleration of the satellite *relative* to earth. The effect of the last term of equation (1.2-17) is to account for the perturbing effects of the moon, sun and planets on a near earth satellite.

To further simplify this equation it is necessary to determine the magnitude of the perturbing effects compared to the force between earth and satellite. Table 1.2-1 lists the relative accelerations (not the perturbative accelerations) for a satellite in a 200 n. mi orbit about the earth. Notice also that the effect of the nonspherical earth (oblateness) is included for comparison.

COMPARISON OF RELATIVE ACCELERATION (IN G's) FOR A 200 NM EARTH SATELLITE

	Acceleration in G's on 200 nm Earth Satellite
Earth	.89
Sun	6×10^{-4}
Mercury	2.6×10^{-10}
Venus	1.9×10^{-8}
Mars	7.1×10^{-10}
Jupiter	3.2×10^{-8}
Saturn	2.3×10^{-9}
Uranus	8×10^{-11}
Neptune	3.6×10^{-11}
Pluto	10^{-12}
Moon	3.3×10^{-6}
Earth Oblateness	10^{-3}

TABLE 1.2-1

1.3 THE TWO-BODY PROBLEM

Now that we have a general expression for the relative motion of two bodies perturbed by other bodies it would be a simple matter to reduce it to an equation for only two bodies. However, to further clarify the derivation of the equation of relative motion, some of the work of the previous section will be repeated considering just two bodies.

1.3.1 Simplifying Assumptions. There are two assumptions we will make with regard to our model:

1. The bodies are spherically symmetric. This enables us to treat the bodies as though their masses were concentrated at their centers.

2. There are no external nor internal forces acting on the system other than the gravitational forces which act along the line joining the centers of the two bodies.

1.3.2 The Equation of Relative Motion. Before we may apply Newton's second law to determine the equation of relative motion of these two bodies, we must find an inertial (unaccelerated and nonrotating) reference frame for the purpose of measuring the motion or the lack of it. Newton described this inertial reference frame by saying that it was fixed in absolute space, which "in its own nature, without relation to anything external, remains always similar and immovable."[5] However, he failed to indicate how one found this frame which was absolutely at rest. For the time being, let us carry on with our investigation of the relative motion by assuming that we have found such an inertial reference frame and then later return to a discussion of the consequences of the fact that in reality all we can ever find is an "almost" inertial reference frame.

Consider the system of two bodies of mass M and m illustrated in Figure 1.3-1. Let (X', Y', Z') be an inertial set of rectangular cartesian coordinates. Let (X, Y, Z) be a set of nonrotating coordinates parallel to (X', Y', Z') and having an origin coincident with the body of mass M. The position vectors of the bodies M and m with respect to the set (X', Y', Z') are $\mathbf{r}_M$ and $\mathbf{r}_m$ respectively. Note that we have defined

$$\mathbf{r} = \mathbf{r}_m - \mathbf{r}_M \,.$$

Now we can apply Newton's laws in the inertial frame (X', Y', Z') and obtain

$$m\ddot{\mathbf{r}}_m = -\frac{GMm}{r^2}\frac{\mathbf{r}}{r}$$

and $$M\ddot{\mathbf{r}}_M = \frac{GMm}{r^2}\frac{\mathbf{r}}{r}$$

The above equations may be written:

$$\ddot{\mathbf{r}}_m = -\frac{GM}{r^3}\mathbf{r} \tag{1.3-1}$$

and $$\ddot{\mathbf{r}}_M = \frac{Gm}{r^3}\mathbf{r}\,. \tag{1.3-2}$$

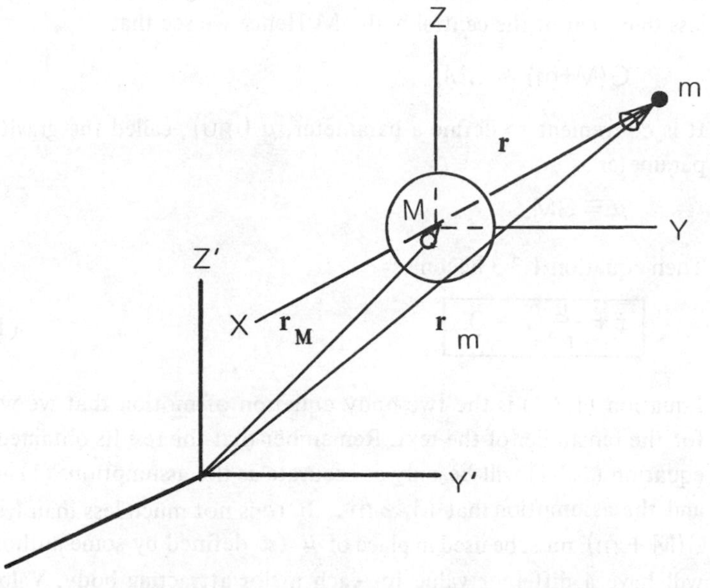

Figure 1.3-1 Relative Motion of Two Bodies

Subtracting equation (1.3-2) from equation (1.3-1) we have

$$\ddot{\mathbf{r}} = -\frac{G(M+m)}{r^3}\,\mathbf{r}\,. \tag{1.3-3}$$

Equation (1.3-3) is the vector differential equation of the relative motion for the two-body problem. Note that this is the same as equation (1.2-17) without perturbing effects and with $\mathbf{r}_{12}$ replaced by $\mathbf{r}$.

Note that since the coordinate set (X, Y, Z) is nonrotating with respect to the coordinate set (X', Y', Z'), the magnitudes and directions of $\mathbf{r}$ and $\ddot{\mathbf{r}}$ as measured in the set (X, Y, Z) will be equal respectively to their magnitudes and directions as measured in the inertial set (X', Y', Z'). Thus having postulated the existence of an inertial reference frame in order to derive equation (1.3-3), we may now discard it and measure the relative position, velocity, and acceleration in a nonrotating, noninertial coordinate system such as the set (X, Y, Z) with its origin in the central body.

Since our efforts in this text will be devoted to studying the motion of artificial satellites, ballistic missiles, or space probes orbiting about some planet or the sun, the mass of the orbiting body, m, will be much less than that of the central body, M. Hence we see that

$$G(M+m) \approx GM.$$

It is convenient to define a parameter, μ (mu), called the gravitational parameter as

$$\mu \equiv GM.$$

Then equation 1.3-3 becomes

$$\boxed{\ddot{\mathbf{r}} + \frac{\mu}{r^3}\mathbf{r} = 0.} \qquad (1.3\text{-}4)$$

Equation (1.3-4) is the two-body equation of motion that we will use for the remainder of the text. Remember that the results obtained from equation (1.3-4) will be only as accurate as the assumptions (1) and (2) and the assumption that $M \gg m$. If m is not much less than M, then $G(M + m)$ must be used in place of μ (so defined by some authors). μ will have a different value for each major attracting body. Values for the earth and sun are listed in the appendix and values for other planets are included in Chapter 8.

1.4 CONSTANTS OF THE MOTION

Before attempting to solve the equation of motion to obtain the trajectory of a satellite we shall derive some useful information about the nature of orbital motion. If you think about the model we have created, namely a small mass moving in a gravitational field whose force is always directed toward the center of a larger mass, you would probably arrive intuitively at the conclusions we will shortly confirm by rigorous mathematical proofs. From your previous knowledge of physics and mechanics you know that a gravitational field is "conservative." That is, an object moving under the influence of gravity alone does not lose or gain mechanical energy but only exchanges one form of energy, "kinetic," for another form called "potential energy." You also know that it takes a tangential component of force to change the angular momentum of a system in rotational motion about some center of rotation. Since the gravitational force is always directed radially toward the center of the large mass we would expect that the angular momentum of the satellite about the center of our reference frame (the large mass) does not change. In the next two sections we will prove these statements.

1.4.1 Conservation of Mechanical Energy. The energy constant of motion can be derived as follows:

1. Dot multiply equation (1.3-4) by $\dot{\mathbf{r}}$

$$\dot{\mathbf{r}} \cdot \ddot{\mathbf{r}} + \dot{\mathbf{r}} \cdot \frac{\mu}{r^3}\,\mathbf{r} = 0\,.$$

2. Since in general $\mathbf{a} \cdot \dot{\mathbf{a}} = a\dot{a}$, $\mathbf{v} = \dot{\mathbf{r}}$ and $\dot{\mathbf{v}} = \ddot{\mathbf{r}}$, then

$$\mathbf{v} \cdot \dot{\mathbf{v}} + \frac{\mu}{r^3}\mathbf{r} \cdot \dot{\mathbf{r}} = 0, \text{ so}$$

$$v\dot{v} + \frac{\mu}{r^3}\, r\dot{r} = 0.$$

3. Noticing that $\frac{d}{dt}\left(\frac{v^2}{2}\right) = v\dot{v}$ and $\frac{d}{dt}\left(-\frac{\mu}{r}\right) = \frac{\mu}{r^2}\dot{r}$

$$\frac{d}{dt}\left(\frac{v^2}{2}\right) + \frac{d}{dt}\left(-\frac{\mu}{r}\right) = 0 \qquad \text{or} \qquad \frac{d}{dt}\left(\frac{v^2}{2} - \frac{\mu}{r}\right) = 0.$$

4. To make step 3 perfectly general we should say that

$$\frac{d}{dt}\left(\frac{v^2}{2} + c - \frac{\mu}{r}\right) = 0$$

where c can be any arbitrary constant since the time derivative of any constant is zero.

5. If the time rate of change of an expression is zero, that expression must be a constant which we will call $\mathcal{E}$. Therefore,

$$\mathcal{E} = \frac{v^2}{2} + \left(c - \frac{\mu}{r}\right) \quad = \text{ a constant called "specific mechanical energy"} \tag{1.4-1}$$

The first term of $\mathcal{E}$ is obviously the kinetic energy per unit mass of the satellite. To convince yourself that the second term is the potential energy per unit mass you need only equate it with the work done in moving a satellite from one point in space to another against the force of gravity. But what about the arbitrary constant, c, which appears in the potential energy term? The value of this constant will depend on the zero reference of potential energy. In other words, at what distance, r, do you want to say the potential energy is zero? This is obviously arbitrary. In your elementary physics courses it was convenient to choose ground level or the surface of the earth as the zero datum for potential energy, in which case an object lying at the bottom of a deep well was found to have a negative potential energy. If we wish to retain the surface of the large mass, e.g. the earth, as our zero reference we would choose $c = \frac{\mu}{r_\oplus}$, where $r_\oplus$ is the radius of the earth. This would be perfectly legitimate but since c is arbitrary, why not set it equal to zero? Setting c equal to zero is equivalent to choosing our zero reference for potential energy at infinity. The price we pay for this simplification is that the potential energy of a satellite (now simply $-\frac{\mu}{r}$) will always be negative.

We conclude, therefore, that the *specific mechanical energy,* $\mathcal{E}$, of a satellite which is the sum of its kinetic energy per unit mass and its potential energy per unit mass remains constant along its orbit, neither increasing nor decreasing as a result of its motion. The expression for $\mathcal{E}$ is

$$\boxed{\mathcal{E} = \frac{v^2}{2} - \frac{\mu}{r}\,.} \tag{1.4-2}$$

1.4.2 Conservation of angular momentum. The angular momentum constant of the motion is obtained as follows:

1. Cross multiply equation (1.3-4) by **r**

$$\mathbf{r} \times \ddot{\mathbf{r}} + \mathbf{r} \times \frac{\mu}{r^3}\mathbf{r} = 0.$$

2. Since in general $\mathbf{a} \times \mathbf{a} = 0$, the second term vanishes and

$$\mathbf{r} \times \ddot{\mathbf{r}} = 0.$$

3. Noticing that $\frac{d}{dt}(\mathbf{r}\times\dot{\mathbf{r}}) = \dot{\mathbf{r}}\times\dot{\mathbf{r}} + \mathbf{r}\times\ddot{\mathbf{r}}$ the equation above becomes

$$\frac{d}{dt}(\mathbf{r} \times \dot{\mathbf{r}}) = 0 \qquad \text{or} \qquad \frac{d}{dt}(\mathbf{r} \times \mathbf{v}) = 0.$$

The expression **r** × **v** which must be a constant of the motion is simply the vector **h**, called *specific angular momentum*. Therefore, we have shown that the specific angular momentum, **h**, of a satellite remains constant along its orbit and that the expression for **h** is

$$\boxed{\mathbf{h} = \mathbf{r} \times \mathbf{v}.} \tag{1.4-3}$$

Since **h** is the vector cross product of **r** and **v** it must always be perpendicular to the plane containing **r** and **v**. But **h** is a constant vector so **r** and **v** must always remain in the same plane. Therefore, we conclude that the satellite's motion must be confined to a plane which is fixed in space. We shall refer to this as the orbital plane.

By looking at the vectors **r** and **v** in the orbital plane and the angle between them (see Figure 1.4-1) we can derive another useful expression for the magnitude of the vector **h**.

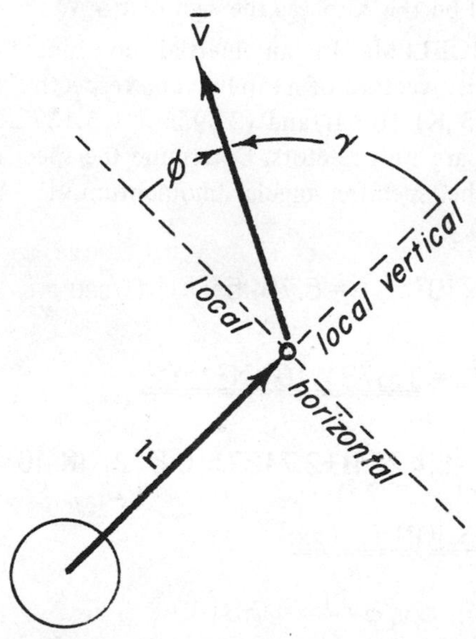

Figure 1.4-1 Flight-path angle, ϕ

No matter where a satellite is located in space it is always possible to define "up and down" and "horizontal." "Up" simply means away from the center of the earth and "down" means toward the center of the earth. So the local vertical at the location of the satellite coincides with the direction of the vector **r**. The local horizontal plane must then be perpendicular to the local vertical. We can now define the direction of the velocity vector, **v**, by specifying the angle it makes with the local vertical as γ (gamma), the zenith angle. The angle between the velocity vector and the local horizontal plane is called ϕ (phi), the flight-path elevation angle or simply "flight-path angle." From the definition of the cross product the magnitude of **h** is

$$h = rv \sin \gamma,$$

We will find it more convenient, however, to express h in terms of the flight-path angle, ϕ. Since γ and ϕ are obviously complementary angles

$$\boxed{h = rv \cos \phi.} \tag{1.4-4}$$

The sign of ϕ will be the same as the sign of $\mathbf{r} \cdot \mathbf{v}$.

EXAMPLE PROBLEM. In an inertial coordinate system, the position and velocity vectors of a satellite are, respectively, $(4.1852\,\mathbf{I} + 6.2778\,\mathbf{J} + 10.463\,\mathbf{K})\,10^7$ ft and $(2.5936\,\mathbf{I} + 5.1872\,\mathbf{J})\,10^4$ ft/sec where **I**, **J** and **K** are unit vectors. Determine the specific mechanical energy, $\mathcal{E}$, and the specific angular momentum, h. Also find the flight-path angle, ϕ.

$$r = 12.899 \times 10^7 \text{ ft}, \quad v = 5.7995 \times 10^4 \text{ ft/sec}$$

$$\mathcal{E} = \frac{v^2}{2} - \frac{\mu}{r} = \underline{\underline{1.573 \times 10^9 \text{ ft}^2/\text{sec}^2}}$$

$$\mathbf{h} = \mathbf{r} \times \mathbf{v} = (-5.4274\mathbf{I} + 2.7137\mathbf{J} + 0.54273\mathbf{K})10^{12} \text{ ft}^2/\text{sec}$$

$$h = \underline{\underline{6.0922 \times 10^{12} \text{ ft}^2/\text{sec}}}$$

$$h = rv \cos \phi, \quad \cos \phi = \frac{h}{rv} = 0.8143$$

$\mathbf{r} \cdot \mathbf{v} > 0$, therefore:

$$\phi = \arccos 0.8143 = \underline{\underline{35.42^\circ}}$$

1.5 THE TRAJECTORY EQUATION

Earlier we wrote the equation of motion for a small mass orbiting a large central body. While this equation (1.3-3) is simple, its complete solution is not. A partial solution which will tell us the size and shape of the orbit is easy to obtain. The more difficult question of how the satellite moves around this orbit as a function of time will be postponed to Chapter 4.

1.5.1 Integration of the Equation of Motion. You recall that the equation of motion for the two-body problem is

$$\ddot{\mathbf{r}} = -\frac{\mu}{r^3}\,\mathbf{r}.$$

Crossing this equation into $\mathbf{h}$ leads toward a form which can be integrated:

$$\ddot{\mathbf{r}} \times \mathbf{h} = \frac{\mu}{r^3}\,(\mathbf{h} \times \mathbf{r}). \tag{1.5-1}$$

The left side of equation (1.5-1) is clearly $d/dt\ (\dot{\mathbf{r}} \times \mathbf{h})$- try it and see. Looking for the right side to also be the time rate of change of some vector quantity, we see that

$$\frac{\mu}{r^3}\,(\mathbf{h} \times \mathbf{r}) = \frac{\mu}{r^3}\,(\mathbf{r} \times \mathbf{v}) \times \mathbf{r} = \frac{\mu}{r^3}\,[\mathbf{v}\,(\mathbf{r} \cdot \mathbf{r}) - \mathbf{r}\,(\mathbf{r} \cdot \mathbf{v})]$$

$$= \frac{\mu}{r}\,\mathbf{v} - \frac{\mu \dot{r}}{r^2}\,\mathbf{r}$$

since $\mathbf{r} \cdot \dot{\mathbf{r}} = r\dot{r}$. Also note that μ times the derivative of the unit vector is also

$$\mu \frac{d}{dt}\left(\frac{\mathbf{r}}{r}\right) = \frac{\mu}{r}\,\mathbf{v} - \frac{\mu \dot{r}}{r^2}\,\mathbf{r}.$$

We can rewrite equation (1.5-1) as

$$\frac{d}{dt}\,(\dot{\mathbf{r}} \times \mathbf{h}) = \mu \frac{d}{dt}\left(\frac{\mathbf{r}}{r}\right).$$

Integrating both sides

$$\dot{\mathbf{r}} \times \mathbf{h} = \mu \frac{\mathbf{r}}{r} + \mathbf{B}. \tag{1.5-2}$$

Where **B** is the vector constant of integration. If we now dot multiply this equation by **r** we get a scalar equation:

$$\mathbf{r} \cdot \dot{\mathbf{r}} \times \mathbf{h} = \mathbf{r} \cdot \mu \frac{\mathbf{r}}{r} + \mathbf{r} \cdot \mathbf{B}.$$

Since, in general, $\mathbf{a} \cdot \mathbf{b} \times \mathbf{c} = \mathbf{a} \times \mathbf{b} \cdot \mathbf{c}$ and $\mathbf{a} \cdot \mathbf{a} = a^2$

$$h^2 = \mu r + rB \cos \nu$$

where ν (nu) is the angle between the constant vector **B** and the radius vector **r**. Solving for r, we obtain

$$\boxed{r = \frac{h^2/\mu}{1 + (B/\mu) \cos \nu}}\,. \tag{1.5-3}$$

1.5.2 The Polar Equation of a Conic Section. Equation (1.5-3) is the trajectory equation expressed in polar coordinates where the polar angle, ν, is measured from the fixed vector **B** to **r**. To determine what kind of a curve it represents we need only compare it to the general equation of a conic section written in polar coordinates with the origin located at a focus and where the polar angle, ν, is the angle between **r** and the point on the conic nearest the focus:

$$\boxed{r = \frac{p}{1 + e \cos \nu}} \tag{1.5-4}$$

In this equation, which is mathematically identical in form to the trajectory equation, p is a geometrical constant of the conic called the "parameter" or "semi-latus rectum." The constant e is called the "eccentricity" and it determines the type of conic section represented by equation (1.5-4).

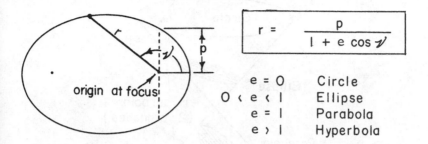

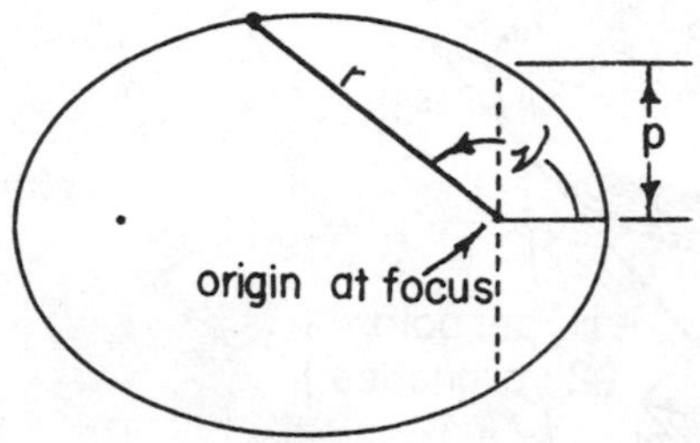

$$r = \frac{p}{1 + e \cos \nu}$$

$e = 0$	Circle
$0 < e < 1$	Ellipse
$e = 1$	Parabola
$e > 1$	Hyperbola

Figure 1.5-1 General equation of any conic section in polar coordinates

The similarity in form between the trajectory equation (1.5-3) and the equation of a conic section (1.5-4) not only verifies Kepler's first law but allows us to extend the law to include orbital motion along any conic section path, not just ellipses.

We can summarize our knowledge concerning orbital motion up to this point as follows:

1. The family of curves called "conic sections" (circle, ellipse, parabola, hyperbola) represent *the only possible paths* for an orbiting object in the two-body problem.

2. The focus of the conic orbit *must* be located at the center of the central body.

3. The mechanical energy of a satellite (which is the sum of kinetic and potential energy) does not change as the satellite moves along its conic orbit. There is, however, an exchange of energy between the two forms, potential and kinetic, which means that the satellite must slow-down as it gains altitude (as r increases) and speed-up as r decreases in such a manner that $\mathcal{E}$ *remains constant.*

4. The orbital motion takes place in a plane which is *fixed in inertial space.*

5. The specific angular momentum of a satellite about the central attracting body *remains constant*. As r and v change along the orbit, the flight-path angle, ϕ, must change so as to keep h constant. (See Figure 1.4-1 and equation (1.4-4)

1.5.3 Geometrical Properties Common to All Conic Sections. Although Figure 1.5-1 illustrates an ellipse, the ellipse is only one of the family of curves called conic sections. Before discussing the factors

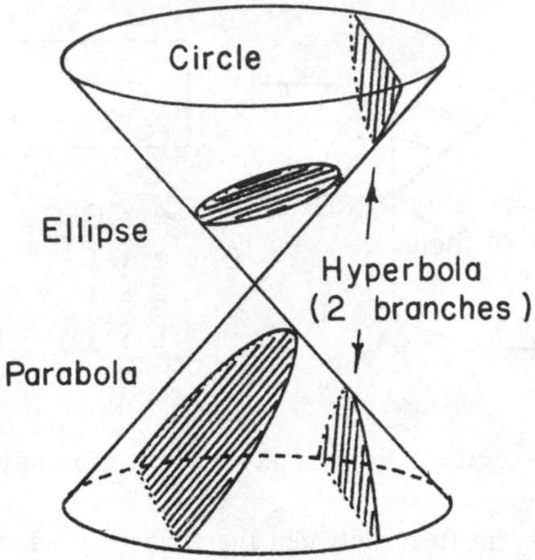

Figure 1.5-2 The conic sections

which determine which of these conic curves a satellite will follow, we need to know a few facts about conic sections in general.

The conic sections have been known and studied for centuries. Many of their most interesting properties were discovered by the early Greeks. The name derives from the fact that a conic section may be defined as the curve of intersection of a plane and a right circular cone. Figure 1.5-2 illustrates this definition. If the plane cuts across one nappe (half-cone), the section is an *ellipse*. A *circle* is just a special case of the ellipse where the plane is parallel to the base of the cone. If, in addition to cutting just one nappe of the cone, the plane is parallel to a line in the surface of the cone, the section is a *parabola*. If the plane cuts both nappes, the section is a *hyperbola* having two branches. There are also *degenerate conics* consisting of one or two straight lines, or a single point, these are produced by planes passing through the apex of the cone.

There is an alternate definition of a conic which is mathematically equivalent to the geometrical definition above:

> A conic is a circle or the locus of a point which moves so that the ratio of its absolute distance from a given point (a focus) to

its absolute distance from a given line (a directrix) is a positive constant e (the eccentricity).

While the directrix has no physical significance as far as orbits are concerned, the focus and eccentricity are indispensable concepts in the understanding of orbital motion. Figure 1.5-3 below illustrates certain other geometrical dimensions and relationships which are common to all conic sections.

Because of their symmetry all conic sections have two foci, F and F'. The prime focus, F, marks the location of the central attracting body in an orbit. The secondary or vacant focus, F', has little significance in

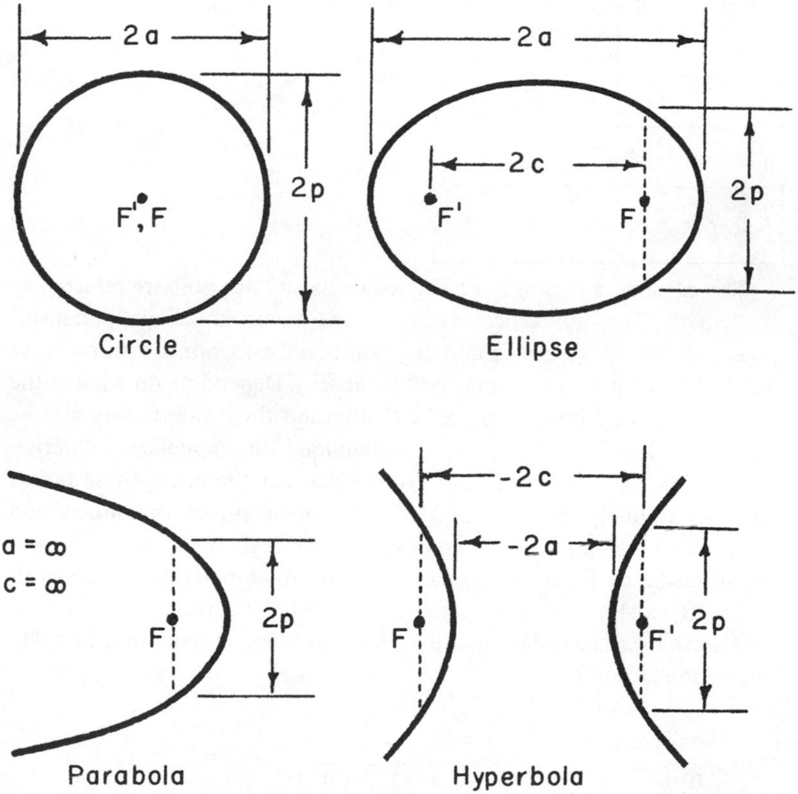

Figure 1.5-3 Geometrical dimensions common to all conic sections

orbital mechanics. In the parabola, which represents the borderline case between the open and closed orbits, the secondary focus is assumed to lie an infinite distance to the left of F. The width of each curve *at the focus* is a positive dimension called the latus rectum and is labeled 2 p in Figure 1.5-3. The length of the chord passing through the foci is called the major axis of the conic and is labeled 2 a. The dimension, a, is called the semi-major axis. Note that for the circle 2 a is simply the diameter, for the parabola 2 a is infinite and for the hyperbola 2 a is taken as negative. The distance between the foci is given the symbol 2 c. For the circle the foci are considered coincident and 2 c is zero, for the parabola 2 c is infinite and for the hyperbola 2 c is taken as a negative. It follows directly from the definition of a conic section given above that for any conic *except a parabola*

$$e = \frac{c}{a} \tag{1.5-5}$$

and

$$p = a\,(1 - e^2). \tag{1.5-6}$$

The extreme end-points of the major axis of an orbit are referred to as "apses". The point nearest the prime focus is called "periapsis" (meaning the "near apse") and the point farthest from the prime focus is called "apoapsis" (meaning the "far apse"). Depending on what is the central attracting body in an orbital situation these points may also be called "perigee" or "apogee," "perihelion" or "aphelion," "periselenium" or "aposelenium," etc. Notice that for the circle these points are not uniquely defined and for the open curves (parabolas and hyperbolas) the apoapsis has no physical meaning.

The distance from the prime focus to either periapsis or apoapsis (where it exists) can be expressed by simply inserting $\nu = 0^\circ$ or $\nu = 180^\circ$ in the general polar equation of a conic section (equation (2.5-4)). Thus, *for any conic*

$$r_{min} = r_{periapsis} = \frac{p}{1 + e\cos 0^\circ}\,.$$

Combining this with equation (1.5-6) gives

$$r_p = \frac{p}{1+e} = a(1-e). \tag{1.5-7}$$

Similarly

$$r_{max} = r_{apoapsis} = \frac{p}{1 + e \cos 180^0}$$

and

$$r_a = \frac{p}{1-e} = a(1+e). \tag{1.5-8}$$

1.5.4 The Eccentricity Vector. In the derivation of equation (1.5-3), the trajectory equation, we encountered the vector constant of integration, **B**, which points toward periapsis. By comparing equations (1.5-3) and (1.5-4) we conclude that $B=\mu e$. Quite obviously, since **e** is also a constant vector pointing toward periapsis.,

$$\mathbf{e} = \mathbf{B}/\mu. \tag{1.5-9}$$

By integrating the two-body equation of motion we obtained the following result

$$\dot{\mathbf{r}} \times \mathbf{h} = \mu \frac{\mathbf{r}}{r} + \mathbf{B} \tag{1.5-2}$$

Solving for **B** and noting that

$$\mathbf{B} = \mathbf{v} \times \mathbf{h} - \mu \frac{\mathbf{r}}{r} \ .$$

Hence

$$\mathbf{e} = \frac{\mathbf{v} \times \mathbf{h}}{\mu} - \frac{\mathbf{r}}{r} \ . \tag{1.5-10}$$

We can eliminate $\mathbf{h}$ from this expression by substituting $\mathbf{h} = \mathbf{r} \times \mathbf{v}$, so

$$\mu \mathbf{e} = \mathbf{v} \times (\mathbf{r} \times \mathbf{v}) - \mu \frac{\mathbf{r}}{r} \cdot$$

Expanding the vector triple product, we get

$$\mu \mathbf{e} = (\mathbf{v} \cdot \mathbf{v})\, \mathbf{r} - (\mathbf{r} \cdot \mathbf{v})\, \mathbf{v} - \mu \frac{\mathbf{r}}{r} \cdot$$

Noting that $(\mathbf{v} \cdot \mathbf{v}) = v^2$ and collecting terms,

$$\boxed{\mu \mathbf{e} = \left(v^2 - \frac{\mu}{r}\right) \mathbf{r} - (\mathbf{r} \cdot \mathbf{v})\mathbf{v}.} \tag{1.5-11}$$

The eccentricity vector will be used in orbit determination in Chapter 2.

1.6 RELATING $\mathcal{E}$ AND h TO THE GEOMETRY OF AN ORBIT

By comparing equation (1.5-3) and equation (1.5-4) we see immediately that the parameter or semi-latus rectum, p, of the orbit depends only on the specific angular momentum, h, of the satellite. By inspection, *for any orbit,*

$$\boxed{p = h^2/\mu.} \tag{1.6-1}$$

In order to see intuitively why an increase in h should result in a larger value for p consider the following argument:

Suppose that a cannon were set up on the top of a high mountain whose summit extends above the sensible atmosphere (so that we may neglect atmospheric drag). If the muzzle of the cannon is aimed horizontally and the cannon is fired, equation (1.4-4) tells us that $h=rv$ since the flight-path angle, ϕ, is zero. Therefore, progressively increasing the muzzle velocity, v, is equivalent to increasing h. Figure 1.6-1 shows the family of curves which represent the trajectory or orbit of the cannonball as the angular momentum of the "cannonball satellite" is progressively increased. Notice that each trajectory is a conic section with the focus located at the center of the earth, and that as h is increased the parameter, p, of the orbit also increases just as equation (1.6-1) predicts.

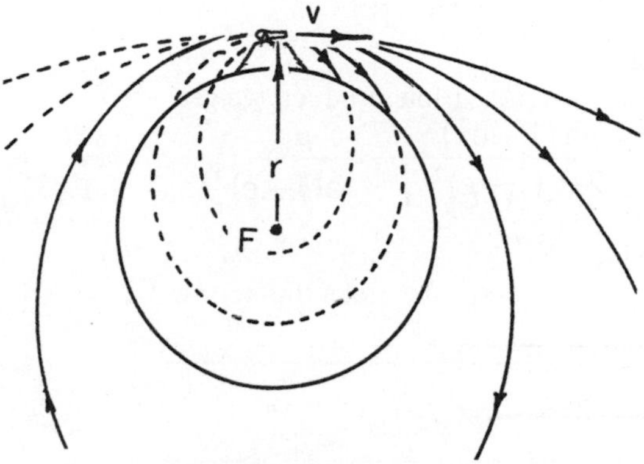

Figure 1.6-1 "Cannonball satellite"

As a by-product of this example we note that at periapsis or apoapsis of any conic orbit the velocity vector (which is always tangent to the orbit) is directed horizontally and the flight-path angle, ν, is zero. We can then write, as a corollary to equation (1.4-4),

$$h = r_p v_p = r_a v_a \,. \tag{1.6-2}$$

If we write the energy equation (1.4-2) for the periapsis point and substitute from equation (1.6-2) we obtain

$$\mathcal{E} = \frac{v^2}{2} - \frac{\mu}{r} = \frac{h^2}{2r_p^2} - \frac{\mu}{r_p} \,.$$

But from equation (1.5-7)

$$r_p = a(1 - e)$$

and from equation (1.5-6) and equation (1.6-1)

$$h^2 = \mu a(1-e^2),$$

therefore

$$\mathcal{E} = \frac{\mu a(1-e^2)}{2a^2(1-e)^2} - \frac{\mu}{a(1-e)}$$

which reduces to

$$\boxed{\mathcal{E} = -\frac{\mu}{2a}\,.} \tag{1.6-3}$$

This simple relationship which is valid *for all conic orbits* tells us that the semi-major axis, a, of an orbit depends only on the specific mechanical energy, $\mathcal{E}$, of the satellite (which in turn depends only on r and v at any point along the orbit). Figure 1.6-1 serves as well to illustrate the intuitive explanation of this fundamental relationship since progressively increasing the muzzle velocity of the cannon also progressively increases $\mathcal{E}$.

Many students find equation (1.6-3) misleading. You should study it together with Figure 1.5-3 which shows that for the circle and ellipse a is positive, while for the parabola a is infinite and for the hyperbola a is negative. This implies that the specific mechanical energy of a satellite in a closed orbit (circle or ellipse) is negative, while $\mathcal{E}$ for a satellite in a parabolic orbit is zero and on a hyperbolic orbit the energy is positive. Thus, the energy of a satellite (negative, zero, or positive) alone determines the type of conic orbit the satellite is in.

Since h alone determines p and since $\mathcal{E}$ alone determines a, the two together determine e (which specifies the exact shape of conic orbit). This can be shown as follows:

$$p = a(1-e^2) \quad \text{therefore} \quad e = \sqrt{1 - \frac{p}{a}}$$

$$p = h^2/\mu \quad \text{and} \quad a = -\mu/2\mathcal{E}$$

so, *for any conic orbit*,

$$e = \sqrt{1 + \frac{2\mathcal{E}h^2}{\mu^2}} \,. \tag{1.6-4}$$

Notice again that if $\mathcal{E}$ is negative, e is positive and less than 1 (an ellipse); if $\mathcal{E}$ is zero, e is exactly 1 (a parabola); if $\mathcal{E}$ is positive, e is greater than 1 (a hyperbola). But what if h is zero regardless of what value $\mathcal{E}$ has? The eccentricity will be exactly 1 but the orbit will not be a parabola! Rather, the orbit will be a degenerate conic (a point or straight line). The student should be aware of this pitfall. Namely, all parabolas have an eccentricity of 1 but an orbit whose eccentricity is 1 need not be a parabola–it could be a degenerate conic.

EXAMPLE PROBLEM. For a given satellite, $\mathcal{E} = -2.0\times10^8$ ft²/sec² and $e = 0.2$ Determine its specific angular momentum, semi-latus rectum, and semi-major axis.

$$a = -\frac{\mu}{2\mathcal{E}} = \underline{\underline{3.5198 \times 10^7 \text{ ft}}}$$

$$p = a(1 - e^2) = \underline{\underline{3.3790 \times 10^7 \text{ ft}}}$$

$$h = \sqrt{p\mu} = \underline{\underline{6.897 \times 10^{11} \text{ ft}^2/\text{sec}}}$$

EXAMPLE PROBLEM. A radar tracking station tells us that a certain decaying weather satellite has $e = 0.1$ and perigee altitude = 200 n. mi. Determine its altitude at apogee, specific mechanical energy, and specific angular momentum.

$$r_p = r_\oplus + 200 = 3643.9 \text{ n.mi.}$$

$$p = r_p(1 + e) = 4008.3 \text{ n.mi.}$$

$$r_a = \frac{p}{1 - e} = 4453.7 \text{ n.mi.}$$

altitude at apogee $= r_a - r_\oplus = \underline{\underline{1009.8 \text{ n.mi.}}}$

$= 6.135 \times 10^6$ ft

$h = \sqrt{p\mu} = \underline{\underline{5.855 \times 10^{11} \text{ ft}^2/\text{sec}}}$

$2a = r_a + r_p = 8097.6$ n.mi.

$\mathcal{E} = -\frac{\mu}{2a} = \underline{\underline{-2.861 \times 10^8 \text{ ft}^2/\text{sec}^2}}$

1.7 THE ELLIPTICAL ORBIT

The orbits of all the planets in the solar system as well as the orbits of all earth satellites are ellipses. Since an ellipse is a closed curve, an object in an elliptical orbit travels the same path over and over. The time for the satellite to go once around its orbit is called the period. We will first look at some geometrical results which apply only to the ellipse and then derive an expression for the period of an elliptical orbit.

1.7.1 Geometry of the Ellipse. An ellipse can be constructed using two pins and a loop of thread. The method is illustrated in Figure 1.7-1. Each pin marks the location of a focus and since the length of the thread is constant, the sum of the distances from any point on an ellipse to each focus ($r + r'$) is a constant. When the pencil is at either end-point of the ellipse it is easy to see that, specifically

$$r + r' = 2a. \tag{1.7-1}$$

By inspection, the radius of periapsis and the radius of apoapsis are related to the major axis of an ellipse as

$$\boxed{r_p + r_a = 2a.} \tag{1.7-2}$$

Also by inspection, the distance between the foci is

$$r_a - r_p = 2c. \tag{1.7-3}$$

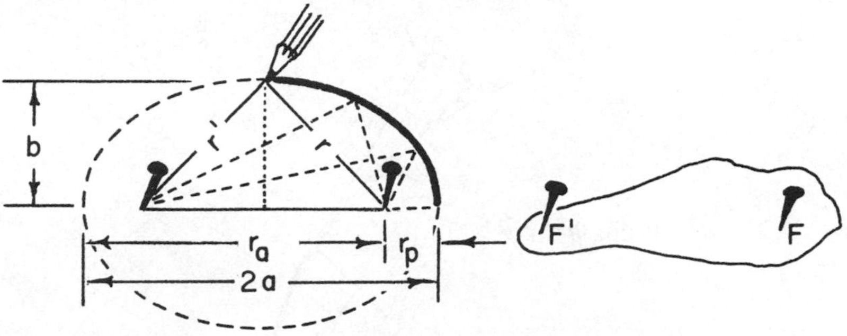

Figure 1.7-1 Simple way to construct an ellipse

Since, in general, e is defined as c/a, equation (1.7-2) and (1.7-3) combine to yield

$$e = \frac{r_a - r_p}{r_a + r_p} \cdot \qquad (1.7\text{-}4)$$

The width of an ellipse at the center is called the minor axis, 2 b. At the end of the minor axis r and r′ are equal as illustrated at the right of Figure 1.7-1. Since $r + r' = 2a$, r and r′ must both be equal to a at this point. Dropping a perpendicular to the major axis (dotted line in Figure 1.7-1) we can form a right triangle from which we conclude that

$$a^2 = b^2 + c^2 . \qquad (1.7\text{-}5)$$

1.7.2 Period of an Elliptical Orbit. If you refer to Figure 1.7-2, you will see that the horizontal component of velocity of a satellite is simply $v \cos \phi$ which can also be expressed as $r \dot{\nu}$. Using equation (1.4-4) we can express the specific angular momentum of the satellite as

$$h = \frac{r^2 d\nu}{dt}$$

which, when rearranged, becomes

$$dt = \frac{r^2}{h} \; d\nu. \qquad (1.7\text{-}6)$$

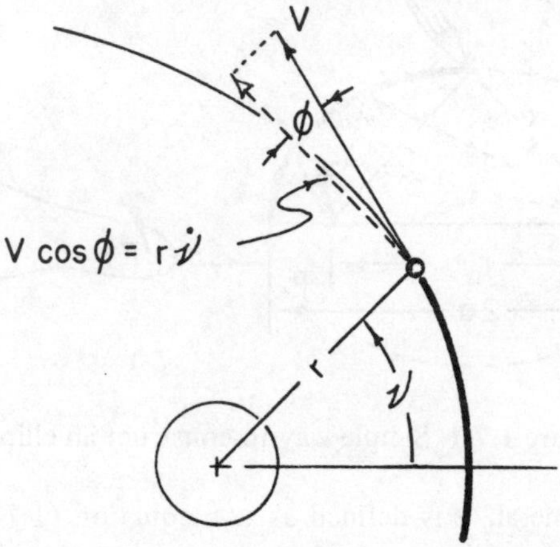

Figure 1.7-2 Horizontal component of v

But from elementary calculus we know that the differential element of area, d A, swept out by the radius vector as it moves through an angle, d ν, is given by the expression

$$dA = \frac{1}{2} r^2 d\nu .$$

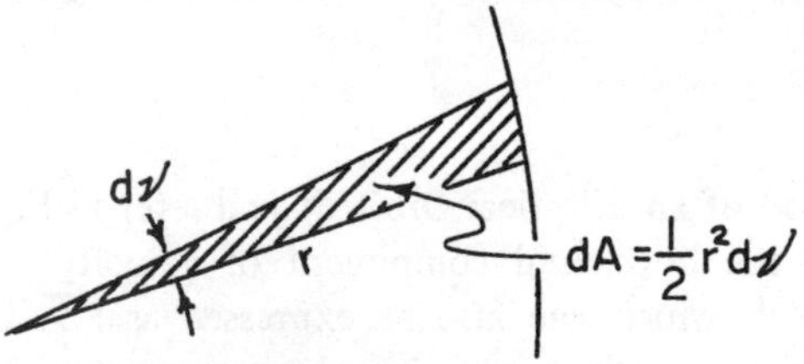

Figure 1.7-3 Differential element of area

So, we can rewrite equation (1.7-6) as

$$dt = \frac{2}{h} dA. \tag{1.7-7}$$

Equation (1.7-7) proves Kepler's second law that "equal areas are swept out by the radius vector in equal time intervals" since h is a constant for an orbit.

During one orbital period the radius vector sweeps out the entire area of the ellipse. Integrating equation (1.7-7) for one period gives us

$$\mathbb{TP} = \frac{2\pi ab}{h} \tag{1.7-8}$$

where π a b is the total area of an ellipse and $\mathbb{TP}$ is the period. From equations (1.7-5), (1.5-5) and (1.5-6)

$$b = \sqrt{a^2 - c^2} = \sqrt{a^2(1 - e^2)} = \sqrt{ap}$$

and, since $h = \sqrt{\mu p}$,

$$\boxed{\mathbb{TP} = \frac{2\pi}{\sqrt{\mu}} a^{3/2}\ .} \tag{1.7-9}$$

Thus, the period of an elliptical orbit *depends only on the size of the semi-major axis,* a. Equation (1.7-9), incidentally, proves Kepler's third law that "the square of the period is proportional to the cube of the mean distance" since a, being the average of the periapsis and apoapsis radii, is the "mean distance" of a satellite from the prime focus.

1.8 THE CIRCULAR ORBIT

The circle is just a special case of an ellipse so all the relationships we just derived for the elliptical orbit including the period are also valid for the circular orbit. Of course, the semi-major axis of a circular orbit is just its radius, so equation (1.7-9) is simply

$$\mathbb{TP}_{cs} = \frac{2\pi}{\sqrt{\mu}} r_{cs}^{3/2} \tag{1.8-1}$$

1.8.1 Circular Satellite Speed. The speed necessary to place a satellite in a circular orbit is called circular speed. Naturally, the satellite must be launched in the horizontal direction at the desired altitude to achieve a circular orbit. The latter condition is called circular velocity and implies both the correct speed and direction. We can calculate the speed required for a circular orbit of radius, r_{cs}, from the energy equation.

$$\mathcal{E} = \frac{v^2}{2} - \frac{\mu}{r} = -\frac{\mu}{2a}$$

If we remember that $r_{cs} = a$, we obtain

$$\frac{v_{cs}^2}{2} - \frac{\mu}{r_{cs}} = -\frac{\mu}{2r_{cs}}$$

which reduces to

$$\boxed{v_{cs} = \sqrt{\frac{\mu}{r_{cs}}}\,.} \qquad (1.8\text{-}2)$$

Notice that the greater the radius of the circular orbit the less speed is required to keep the satellite in this orbit. For a low altitude earth orbit, circular speed is about 26,000 ft/sec while the speed required to keep the moon in its orbit around the earth is only about 3,000 ft/sec.

1.9 THE PARABOLIC ORBIT

The parabolic orbit is rarely found in nature although the orbits of some comets approximate a parabola. The parabola is interesting because it represents the borderline case between the open and closed orbits. An object traveling a parabolic path is on a one-way trip to infinity and will never retrace the same path again.

1.9.1 Geometry of the Parabola. There are only a few geometrical properties peculiar to the parabola which you should know. One is that the two arms of a parabola become more and more nearly parallel as one extends them further and further to the left of the focus in Figure 1.9-1. Another is that, since the eccentricity of a parabola is exactly 1, the periapsis radius is just

$$r_p = \frac{p}{2} \qquad (1.9\text{-}1)$$

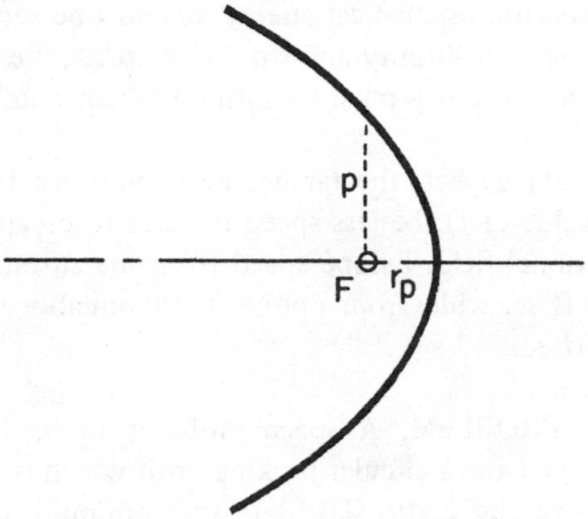

Figure 1.9-1 Geometry of the parabola

which follows from equation (1.5-7). Of course, there is no apoapsis for a parabola and it may be thought of as an "infinitely long ellipse."

1.9.2 Escape Speed. Even though the gravitational field of the sun or a planet theoretically extends to infinity, its strength decreases so rapidly with distance that only a finite amount of kinetic energy is needed to overcome the effects of gravity and allow an object to coast to an infinite distance without "falling back." The speed which is just sufficient to do this is called *escape speed*. A space probe which is given escape speed in any direction will travel on a parabolic *escape trajectory*. Theoretically, as its distance from the central body approaches infinity its speed approaches zero. We can calculate the speed necessary to escape by writing the energy equation for two points along the escape trajectory; first at a general point a distance, r, from the center where the "local escape speed" is v_{esc}, and then at infinity where the speed will be zero:

$$\mathcal{E} = \frac{v_{esc}^2}{2} - \frac{\mu}{r} = \frac{v_\infty^2}{2} \nearrow^0 - \frac{\mu}{r_\infty} \nearrow^0 = 0$$

from which

$$\boxed{v_{esc} = \sqrt{\frac{2\mu}{r}}\,.} \tag{1.9-2}$$

Since the specific mechanical energy, $\mathcal{E}$, must be zero if the probe is to have zero speed at infinity and since $\mathcal{E} = -\mu/2a$, the semi-major axis, a, of the escape trajectory must be infinite which confirms that it is a parabola.

As you would expect, the farther away you are from the central body (larger value of r) the less speed it takes to escape the remainder of the gravitational field. Escape speed from the surface of the earth is about 36,700 ft/sec while from a point 3,400 nm above the surface it is only 26,000 ft/sec.

EXAMPLE PROBLEM. A space probe is to be launched on an escape trajectory from a circular parking orbit which is at an altitude of 100 n mi above the earth. Calculate the minimum escape speed required to escape from the parking orbit altitude. (Ignore the gravitational forces of the sun and other planets.) Sketch the escape trajectory and the circular parking orbit.

a. Escape Speed:

Earth gravitational parameter is

$$\mu = 1.407654 \times 10^{16} \text{ ft}^3/\text{sec}^2$$

Radius of circular orbit is

$$r = r_{earth} + \text{Altitude Circular Orbit}$$

$$= 21.53374 \times 10^6 \text{ ft}$$

From equation (1.9-2)

$$v_{esc} = \sqrt{\frac{2\mu}{r}} = 36{,}157.9 \text{ ft/sec}$$

b. Sketch of escape trajectory and circular parking orbit:

From the definition of escape speed the energy constant is zero on the escape trajectory which is therefore parabolic. The parameter p is determined by equation (1.5-7).

$$p = r_p\,(1 + e)$$
$$= 21.53374 \times 10^6 \text{ ft} \times 2$$
$$= 43.06748 \times 10^6 \text{ ft}$$
$$= 7087.8 \text{n.mi.}$$

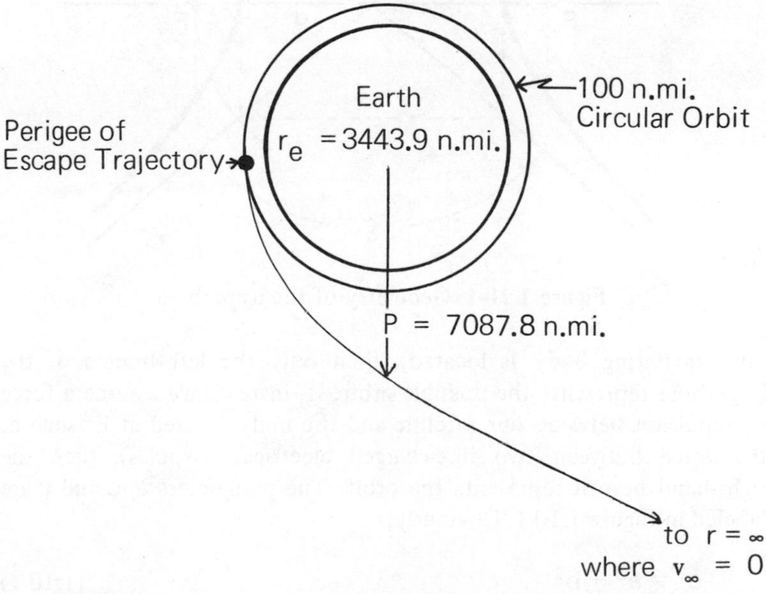

Figure 1.9-2 Escape trajectory for example problem

1.10 THE HYPERBOLIC ORBIT

Meteors which strike the earth and interplanetary probes sent from the earth travel hyperbolic paths relative to the earth. A hyperbolic orbit is necessary if we want the probe to have some speed left over after it escapes the earth's gravitational field. The hyperbola is an unusual and interesting conic section because it has two branches. Its geometry is worth a few moments of study.

1.10.1 Geometry of the Hyperbola. The arms of a hyperbola are asymptotic to two intersecting straight lines (the asymptotes). If we consider the left-hand focus, F, as the prime focus (where the center of

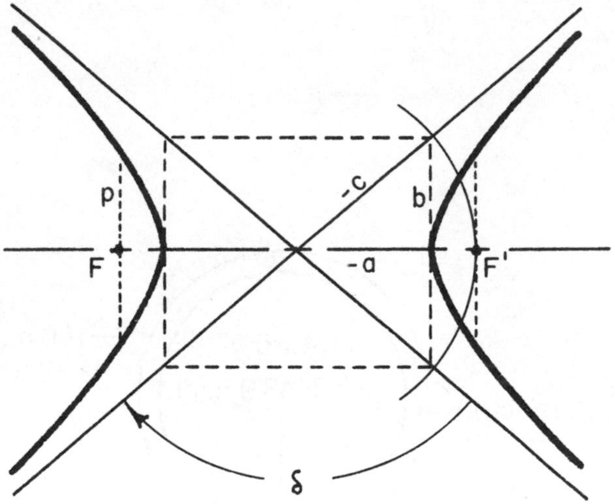

Figure 1.10-1 Geometry of the hyperbola

our gravitating body is located), then only the left branch of the hyperbola represents the possible orbit. If, instead, we assume a force of repulsion between our satellite and the body located at F (such as the force between two like-charged electrical particles), then the right-hand branch represents the orbit. The parameters a, b and c are labeled in Figure 1.10-1. Obviously,

$$c^2 = a^2 + b^2 \tag{1.10-1}$$

for the hyperbola. The angle between the asymptotes, which represents the angle through which the path of a space probe is turned by its encounter with a planet, is labeled δ (delta) in Figure 1.10-1. The turning angle, δ, is related to the geometry of the hyperbola as follows:

$$\sin \frac{\delta}{2} = \frac{a}{c} \tag{1.10-2}$$

but since $e = c/a$ equation (1.10-2) becomes

$$\boxed{\sin \frac{\delta}{2} = \frac{1}{e} \quad .} \tag{1.10-3}$$

The greater the eccentricity of the hyperbola, the smaller will be the turning angle, δ.

1.10.2 Hyperbolic Excess Speed. If you give a space probe exactly escape speed, it will just barely escape the gravitational field which

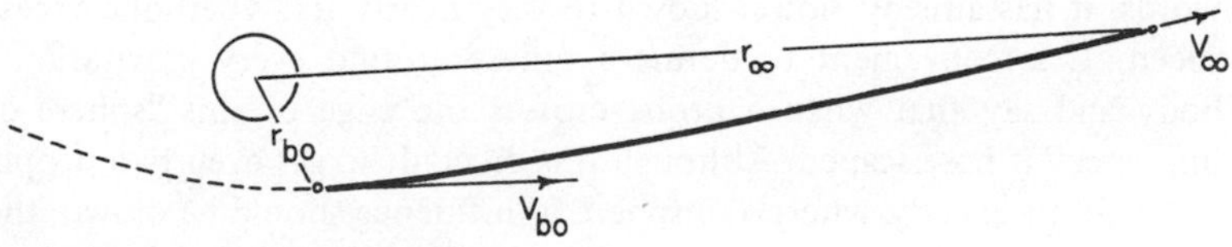

Figure 1.10-2 Hyperbolic excess speed

means that its speed will be approaching zero as its distance from the force center approaches infinity. If, on the other hand, we give our probe more than escape speed at a point near the earth, we would expect the speed at a great distance from the earth to be approaching some finite constant value. This residual speed which the probe would have left over even at infinity is called "hyperbolic excess speed." We can calculate this speed from the energy equation written for two points on the hyperbolic escape trajectory—a point near the earth called the "burnout point" and a point an infinite distance from the earth where the speed will be the hyperbolic excess speed, v_∞.

Since specific mechanical energy does not change along an orbit, we may equate $\mathcal{E}$ at the burnout point and $\mathcal{E}$ at infinity:

$$\mathcal{E} = \frac{v_{bo}^2}{2} - \frac{\mu}{r_{bo}} = \frac{v_\infty^2}{2} - \frac{\mu}{r_\infty}^{\nearrow 0} \tag{1.10-4}$$

from which we conclude that

$$v_\infty^2 = v_{bo}^2 - \frac{2\mu}{r_{bo}} = v_{bo}^2 - v_{esc}^2 \,. \tag{1.10-5}$$

Note that if v_∞ is zero (as it is on a parabolic trajectory) v_{bo} becomes simply the escape speed.

1.10.3 Sphere of Influence. It is, of course, absurd to talk about a space probe "reaching infinity" and in this sense it is meaningless to talk about escaping a gravitational field completely. It is a fact, however, that once a space probe is a great distance (say, a million miles) from earth, for all practical purposes it has escaped. In other words, it has already slowed down to very nearly its hyperbolic excess speed. It is convenient to define a sphere around every gravitational body and say that when a probe crosses the edge of this "sphere of influence" it has escaped. Although it is difficult to get even two people to agree on exactly where the sphere of influence should be drawn, the concept is convenient and is widely used, especially in lunar and interplanetary trajectories.

1.11 CANONICAL UNITS

Astronomers are as yet unable to determine the precise distance and mass of objects in space. Such fundamental quantities as the mean distance from the earth to the sun, the mass and mean distance of the moon and the mass of the sun are not accurately known. This dilemma is avoided in mathematical calculations if we assume the mass of the sun to be 1 "mass unit" and the mean distance from the earth to the sun to be our unit of distance which is called an "astronomical unit." All other masses and distances can then be given in terms of these assumed units even though we do not know precisely the absolute value of the sun's mass and distance in pounds or miles. Astronomers call this normalized system of units "canonical units."

We will adopt a similar system of normalized units in this text primarily for the purpose of simplifying the arithmetic of our orbit calculations.

1.11.1 The Reference Orbit. We will use a system of units based on a hypothetical circular reference orbit. In a two-body problem where the sun is the central body the reference orbit will be a circular orbit whose radius is one astronomical unit (AU). For other problems where the earth, moon, or some other planet is the central body the reference orbit will be a minimum altitude circular orbit just grazing the surface of the planet.

We will define our distance unit (DU) to be the radius of the reference orbit. If we now define our time unit (TU) such that the speed of a satellite in the hypothetical reference orbit is 1 DU/TU, then the value of the gravitational parameter, μ, will turn out to be 1 DU^3/TU^2.

Unless it is perfectly clear which reference orbit the units in your problem are based on you will have to indicate this by means of a subscript on the symbol DU and TU. This is most easily done by annexing as a subscript the astronomer's symbol for the sun, earth, or other planet. The most commonly used symbols are:

☉ The Sun
☾ The Moon
☿ Mercury
♀ Venus
⊕ The Earth
♂ Mars
♃ Jupiter
♄ Saturn
♅ Uranus
♆ Neptune
♇ Pluto

The concept of the reference orbit is illustrated in Figure 1.11-1.

Values for the commonly used astrodynamic constants and their relationship to canonical units are listed in the appendices.

EXAMPLE PROBLEM. A space object is sighted at an altitude of 1.046284×10^7 ft above the earth traveling at 2.593625×10^4 ft/sec and a flight path angle of 0° at the time of sighting. Using canonical units determine $\mathcal{E}$, h, p, e, r_a, r_p.

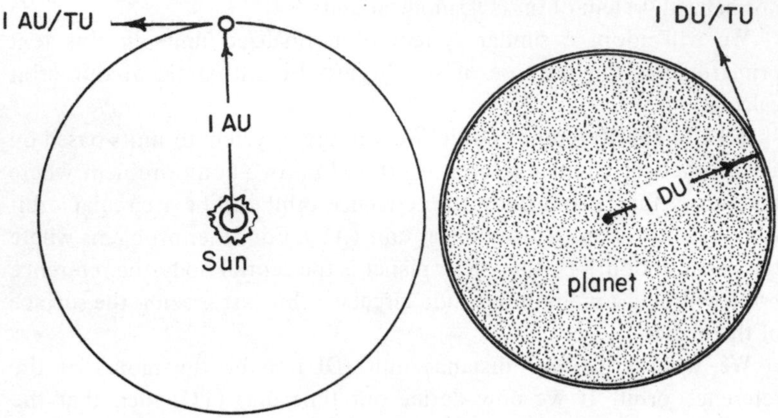

Figure 1.11-1 Reference circular orbits

Convert altitude and speed to earth canonical units.

$$\text{Alt} = .5\ DU_{\oplus}$$

$$v = 1\ DU_{\oplus}/TU_{\oplus}$$

The gravitational parameter and earth radius are:

$$\mu_{\oplus} = 1.407647 \times 10^{16}\ ft^3/sec^2 = 1 DU_{\oplus}{}^3/TU_{\oplus}{}^2$$

$$r_{\oplus} = 1\ DU_{\oplus}$$

The radius of the object from the center of the earth is:

$$r = r_{\oplus} + \text{Alt} = 1.5\ DU_{\oplus}$$

Find $\mathcal{E}$ from equation (1.4-2).

$$\mathcal{E} = \frac{v^2}{2} - \frac{\mu_{\oplus}}{r} = -.167 DU_{\oplus}^2/TU_{\oplus}^2 = -\ 1.12339 \times 10^8\ ft^2/sec^2$$

Find h from equation (1.4-4)

$$h = rv\cos\phi = 1.5DU_{\oplus}^2/TU_{\oplus} = 8.141 \times 10^{11}\ ft^2/sec$$

Find p from equation (1.6-1)

$$p = \frac{h^2}{\mu_{\oplus}} = 2.25DU_{\oplus} = 4.7082763 \times 10^7\ ft$$

Find e from equation (1.6-4)

$$e = \sqrt{1 + \frac{2\mathcal{E}h^2}{\mu^2}} = .5$$

Find r_a from equation (1.5-8)

$$r_a = \frac{p}{1-e} = 4.5DU_{\oplus} = 9.416553 \times 10^7\ ft$$

Find r_p from equation (1.5-7)

$$r_p = \frac{p}{1+e} = 1.5DU_{\oplus} = 3.138851 \times 10^7\ ft$$

EXERCISES

1.1 The position and velocity of a satellite at a given instant are described by

$$\mathbf{r} = 2\mathbf{I} + 2\mathbf{J} + 2\mathbf{K} \qquad \text{(Distance Units)}$$

$$\mathbf{v} = -.4\mathbf{I} + .2\mathbf{J} + .4\mathbf{K} \qquad \text{(Distance Units per Time Unit)}$$

where **I J K** is a nonrotating geocentric coordinate system. Find the specific angular momentum and total specific mechanical energy of the satellite.
(Answer: $\mathbf{h} = .4\mathbf{I} - 1.6\mathbf{J} + 1.2\mathbf{K}$ DU^2/TU,
$\mathcal{E} = -.1087$ DU^2/TU^2)

1.2 For a certain satellite the observed velocity and radius at $\nu = 90^\circ$ is observed to be 45,000 ft/sec and 4,000 n mi, respectively. Find the eccentricity of the orbit.
(Answer: $e = 1.581$)

1.3 An earth satellite is observed to have a height of perigee of 100 n mi and a height of apogee of 600 n mi. Find the period of the orbit.

1.4 Six constants of integration (or effectively, 6 orbital elements) are required for a complete solution to the two-body problem. Why, in general, is a completely determined closed solution of the N-body problem an impossibility if $N \geqslant 3$?

1.5 For a certain earth satellite it is known that the semi-major axis, a, is 30×10^6 ft. The orbit eccentricity is 0.2.

a. Find its perigee and apogee distances from the center of the earth.

b. Find the specific energy of the trajectory.

c. Find the semi-latus rectum or parameter (p) of the orbit.

d. Find the length of the position vector at a true anomaly of 135°.
(Ans. $r = 3.354 \times 10^{7}$ ft)

1.6 Find an equation for the velocity of a satellite as a function of total specific mechanical energy and distance from the center of the earth.

1.7 Prove that $r_{apoapsis} = a(1+e)$.

1.8 Identify each of the following trajectories as either circular, elliptical, hyperbolic, or parabolic:

a. $r = 3$ DU

$v = 1.5$ DU/TU

b. $r_{perigee} = 1.5$ DU

$p = 3$ DU

c. $\mathcal{E} = -1/3\ DU^2/TU^2$

$p = 1.5$ DU

d. $\mathbf{r} = \mathbf{J} + .2\mathbf{K}$

$\mathbf{v} = .9\mathbf{I} + .123\mathbf{K}$

e. $\mathbf{r} = 1.01\mathbf{K}$

$\mathbf{v} = \mathbf{I} + 1.4\mathbf{K}$

1.9 A space vehicle enters the sensible atmosphere of the earth (300,000 ft) with a velocity of 25,000 ft/sec at a flight-path angle of -60°. What was its velocity and flight-path angle at an altitude of 100 n. mi during descent?
(Ans. $v = 24{,}618$ ft/sec, $\phi = -59^{\circ}\ 58'$)

1.10 Show that two-body motion is confined to a plane fixed in space.

1.11 A sounding rocket is fired vertically. It achieves a burnout speed of 10,000 ft/sec at an altitude of 100,000 ft. Determine the maximum altitude attained. (Neglect atmospheric drag.)

1.12 Given that $e = \frac{c}{a}$, derive values for e for circles, ellipses and hyperbolas.

1.13 Show by means of the differential calculus that the position vector is an extremum (maximum or minimum) at the apses of the orbit.

1.14 Given the equation $r = \frac{p}{1+e\cos\nu}$ plot at least four points, sketch and identify the locus and label the major dimensions for the following conic sections:

a. $p=2,\qquad e=0$

b. $p=6,\qquad e=.2$

c. $p=6,\qquad e=.6$

d. $p=3,\qquad e=1$

e. $p=2,\qquad e=2$

(HINT: Polar graph paper would be of help here!)

1.15 Starting with:

$$m_k\ddot{\mathbf{r}}_k = \sum_{\substack{j=1\\ j\neq k}}^{m} \frac{Gm_jm_k}{r_{jk}^2}\,\frac{\mathbf{r}_j-\mathbf{r}_k}{r_{jk}}$$

where $\mathbf{r}_j$ is the vector from the origin of an inertial frame to any j^{th}

body and r_{jk} is the scalar distance between the jth and k^{th} bodies ($r_{jk} \equiv r_{kj}$)

a. Show that:

$$\sum_{k=1}^{m} m_k \ddot{\mathbf{r}}_k = 0$$

b. Using the definition of a system's mass center show:

$$\mathbf{r}_c = \mathbf{a}t + \mathbf{b}$$

where $\mathbf{r}_c$ is the vector from the inertial origin to the system mass center and **a** and **b** are constant vectors.

c. What is the significance of the equation derived in part b above?

1.16 A satellite is injected into an elliptical orbit with a semi-major axis equal to $4\,DU_\oplus$. When it is precisely at the end of the semi-minor axis it receives an impulsive velocity change just sufficient to place it into an escape trajectory. What was the magnitude of the velocity change?
(Ans. $\Delta v = 0.207\ DU_\oplus/TU_\oplus$)

1.17 Show that the speed of a satellite on an elliptical orbit at either end of the *minor* axis is the same as local circular satellite speed at that point.

1.18 Show that when an object is located at the intersection of the semi-minor axis of an elliptical orbit the eccentricity of the orbit can be expressed as $e = -\cos\nu$.

* **1.19** BMEWS (Ballistic Missile Early Warning System) detects an unidentified object with the following parameters:

altitude = .5 DU

speed = $\sqrt{2/3}$ DU/TU

flight-path angle = 30°

Is it possible that this object is a space probe intended to escape the earth, an earth satellite or a ballistic missile?

* **1.20** Prove that the flight-path angle is equal to 45° when $\nu = 90^{\circ}$ on all parabolic trajectories.

* **1.21** Given two spherically symmetric bodies of considerable mass, assume that the only force that acts is a repulsive force, proportional to the product of the masses and inversely proportional to the cube of the distance between the masses that acts along the line connecting the centers of the bodies. Assume that Newton's second law holds ($\Sigma \mathbf{F} = m\mathbf{a}$) and derive a differential equation of motion for these bodies.

* **1.22** A space vehicle destined for Mars was first launched into a 100 n mi circular parking orbit.

a. What was speed of vehicle at injection into parking orbit?

The vehicle coasted in orbit for a period of time to allow system checks to be made and then was restarted to increase its velocity 37,600 ft/sec which placed it on an interplanetary trajectory toward Mars.

b. Find e, h and $\mathcal{E}$ relative to the earth for the escape orbit. What kind of orbit is it?

c. Compare the velocity at 1,000,000 n mi from the earth with the hyperbolic excess velocity, v_∞. Why are the two so nearly alike?

LIST OF REFERENCES

1. Newman, James R. "Commentary on Isaac Newton," in *The World of Mathematics*. Vol 1. New York, NY, Simon and Schuster, 1956.

2. Lodge, Sir Oliver. "Johann Kepler," in *The World of Mathematics*. Vol 1. New York, NY, Simon and Schuster, 1956.

3. Koestler, Arthur. *The Watershed*. Garden City, NY, Anchor Books, Doubleday & Company, Inc, 1960.

4. Turnbull, Herbert Westren. *The Great Mathematicians.* 4th ed. London, Methuen & Co, Ltd, 1951.

5. Newton, Sir Isaac. *Principia*. Motte's translation revised by Cajori. Vol 1. Berkeley and Los Angeles, University of California Press, 1962.

CHAPTER 2

ORBIT DETERMINATION FROM OBSERVATIONS

Finally, as all our observations, on account of the imperfection of the instruments and of the senses, are only approximations to the truth, an orbit based only on the six absolutely necessary data may be still liable to considerable errors. In order to diminish these as much as possible, and thus to reach the greatest precision attainable, no other method will be given except to, accumulate the greatest number of the most perfect observations, and to adjust the elements, not so as to satisfy this or that set of observations with absolute exactness, but so as to agree with all in the best possible manner.

—Carl Friedrich Gauss[1]

2.1 HISTORICAL BACKGROUND

The first method of finding the orbit of a body from three observations was devised by Newton and is given in the *Principia*. The most publicized result of applying Newton's method of orbit determination belongs to our old friend Sir Edmond Halley.

In 1705, shortly after the publication of the *Principia* (1687), Halley set to work calculating the orbits of 24 comets which had been sighted between 1337 and 1698, using the method which Newton had described. Newton's arguments, presented in a condensed form, were so difficult to understand that of all his contemporaries only Halley was able to master his technique and apply it to the calculation of the orbits of those comets for which there was a sufficient number of observations.

In a report on his findings published in 1705 he states: "Many considerations incline me to believe the comet of 1531 observed by Apianus to have been the same as that described by Kepler and Longomontanus in 1607 and which I again observed when it returned in 1682. All the elements agree; .... whence I would venture confidently to predict its return, namely in the year 1758."[2]

In a later edition of the same work published in 1752, Halley confidently identified his comet with those that had appeared in 1305, 1380 and 1456 and remarked, "wherefore if according to what we have already said it should return again about the year 1758, candid posterity will not refuse to acknowledge that this was first discovered by an Englishman."[2]

No one could be certain he was right, much less that it might not succumb to a churlish whim and bump into the earth. The realization that the stability of the earth's orbit depends "on a nice balance between the velocity with which the earth is falling toward the sun and its tangential velocity at right angles to that fall" did not unduly disturb the handful of mathematicians and astronomers who understood what that equilibrium meant. But there were many others, of weaker faith in arithmetic and geometry, who would have preferred a less precarious arrangement.[3]

On Christmas day, 1758, Halley's comet reappeared, just as he said it would. Recent investigations have revealed, in ancient Chinese chronicles, an entire series of earlier appearances of Halley's comet–the earliest in 467 BC. The comet has appeared since in 1835 and 1910; its next visit is expected in 1986.

Newton's method of determining a parabolic orbit from observations depended on a graphical construction which, by successive approximations, led to the elements. The first completely analytical method for solving the same problem was given by Euler in 1744 in his *Theory of the Motion of Planets and Comets.* To Euler belongs the discovery of the equation connecting two radius vectors and the subtended chord of the parabola with the interval of time during which the comet describes the corresponding arc.

Lambert, in his works of 1761 to 1771, gave a generalized formulation of the theorem of Euler for the case of elliptical and hyperbolic orbits. But Lambert was a geometrician at heart and was not inclined to analytical development of his method. Nevertheless, he had an unusual grasp of the physics of the problem and actually anticipated many of the ideas that were ultimately carried out by his successors in better and more convenient ways.

Lagrange, who at age 16 was made Professor of Mathematics at Turin, published three memoirs on the theory of orbits, two in 1778 and one in 1783. It is interesting to note that the mathematical spark was kindled in the young Lagrange by reading a memoir of Halley. From the very first his writings were elegance itself; he would set to mathematics all the problems which his friends brought him, much as Schubert would set to music any stray rhyme that took his fancy. As one would expect, Lagrange brought to the incomplete theories of Euler and Lambert generality, precision, and mathematical elegance.

In 1780 Laplace published an entirely new method of orbit determination. The basic ideas of his method are given later in this chapter and are still of fundamental importance.

The theory of orbit determination was really brought to fruition through the efforts of the brilliant young German mathematician, Gauss, whose work led to the rediscovery of the asteroid Ceres in 1801 after it had become "lost." Gauss also invented the "method of least squares" to deal with the problem of fitting the best possible orbit to a large number of observations.

Today, with the availability of radar, the problem of orbit determination is much simpler. Before showing you how it is done, however, we must digress a moment to explain how the size, shape and orientation of an orbit in three dimensional space can be described by six quantities called "orbital elements."

2.2 COORDINATE SYSTEMS

Our first requirement for describing an orbit is a suitable inertial reference frame. In the case of orbits around the sun such as planets, asteroids, comets and some deep-space probes describe, the heliocentric-ecliptic coordinate system is convenient. For satellites of the earth we will want to use the geocentric-equatorial system. In order to describe these coordinate systems we will give the position of the origin, the orientation of the fundamental plane (i.e. the X-Y plane), the principal direction (i.e. the direction of the X-axis), and the direction of the Z-axis. Since the Z-axis must be perpendicular to the fundamental plane it is only necessary to specify which direction is positive. The Y-axis is always chosen so as to form a right-handed set of

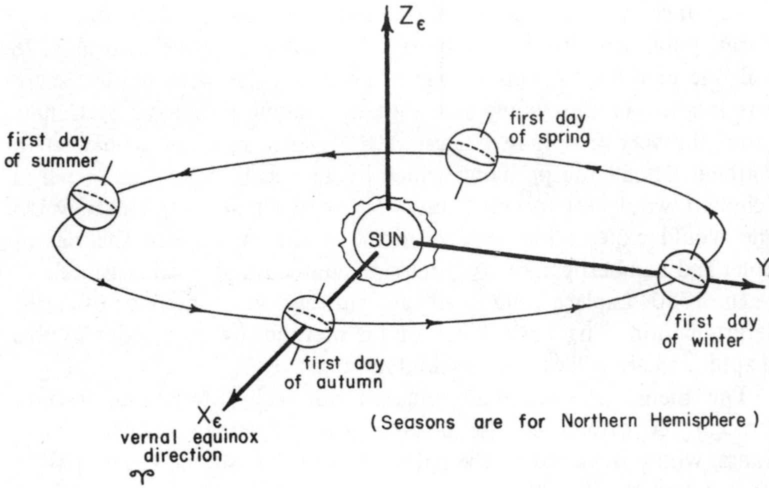

Figure 2.2-1 Heliocentric–ecliptic coordinate system

coordinate axes.

2.2.1 The Heliocentric-Ecliptic Coordinate System. As the name implies, the heliocentric-ecliptic system has its origin at the center of the sun. The X_ϵ-Y_ϵ or fundamental plane coincides with the "ecliptic" which is the plane of the earth's revolution around the sun. The line-of-intersection of the ecliptic plane and the earth's equatorial plane defines the direction of the X_ϵ-axis as shown in Figure 2.2-1. On the first day of Spring a line joining the center of the earth and the center of the sun points in the direction of the positive X_ϵ-axis. This is called the vernal equinox direction and is given the symbol ♈ by astronomers because it used to point in the direction of the constellation Aries (the ram). As you know, the earth wobbles slightly and its axis of rotation shifts in direction slowly over the centuries. This effect is known as precession and causes the line-of-intersection of the earth's equator and the ecliptic to shift slowly. As a result the heliocentric-ecliptic system is

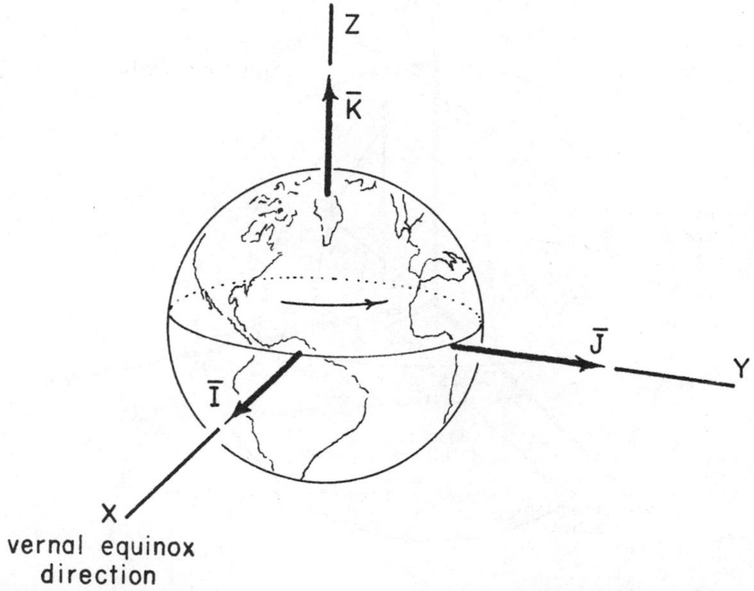

Figure 2.2-2 Geocentric-equatorial coordinate system

not really an inertial reference frame and where extreme precision is required it would be necessary to specify that the XYZ_{ϵ} coordinates of an object were based on the vernal equinox direction of a particular year or "epoch."[4]

2.2.2 The Geocentric-Equatorial Coordinate System. The geocentric-equatorial system has its origin at the earth's center. The fundamental plane is the equator and the positive X-axis points in the vernal equinox direction. The Z-axis points in the direction of the north pole. It is important to keep in mind when looking at Figure 2.2-2 that the XYZ system is not fixed to the earth and turning with it; rather, the geocentric-equatorial frame is nonrotating with respect to the stars (except for precession of the equinoxes) and the earth turns relative to it.

Unit vectors, **I**, **J** and **K**, shown in Figure 2.2-2, lie along the X, Y and Z axes respectively and will be useful in describing vectors in the geocentric-equatorial system. Note that vectors in this figure and in many others are designated by a bar over the letter, whereas they are

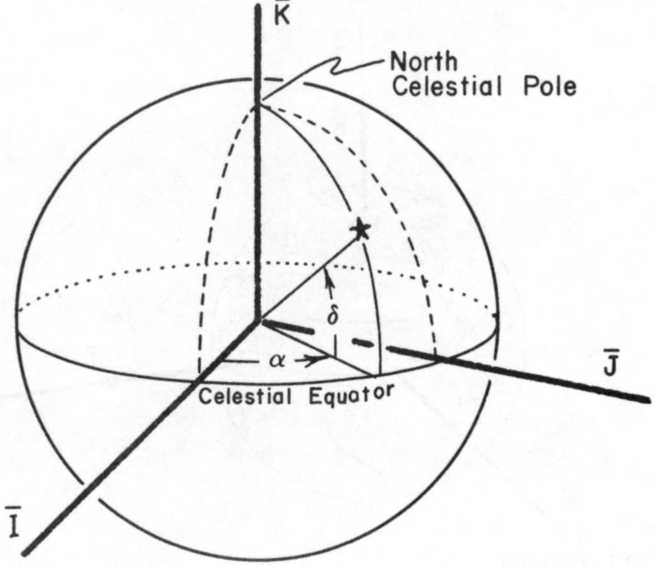

Figure 2.2-3 Right ascension–declination coordinate system

boldface in the text material.

2.2.3 The Right Ascension-Declination System. A coordinate system closely related to the geocentric-equatorial frame is the right ascension-declination system. The fundamental plane is the "celestial equator"–the extension of the earth's equatorial plane to a fictitious sphere of infinite radius called the "celestial sphere." The position of an object projected against the celestial sphere is described by two angles called right ascension and declination.

As shown in Figure 2.2-3, the right ascension, α, is measured eastward in the plane of the celestial equator from the vernal equinox direction. The declination, δ, is measured northward from the celestial equator to the line-of-sight.

The origin of the system may be at the center of the earth (geocenter) or a point on the surface of the earth (topocenter) or anywhere else. For all intents and purposes any point may be considered the center of the infinite celestial sphere.

Astronomers use the right ascension-declination system to catalog star positions accurately. Because of the enormous distances to the stars, their coordinates remain essentially unchanged even when viewed from opposite sides of the earth's orbit around the sun. Only a few stars are close enough to show a measurable parallax between observations made 6 months apart.

Because star positions are known accurately to fractions of an arc-second, a photograph of an earth satellite against a star background can be used to determine the topocentric right ascension and declination of the satellite at the time of observation. Orbit determination from such optical sightings of a satellite will be discussed in a later section.

2.2.4 The Perifocal Coordinate System. One of the most convenient coordinate frames for describing the motion of a satellite is the perifocal coordinate system. Here the fundamental plane is the plane of the satellite's orbit. The coordinate axes are named x_ω, y_ω and z_ω. The x_ω axis points toward the periapsis; the y_ω axis is rotated 90° in the direction of orbital motion and lies in the orbital plane; the z_ω axis along **h** completes the right-handed perifocal system. Unit vectors in the direction of x_ω, y_ω and z_ω are called **P**, **Q** and **W** respectively.

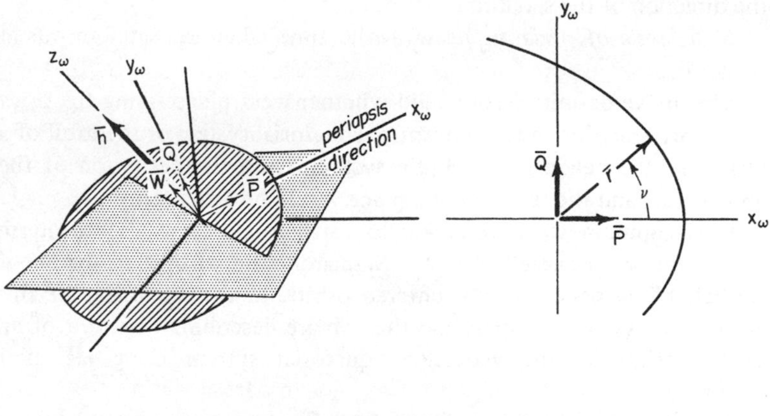

Figure 2.2-4 Perifocal coordinate system

In the next section we will define orbital elements in relation to the **IJK** vectors. When describing a heliocentric orbit, you may use the same definitions applied to the XYZ_ϵ axes. Another system, the topocentric-horizon coordinate system will be introduced in a later section.

2.3 CLASSICAL ORBITAL ELEMENTS

Five independent quantities called "orbital elements" are sufficient to completely describe the size, shape and orientation of an orbit. A sixth element is required to pinpoint the position of the satellite along the orbit at a particular time. The classical set of six orbital elements are defined with the help of Figure 2.3-1 as follows:

1. a, *semi-major axis*—a constant defining the size of the conic orbit.
2. e, *eccentricity*—a constant defining the shape of the conic orbit.
3. i, *inclination*—the angle between the **K** unit vector and the angular momentum vector, **h**.
4. Ω, *longitude of the ascending node*—the angle, in the fundamental plane, between the **I** unit vector and the point where the satellite crosses through the fundamental plane in a northerly direction (ascending node) measured counterclockwise when viewed from the north side of the fundamental plane.
5. ω, *argument of periapsis*—the angle, in the plane of the satellite's orbit, between the ascending node and the periapsis point, measured in the direction of the satellite's motion.
6. T, *time of periapsis passage*—the time when the satellite was at periapsis.

The above definitions are valid whether we are describing the orbit of an earth satellite in the geocentric-equatorial system or the orbit of a planet in the heliocentric-ecliptic system. Only the definition of the unit vectors and the fundamental plane would be different.

It is common when referring to earth satellites to use the term "argument of perigee" for ω. Similarly, the term "argument of perihelion" is used for sun-centered orbits. In the remainder of this chapter we shall tacitly assume that we are describing the orbit of an earth satellite in the geocentric-equatorial system using **IJK** unit vectors.

The list of six orbital elements defined above is by no means exhaustive. Frequently the semi-latus rectum, p, is substituted for a in the above list. Obviously, if you know a and e you can compute p.

Instead of argument of periapsis, the following is sometimes used:

Π, *longitude of periapsis*—the angle from **I** to periapsis measured

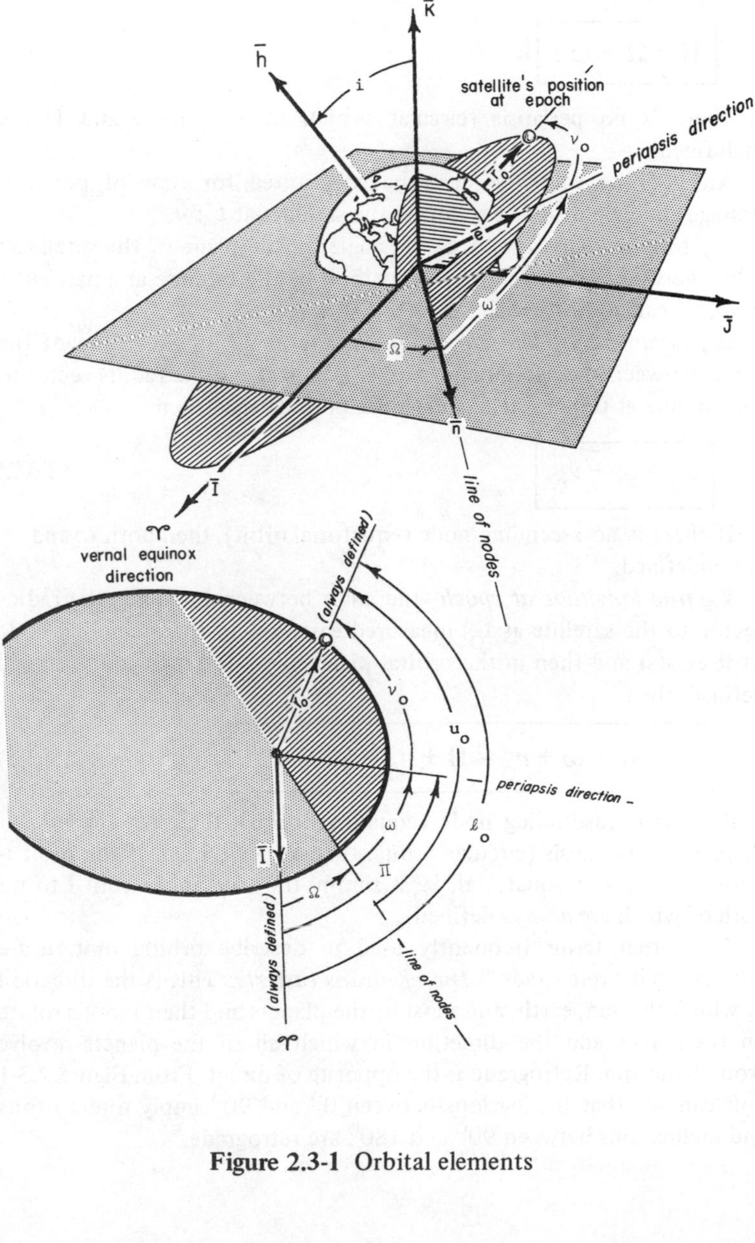

Figure 2.3-1 Orbital elements

eastward to the ascending node (if it exists) and then in the orbital plane to periapsis. If both Ω and ω are defined then

$$\boxed{\Pi = \Omega + \omega\,.} \tag{2.3-1}$$

If there is no periapsis (circular orbit), then both ω and Π are undefined.

Any of the following may be substituted for time of periapsis passage and would suffice to locate the satellite at t_o:

ν_o, *true anomaly at epoch*–the angle, in the plane of the satellite's orbit, between periapsis and the position of the satellite at a particular time, t_o, called the "epoch."

u_o, *argument of latitude at epoch*–the angle, in the plane of the orbit, between the ascending node (if it exists) and the radius vector to the satellite at time t_o. If ω and ν_o are both defined then

$$\boxed{u_o = \omega + \nu_o.} \tag{2.3-2}$$

If there is no ascending node (equatorial orbit), then both ω and u_o are undefined.

ℓ_o, *true longitude at epoch*–the angle between $\mathbf{I}$ and $\mathbf{r}_o$ (the radius vector to the satellite at t_o) measured eastward to the ascending node (if it exists) and then in the orbital plane to $\mathbf{r}_o$. If Ω, ω and ν_o are all defined, then

$$\boxed{\ell_o = \Omega + \omega + \nu_o = \Pi + \nu_o = \Omega + u_o\,.} \tag{2.3-3}$$

If there is no ascending node (equatorial orbit), then $\ell_o = \Pi + \nu_o$. If there is no periapsis (circular orbit), then $\ell_o = \Omega + u_o$. If the orbit is both circular and equatorial, ℓ_o is simply the true angle from $\mathbf{I}$ to $\mathbf{r}_o$, both of which are *always* defined.

Two other terms frequently used to describe orbital motion are "direct" and "retrograde." *Direct means easterly.* This is the direction in which the sun, earth, and most of the planets and their moons rotate on their axes and the direction in which all of the planets revolve around the sun. Retrograde is the opposite of direct. From Figure 2.3-1 you can see that inclinations between 0° and 90° imply direct orbits and inclinations between 90° and 180° are retrograde.

2.4 DETERMINING THE ORBITAL ELEMENTS FROM r AND v

Let us assume that a radar site on the earth is able to provide us with the vectors **r** and **v** representing the position and velocity of a satellite relative to the geocentric-equatorial reference frame at a particular time, t_O. How do we find the six orbital elements which describe the motion of the satellite? The first step is to form the three vectors, **h**, **n** and **e**.

2.4.1 Three Fundamental Vectors–h, n and e. We have already encountered the angular momentum vector, **h**.

$$\boxed{\mathbf{h} = \mathbf{r} \times \mathbf{v}} \qquad (2.4\text{-}1)$$

Thus

$$\mathbf{h} = \begin{vmatrix} \mathbf{I} & \mathbf{J} & \mathbf{K} \\ r_I & r_J & r_K \\ v_I & v_J & v_K \end{vmatrix} = h_I\mathbf{I} + h_J\mathbf{J} + h_K\mathbf{K}\,. \qquad (2.4\text{-}2)$$

An important thing to remember is that **h** is a vector *perpendicular to the plane of the orbit.*

The node vector, **n**, is defined as

$$\boxed{\mathbf{n} \equiv \mathbf{K} \times \mathbf{h}\,.} \qquad (2.4\text{-}3)$$

Thus

$$\mathbf{n} = \begin{vmatrix} \mathbf{I} & \mathbf{J} & \mathbf{K} \\ 0 & 0 & 1 \\ h_I & h_J & h_K \end{vmatrix} = n_I\mathbf{I} + n_J\mathbf{J} + n_K\mathbf{K} = -h_J\mathbf{I} + h_I\mathbf{J}\,. \qquad (2.4\text{-}4)$$

From the definition of a vector cross product, **n** must be perpendicular to both **K** and **h**. To be perpendicular to **K**, **n** would have to lie in the equatorial plane. To be perpendicular to **h**, **n** would have to lie in the orbital plane. Therefore, **n** must lie in both the equatorial and orbital planes, or in their intersection which is called the "line of nodes." Specifically, **n** *is a vector pointing along the line of nodes in the direction of the ascending node.* The magnitude of **n** is of no consequence to us. We are only interested in its direction.

The vector, **e**, is obtained from

$$\boxed{\mathbf{e} = \frac{1}{\mu}\left[\left(v^2 - \frac{\mu}{r}\right)\mathbf{r} - (\mathbf{r}\cdot\mathbf{v})\,\mathbf{v}\right]} \qquad (2.4\text{-}5)$$

and is derived in Chapter 1 (equation (1.5-11)). The vector **e** points *from the center of the earth (focus of the orbit) toward perigee with a magnitude exactly equal to the eccentricity of the orbit.*

All three vectors, **h**, **n** and **e** are illustrated in Figure 2.3-1. Study this figure carefully. An understanding of it is essential to what follows.

2.4.2 Solving for the Orbital Elements. Now that we have **h**, **n** and **e** we can proceed rather easily to obtain the orbital elements. The parameter, p, and eccentricity follow directly from **h** and **e** while all the remaining orbital elements are simply angles between two vectors whose components are now known. If we know how to find the angle between two vectors the problem is solved. In general, the cosine of the angle, α, between two vectors **A** and **B** is found by dotting the two vectors together and dividing by the product of their magnitudes. Since

$$\mathbf{A}\cdot\mathbf{B} = AB\cos\alpha$$

then

$$\cos\alpha = \frac{\mathbf{A}\cdot\mathbf{B}}{AB} \; . \qquad \text{(see Figure (2.4-1)} \qquad (2.4\text{-}6)$$

Of course, being able to evaluate the cosine of an angle does not mean that you know the angle. You still have to decide whether the angle is smaller or greater than 180°. The answer to this quadrant resolution problem must come from other information in the problem as we shall see.

We can outline the method of finding the orbital elements as follows:

1. $p = h^2/\mu$

2. $e = |\mathbf{e}|$

3. Since i is the angle between **K** and **h**,

$$\boxed{\cos i = \frac{h_K}{h}}. \tag{2.4-7}$$

(Inclination is always less than 180°.)

4. Since Ω is the angle between **I** and **n**,

$$\boxed{\cos \Omega = \frac{n_I}{n}}. \tag{2.4-8}$$

(If $n_J > 0$ then Ω is less than 180°.)

5. Since ω is the angle between **n** and **e**,

$$\boxed{\cos \omega = \frac{\mathbf{n} \cdot \mathbf{e}}{ne}}. \tag{2.4-9}$$

(If $e_K > 0$ then ω is less than 180°.)

6. Since ν_0 is the angle between **e** and **r**,

$$\boxed{\cos \nu_0 = \frac{\mathbf{e} \cdot \mathbf{r}}{er}} \tag{2.4-10}$$

(If $\mathbf{r} \cdot \mathbf{v} > 0$ then ν_0 is less than 180°.)

7. Since u_0 is the angle between **n** and **r**,

$$\boxed{\cos u_0 = \frac{\mathbf{n} \cdot \mathbf{r}}{nr}}. \tag{2.4-11}$$

(If $r_K > 0$ then u_0 is less than 180°.)

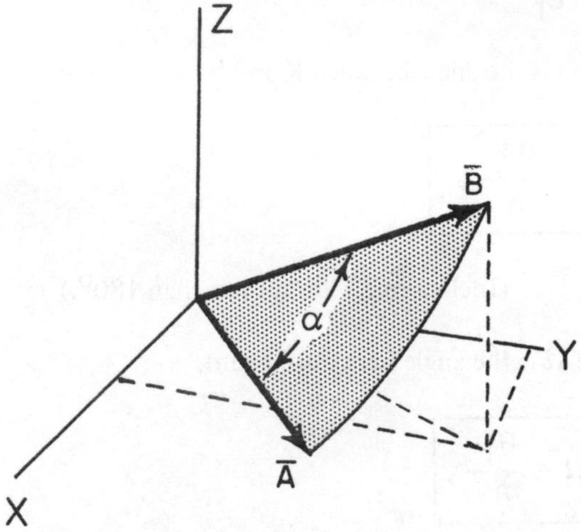

Figure 2.4-1 Angle between vectors

8. $\ell_0 = \Omega + \omega + \nu_0 = \Omega + u_0$

All of the quadrant checks in parentheses make physical sense. If they don't make sense to you, look at the geometry of Figure 2.3-1 and study it until they do. The quadrant check for ν_0 is nothing more than a method of determining whether the satellite is between periapsis and apoapsis (where flight-path angle is always positive) or between apoapsis and periapsis (where ϕ is always negative). With this hint see if you can fathom the logic of checking $\mathbf{r} \cdot \mathbf{v}$.

The three orbits illustrated in the following example problem should help you visualize what we have been talking about up to now.

Find the orbital elements of the three orbits in the following illustrations.

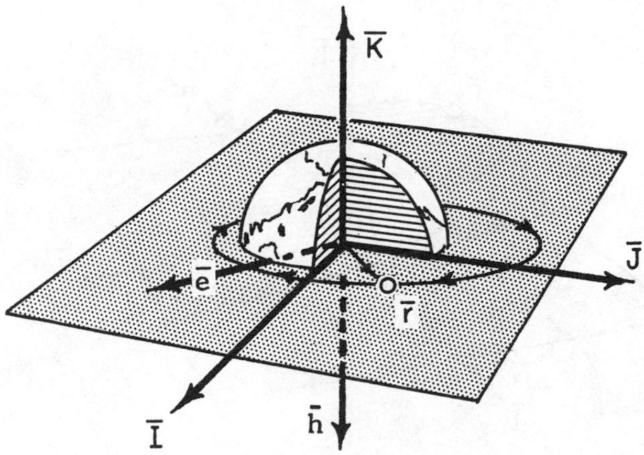

Figure 2.4-2 Orbit 1

ORBIT 1 (retrograde equatorial)

$p = 1.5$ DU	$\Pi = 45^0$
$e = .2$	$\nu_0 = 270^0$
$i = 180^0$	$u_0 =$ undefined
$\Omega =$ undefined	$\ell_0 = 315^0$

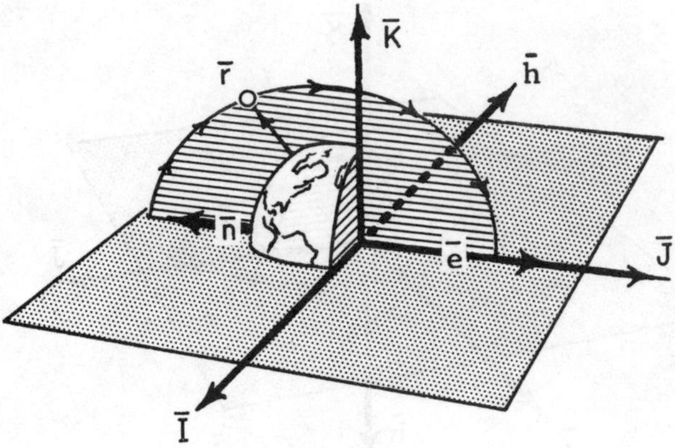

Figure 2.4-3 Orbit 2

ORBIT 2 (polar)

$p = 1.5$ DU	$\omega = 180^0$
$e = .2$	$\nu_0 = 225^0$
$i = 90^0$	$u_0 = 45^0$
$\Omega = 270^0$	$\ell_0 = 315^0$

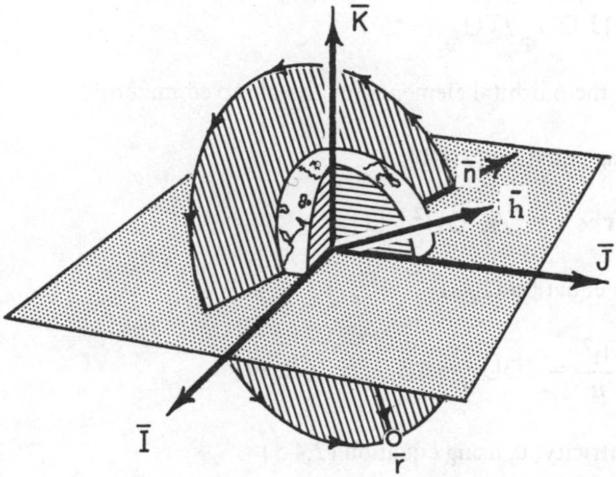

Figure 2.4-4 Orbit 3

ORBIT 3 (direct circular)

$p = 1.5$ DU	ω = undefined
$e = 0$	ν_o = undefined
$i = 60^0$	$u_o = 270^0$
$\Omega = 150^0$	$\ell_o = 420^0$

EXAMPLE PROBLEM. A radar tracks a meteoroid and from the tracking data the following inertial position and velocity vectors are found (expressed in the geocentric-equatorial coordinate system).

$$\mathbf{r} = 2\mathbf{I}\ \mathrm{DU}_\oplus$$

$$\mathbf{v} = 1\mathbf{J}\ \mathrm{DU}_\oplus/\mathrm{TU}_\oplus$$

Determine the 6 orbital elements for the observed meteroid.

Find p using equation (2.4-1)

$$\mathbf{h} = \mathbf{r} \times \mathbf{v} = 2\mathbf{K}\ \mathrm{DU}_\oplus^2/\mathrm{TU}_\oplus .$$

Then from equation (1.6-1)

$$p = \frac{h^2}{\mu} = 4\mathrm{DU}_\oplus = 13{,}775.74 \text{n.mi.}$$

Find eccentricity, e, using equation (2.4-5)

$$\mathbf{e} = \frac{1}{\mu}\left[\left(v^2 - \frac{\mu}{r}\right)\mathbf{r} - (\mathbf{r} \cdot \mathbf{v})\mathbf{v}\right] = 1\mathbf{I}$$

$$e = |\mathbf{e}| = 1$$

Since $h \neq 0$ and $e = 1$ the path of the meteoroid is parabolic with respect to the earth.

Find inclination, i. Using equation (2.4-7)

$$i = \cos^{-1}\left(\frac{\mathbf{h} \cdot \mathbf{k}}{h}\right) = 0^0$$

Therefore the meteoroid is traveling in the equatorial plane.

Find longitude of ascending node, Ω.

From equation (2.4-8)

$$\Omega = \cos^{-1}\left(\frac{\mathbf{n} \cdot \mathbf{I}}{n}\right)$$

However, since the meteoroid is located in the equatorial plane there is no ascending node because its trajectory does not cross the equatorial plane and therefore, for this case Ω, is undefined.

Find argument of periapsis, ω.

From equation (2.4-9)

$$\omega = \cos^{-1}\left(\frac{\mathbf{n} \cdot \mathbf{e}}{ne}\right)$$

Again, since there is no ascending node ω is also undefined. In lieu of the ω the longitude of periapsis, Π, can be determined in this case.

Find longitude of periapsis, Π.

Since the orbital plane is coincident with the equatorial plane ($i = 0^0$) Π is measured from the **I**-axis to periapsis which is colocated with the **e** vector. Therefore from the definition of the dot product

$$\Pi = \cos^{-1}\left(\frac{\mathbf{e} \cdot \mathbf{I}}{e}\right) = 0^0$$

it is determined that perigee is located along the **I** axis.

Find true anomally, ν_0.

From equation (2.4-10)

$$\nu_0 = \cos^{-1}\left(\frac{\mathbf{e} \cdot \mathbf{r}}{er}\right) = 0^0$$

and it is found that the meteoroid is presently at periapsis.

Find true longitude of epoch, ℓ_o.

From equation (2.3-3)

$$\ell_o = \Pi + \nu_o = 0^0$$

EXAMPLE PROBLEM. The following inertial position and velocity vectors are expressed in a geocentric equatorial coordinate system for an observed space object:

$$\mathbf{r} = \frac{3\sqrt{3}}{4}\mathbf{I} + \frac{3}{4}\mathbf{J} \text{ DU}_\oplus$$

$$\mathbf{v} = -\frac{1}{2\sqrt{2}}\mathbf{I} + \frac{\sqrt{3}}{2\sqrt{2}}\mathbf{J} + \frac{1}{\sqrt{2}}\mathbf{K} \text{ DU}_\oplus/\text{TU}_\oplus$$

Determine the 6 orbital elements for the observed object.

Find p.

First find $\mathbf{h}$ by equation (2.4-1)

$$\mathbf{h} = \mathbf{r} \times \mathbf{v} = \frac{1}{\sqrt{2}}\left(\frac{6}{8}\mathbf{I} - \frac{6\sqrt{3}}{8}\mathbf{J} + \frac{12}{8}\mathbf{K}\right) \text{DU}_\oplus^2/\text{TU}_\oplus$$

then from equation (1.6-1)

$$p = \frac{h^2}{\mu} = 2.25 \text{ DU}_\oplus = 7748.85 \text{ n.mi.}$$

Find e.

From equation (2.4-5)

$$\mathbf{e} = \frac{1}{\mu}\left[\left(v^2 - \frac{\mu}{r}\right)\mathbf{r} - (\mathbf{r} \cdot \mathbf{v})\,\mathbf{v}\right] = \frac{\sqrt{3}}{4}\mathbf{I} + \frac{1}{4}\mathbf{J}$$

and the magnitude of the $\mathbf{e}$ vector is:

$$e = .5$$

Find the inclination, i:

From equation (2.4-7)

$$i = \cos^{-1}\left(\frac{\mathbf{h}\cdot\mathbf{k}}{h}\right) = 45^0$$

Find the longitude of ascending node, Ω:

First we must determine the node vector **n**.

From equation (2.4-3)

$$\mathbf{n} = \mathbf{k}\times\mathbf{h} = \frac{3}{4\sqrt{2}}(\sqrt{3}\mathbf{I} + \mathbf{J})$$

From equation (2.4-8)

$$\Omega = \cos^{-1}\left(\frac{\mathbf{n}\cdot\mathbf{I}}{n}\right) = 30^0$$

Find the argument of periapsis, ω:

From equation (2.4-9)

$$\omega = \cos^{-1}\left(\frac{\mathbf{n}\cdot\mathbf{e}}{ne}\right) = 0^0$$

Find the true anomaly, ν_0:

From equation (2.4-10)

$$\nu_0 = \cos^{-1}\left(\frac{\mathbf{e}\cdot\mathbf{r}}{er}\right) = 0^0$$

2.5 DETERMINING r AND v FROM THE ORBITAL ELEMENTS

In the last section we saw how to determine a set of classical orbital elements from the vectors **r** and **v** at some epoch. Now we will look at the inverse problem of determining **r** and **v** when the six classical elements are given.

This is both an interesting and a practical exercise since it represents one way of solving a basic problem of astrodynamics–that of updating the position and velocity of a satellite to some future time. Suppose you know $\mathbf{r}_o$ and $\mathbf{v}_o$ at some time t_o. Using the techniques presented in the last section you could determine the elements p, e, i, Ω, ω and ν_o. Of these six elements, the first five are constant (if we accept the assumptions of the restricted two-body problem) and only the true anomaly, ν, changes with time. In Chapter 4 you will learn how to determine the change in true anomaly, $\Delta\nu$, that occurs in a given time, $t - t_o$. This will enable you to construct a new set of orbital elements and the only step remaining is to determine the new **r** and **v** from this "updated" set.

The method, shown below, consists of two steps; first we must express **r** and **v** in perifocal coordinates and then transform **r** and **v** to geocentric-equatorial components.

2.5.1 Expressing r and v in the Perifocal System. Let's assume that we know p, e, i, Ω, ω and ν. We can immediately write an expression for **r** in terms of the perifocal system (see Figure 2.2-4):

$$\boxed{\mathbf{r} = r\cos\nu\,\mathbf{P} + r\sin\nu\,\mathbf{Q}} \tag{2.5-1}$$

where the scalar magnitude r can be determined from the polar equation of a conic:

$$r = \frac{p}{1 + e\cos\nu} \tag{1.5-4}$$

To obtain **v** we only need to differentiate **r** in equation (2.5-1) keeping in mind that the perifocal coordinate frame is "inertial" and so $\dot{\mathbf{P}} = \dot{\mathbf{Q}} = 0$ and

$$\mathbf{v} = \dot{\mathbf{r}} = (\dot{r}\cos\nu - r\dot{\nu}\sin\nu)\mathbf{P} + (\dot{r}\sin\nu + r\dot{\nu}\cos\nu)\mathbf{Q}$$

This expression for **v** can be simplified by recognizing that $h = r^2\dot{\nu}$ (see Figure 1.7-2), $p = h^2/\mu$ and differentiating equation (1.5-4) above to obtain

$$\dot{r} = \sqrt{\frac{\mu}{p}}\,e\sin\nu \tag{2.5-2}$$

and

$$r\dot{\nu} = \sqrt{\frac{\mu}{p}}\,(1 + e\cos\nu). \tag{2.5-3}$$

Making these substitutions in the expression for **v** and simplifying yields

$$\boxed{\mathbf{v} = \sqrt{\frac{\mu}{p}}\,[-\sin\nu\,\mathbf{P} + (e + \cos\nu)\,\mathbf{Q}]} \tag{2.5-4}$$

EXAMPLE PROBLEM. A space object has the following orbital elements as determined by NORAD space track system:

$p = 2.25\ DU_{\oplus}$ $\quad\Omega = 30^0$

$e = .5$ $\quad\omega = 0^0$

$i = 45^0$ $\quad\nu_0 = 0^0$

Express the **r** and **v** vectors for the space object in the perifocal coordinate system.

Before equation (2.5-1) can be applied the magnitude of the r vector must be determined first by equation (1.5-4)

$$r = \frac{P}{1 + e\cos\nu} = 1.5\ DU_{\oplus}$$

from equation (2.4-2)

$$\mathbf{r} = r\cos\nu\mathbf{P} + r\sin\nu\mathbf{Q} = 1.5\mathbf{P}\ DU_{\oplus}$$

from equation (2.5-4)

$$\mathbf{v} = \sqrt{\frac{\mu}{P}}\,[-\sin\nu\mathbf{P} + (e + \cos\nu)\mathbf{Q}]$$

$$= 1\mathbf{Q}\ DU_{\oplus}/TU_{\oplus}$$

2.6 COORDINATE TRANSFORMATIONS

Before we discuss the transformation of **r** and **v** to the geocentric-equatorial frame we will review coordinate transformations in general.

A vector may be expressed in any coordinate frame. It is common in astrodynamics to use rectangular coordinates although occasionally spherical polar coordinates are more convenient. A rectangular coordinate frame is usually defined by specifying its origin, its fundamental (x-y) plane, the direction of the positive z-axis, and the principal (x) direction in the fundamental plane. Three unit vectors are then defined to indicate the directions of the three mutually perpendicular axes. Any other vector can be expressed as a linear combination of these three unit vectors. This collection of unit vectors is commonly referred to as a "basis."

2.6.1 What a Coordinate Transformation Does. A coordinate transformation merely changes the basis of a vector–nothing else. The vector still has the same length and direction after the coordinate transformation, and it still represents the same thing. For example, suppose you know the south, east, and zenith components of the "velocity of a satellite relative to the topocentric-horizon frame." The phrase in quotes describes what the vector represents. The "basis" of the vector is obviously the set of unit vectors pointing south, east and up.

Now suppose you want to express this vector in terms of a different basis, say the IJK unit vectors of the geocentric-equatorial frame (perhaps because you want to add another vector to it and this other vector is expressed in IJK components). A simple coordinate transformation will do the trick. The vector will still have the same magnitude and direction and will represent the same thing, namely, the "velocity of a satellite relative to the topocentric-horizon frame" even though you now express it in terms of geocentric-equatorial coordinates. In other words, changing the basis of a vector does not change its magnitude, direction, or what it represents.

A vector has only two properties that can be expressed mathematically–magnitude and direction. Certain vectors, such as position vectors, have a definite starting point, but this point of origin cannot be expressed mathematically and does not change in a coordinate transformation. For example, suppose you know the south, east, and zenith components of "the vector from a radar

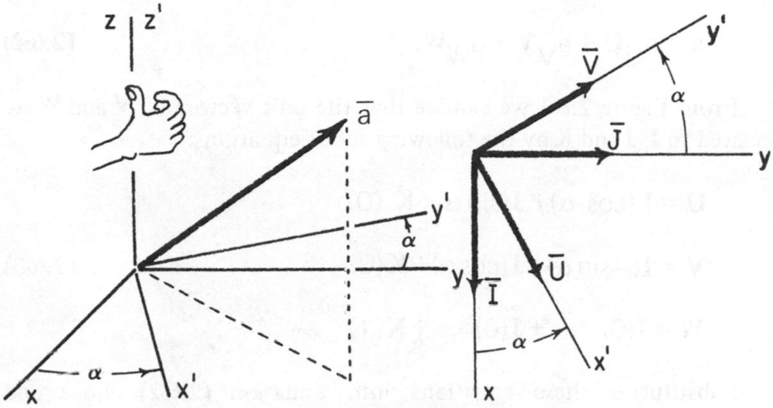

Figure 2.6-1 Rotation about z axis

site on the surface of the earth to a satellite." A simple change of basis will enable you to express this vector in terms of IJK components:

$$\boldsymbol{\rho} = \rho_I \mathbf{I} + \rho_J \mathbf{J} + \rho_K \mathbf{K}.$$

The transformation did not change what the vector represents so it is still the vector "*from* the radar site *to* the satellite." In other words, expressing a vector in coordinates of a particular frame does not imply that the vector has its tail at the origin of that frame.

2.6.2 Change of Basis Using Matrices. Changing from one basis to another can be streamlined by using matrix methods. Suppose we have two coordinate frames xyz and x′ y′ z′ related by a simple rotation through a positive angle α about the z axis. Figure 2.6-1 illustrates the two frames. We will define a positive rotation about any axis by means of the "right-hand rule"–if the thumb of the right hand is extended in the direction of the positive coordinate axis, the fingers curl in the sense of a positive rotation.

Let us imagine three unit vectors **I, J** and **K** extending along the x, y and z axes respectively and another set of unit vectors **U, V** and **W** along the x′, y′ and z′ axes. Now suppose we have a vector **a** which may be expressed in terms of the IJK basis as

$$\mathbf{a} = a_I \mathbf{I} + a_J \mathbf{J} + a_K \mathbf{K} \tag{2.6-1}$$

or in terms of the UVW basis as

$$\mathbf{a} = a_U\mathbf{U} + a_V\mathbf{V} + a_W\mathbf{W}. \tag{2.6-2}$$

From Figure 2.6-1 we can see that the unit vectors **U**, **V** and **W** are related to **I**, **J** and **K** by the following set of equations:

$$\begin{aligned}
\mathbf{U} &= \mathbf{I}(\cos\alpha) + \mathbf{J}(\sin\alpha) + \mathbf{K}(0) \\
\mathbf{V} &= \mathbf{I}(-\sin\alpha) + \mathbf{J}(\cos\alpha) + \mathbf{K}(0) \\
\mathbf{W} &= \mathbf{I}(0) \quad + \mathbf{J}(0) \quad + \mathbf{K}(1)
\end{aligned} \tag{2.6-3}$$

Substituting these equations into equation (2.6-2) above and equating it with (2.6-1) yields

$$\begin{aligned}
a_U &= a_I(\cos\alpha) + a_J(\sin\alpha) + a_K(0) \\
a_V &= a_I(-\sin\alpha) + a_J(\cos\alpha) + a_K(0) \\
a_W &= a_I(0) \quad + a_J(0) \quad + a_K(1).
\end{aligned} \tag{2.6-4}$$

We can express this last set of equations very compactly if we use matrix notation and think of the vector **a** as a triplet of numbers representing a column matrix. We will use subscripts to identify the basis, thus

$$\mathbf{a}_{IJK} = \begin{bmatrix} a_I \\ a_J \\ a_K \end{bmatrix} \text{ and } \mathbf{a}_{UVW} = \begin{bmatrix} a_U \\ a_V \\ a_W \end{bmatrix}.$$

The coefficients of a_I, a_J and a_K in equations (2.6-4) should be taken as the elements of a three by three "transformation matrix" which we can call $\widetilde{\mathbf{A}}$

$$\widetilde{\mathbf{A}} = \begin{bmatrix} \cos\alpha & \sin\alpha & 0 \\ -\sin\alpha & \cos\alpha & 0 \\ 0 & 0 & 1 \end{bmatrix}. \tag{2.6-5}$$

The set of equations (2.6-4) can then be represented as

$$\mathbf{a}_{UVW} = \widetilde{\mathbf{A}}\, \mathbf{a}_{IJK} \tag{2.6-6}$$

which is really matrix shorthand for

$$\begin{bmatrix} a_U \\ a_V \\ a_W \end{bmatrix} = \begin{bmatrix} \cos\alpha & \sin\alpha & 0 \\ -\sin\alpha & \cos\alpha & 0 \\ 0 & 0 & 1 \end{bmatrix} \begin{bmatrix} a_I \\ a_J \\ a_K \end{bmatrix} . \tag{2.6-7}$$

Applying the rules of matrix multiplication to equation (2.6-7) yields the set of equations (2.6-4) above.

2.6.3 Summary of Transformation Matrices for Single Rotation of Coordinate Frame. Arguments similar to those above may be used to derive transformation matrices that will represent rotations of the coordinate frame about the x or y axes. These are summarized below.

Rotation about the x-axis. The transformation matrix $\widetilde{\mathbf{A}}$ corresponding to a single rotation of the coordinate frame about the positive x axis through a positive angle α is

$$\widetilde{\mathbf{A}} = \begin{bmatrix} 1 & 0 & 0 \\ 0 & \cos\alpha & \sin\alpha \\ 0 & -\sin\alpha & \cos\alpha \end{bmatrix} . \tag{2.6-8}$$

Rotation about the y axis. The transformation matrix $\widetilde{\mathbf{B}}$ corresponding to a single rotation of the coordinate frame about the positive y axis through a positive angle β is

$$\widetilde{\mathbf{B}} = \begin{bmatrix} \cos\beta & 0 & -\sin\beta \\ 0 & 1 & 0 \\ \sin\beta & 0 & \cos\beta \end{bmatrix} . \tag{2.6-9}$$

Rotation about the z axis. The transformation matrix $\widetilde{\mathbf{C}}$ corresponding to a single rotation of the coordinate frame about the positive z axis through a positive angle γ is

$$\widetilde{C} = \begin{bmatrix} \cos\gamma & \sin\gamma & 0 \\ -\sin\gamma & \cos\gamma & 0 \\ 0 & 0 & 1 \end{bmatrix} \tag{2.6-10}$$

2.6.4 Successive Rotations About Several Axes. So far we have learned how to use matrices to perform a simple change of basis where the new set of unit vectors is related to the old by a simple rotation about one of the coordinate axes. Let us now look at a more complicated transformation involving more than one rotation.

Suppose we know the IJK components of some general vector, **a**, in the geocentric-equatorial frame and we wish to find its SEZ components in the topocentric-horizon frame (see section 2.7.1).

Figure 2.8-4 shows the angular relationship between two frames. Starting with the IJK frame we can first rotate it through a positive angle θ about the Z (**K**) axis and then rotate it through a positive angle $(90^{\circ} - L)$ about the Y axis to bring it into angular alignment with the SEZ frame.

The three components of **a** after the first rotation can be found from

$$\begin{bmatrix} \cos\theta & \sin\theta & 0 \\ -\sin\theta & \cos\theta & 0 \\ 0 & 0 & 1 \end{bmatrix} \begin{bmatrix} a_I \\ a_J \\ a_K \end{bmatrix}$$

The above expression is actually a column matrix and represents the three components of **a** in the intermediate frame. We can now multiply this column matrix by the appropriate matrix corresponding to the second rotation and obtain

$$\begin{bmatrix} a_S \\ a_E \\ a_Z \end{bmatrix} = \begin{bmatrix} \sin L & 0 & -\cos L \\ 0 & 1 & 0 \\ \cos L & 0 & \sin L \end{bmatrix} \begin{bmatrix} \cos\theta & \sin\theta & 0 \\ -\sin\theta & \cos\theta & 0 \\ 0 & 0 & 1 \end{bmatrix} \begin{bmatrix} a_I \\ a_J \\ a_K \end{bmatrix}$$

Since matrix multiplication is associative we can multiply the two simple rotation matrices together to form a single transformation matrix and write

$$\begin{bmatrix} a_S \\ a_E \\ a_Z \end{bmatrix} = \begin{bmatrix} \sin L \cos\theta & \sin L \sin\theta & -\cos L \\ -\sin\theta & \cos\theta & 0 \\ \cos L \cos\theta & \cos L \sin\theta & \sin L \end{bmatrix} \begin{bmatrix} a_I \\ a_J \\ a_K \end{bmatrix} \tag{2.6-11}$$

Keep in mind that the order in which you multiply the two rotation matrices is important since matrix multiplication is not commutative, (i.e., $\widetilde{A}\widetilde{B} \neq \widetilde{B}\widetilde{A}$). Since matrix multiplication can represent rotation of an axis system we can infer from this that the order in which rotations are performed is not irrelevant. For example if you are in an airplane and you rotate in pitch 45° nose-up and then roll 90° right you will be in a different attitude than if you first roll 90° right and then pitch up 45°.

Equation (2.6-11) represents the transformation from geocentric-equatorial to topocentric coordinates. We can write equation (2.6-11) more compactly as

$$\mathbf{a}_{SEZ} = \widetilde{\mathbf{D}}\mathbf{a}_{IJK}$$

where $\widetilde{\mathbf{D}}$ is the overall transformation matrix.

Now, if we want to perform the inverse transformation (from topocentric to geocentric) we need to find the inverse of matrix $\widetilde{\mathbf{D}}$ which we will call $\widetilde{\mathbf{D}}^{-1}$.

$$\mathbf{a}_{IJK} = \widetilde{\mathbf{D}}^{-1}\, \mathbf{a}_{SEZ}$$

In general the inverse of a matrix is difficult to calculate. Fortunately, all transformation matrices between rectangular frames have the unique property that they are orthogonal.

A three-by-three matrix is called "orthogonal" if the rows and the columns are scalar components of mutually perpendicular unit vectors. The inverse of any orthogonal matrix is equal to its transpose. The transpose of the matrix $\widetilde{\mathbf{D}}$, denoted by $\widetilde{\mathbf{D}}^{T}$, is the new matrix whose rows are the old columns and whose columns are the old rows (in the same order). Hence

$$\begin{bmatrix} a_I \\ a_J \\ a_K \end{bmatrix} = \begin{bmatrix} \sin L \cos\theta & -\sin\theta & \cos L \cos\theta \\ \sin L \sin\theta & \cos\theta & \cos L \sin\theta \\ -\cos L & 0 & \sin L \end{bmatrix} \begin{bmatrix} a_S \\ a_E \\ a_Z \end{bmatrix} \tag{2.6-12}$$

The angles through which one frame must be rotated to bring its axes into coincidence with another frame are commonly referred to as "Euler angles." A maximum of three Euler angle rotations is sufficient to bring any two frames into coincidence.

By memorizing the three basic rotation matrices and the rules for matrix multiplication you will be able to perform any desired change of basis no matter how complicated.

2.6.5 Transformation from the Perifocal to the Geocentric-Equatorial Frame. The perifocal coordinate system is related geometrically to the IJK frame through the angles Ω, i and ω as shown in Figure 2.6-2.

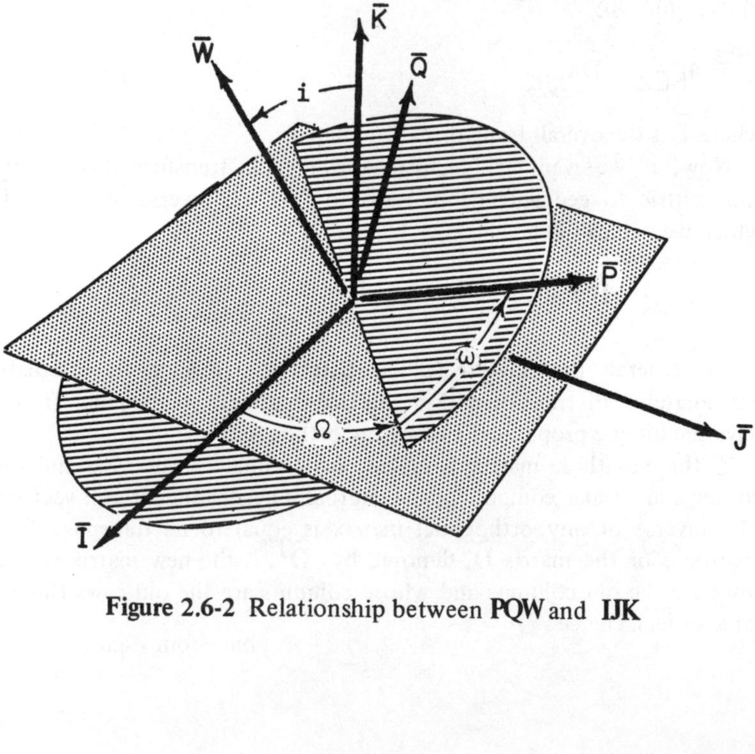

Figure 2.6-2 Relationship between **PQW** and **IJK**

The transformation of coordinates between the **P Q W** and **IJK** systems can be accomplished by means of the rotation matrix, $\widetilde{\mathbf{R}}$.

Thus if a_I, a_J, a_K and a_P, a_Q, a_W are the components of a vector **a** in each of the two systems, then

$$\begin{bmatrix} a_I \\ a_J \\ a_K \end{bmatrix} = \widetilde{R} \begin{bmatrix} a_P \\ a_Q \\ a_W \end{bmatrix}.$$

Since we will often know the elements of the orbit, it may be convenient to use those angles in a transformation to the IJK frame. To

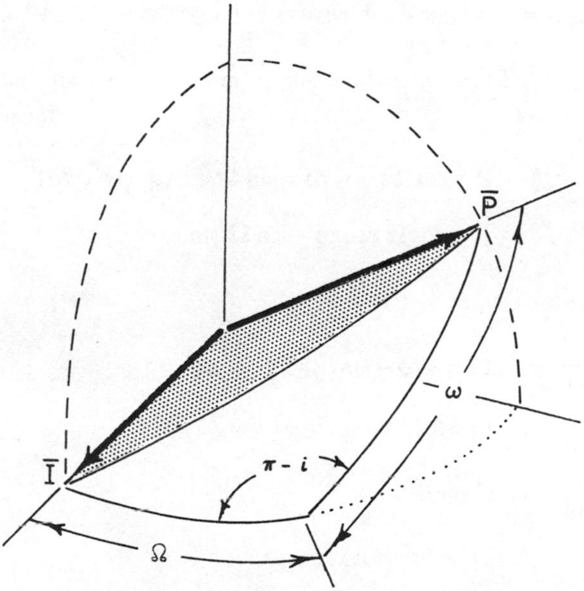

Figure 2.6-3 Angle between **I** and **P**

do this we can use direction cosines which can be found using the cosine law of spherical trigonometry. For example, from Figure 2.6-3, we see that the cosine of the angle between **I** and **P** can be calculated from their dot product, $\mathbf{I} \cdot \mathbf{P}$, since **I** and **P** are unit vectors. Recall from vector theory that the dot product simply gives the projection of one vector upon another. **P** can then be projected into the IJK frame by simply taking its dot product with **I**, **J** and **K**. Thus

$$\widetilde{R} = \begin{bmatrix} \mathbf{I}\cdot\mathbf{P} & \mathbf{I}\cdot\mathbf{Q} & \mathbf{I}\cdot\mathbf{W} \\ \mathbf{J}\cdot\mathbf{P} & \mathbf{J}\cdot\mathbf{Q} & \mathbf{J}\cdot\mathbf{W} \\ \mathbf{K}\cdot\mathbf{P} & \mathbf{K}\cdot\mathbf{Q} & \mathbf{K}\cdot\mathbf{W} \end{bmatrix} = \begin{bmatrix} R_{11} & R_{12} & R_{13} \\ R_{21} & R_{22} & R_{23} \\ R_{31} & R_{32} & R_{33} \end{bmatrix} \tag{2.6-13}$$

In Figure 2.6-3, the angle between **I** and **P** forms one side of a spherical triangle whose other two sides are Ω and ω. The included angle is π - i. The law of cosines for spherical triangles is

$$\cos a = \cos b \cos c + \sin b \sin c \cos A$$

where A, B and C are the three angles and a, b and c are the opposite sides. Now

$$R_{11} = \mathbf{I} \cdot \mathbf{P} = \cos \Omega \cos \omega + \sin \Omega \sin \omega \cos (\pi\text{-i})$$

$$= \cos \Omega \cos \omega - \sin \Omega \sin \omega \cos i\,.$$

Similarly, the elements of $\widetilde{\mathbf{R}}$ are

$$R_{11} = \cos \Omega \cos \omega - \sin \Omega \sin \omega \cos i$$

$$R_{12} = -\cos \Omega \sin \omega - \sin \Omega \cos \omega \cos i$$

$$R_{13} = \sin \Omega \sin i$$

$$R_{21} = \sin \Omega \cos \omega + \cos \Omega \sin \omega \cos i$$

$$R_{22} = -\sin \Omega \sin \omega + \cos \Omega \cos \omega \cos i \tag{2.6-14}$$

$$R_{23} = -\cos \Omega \sin i$$

$$R_{31} = \sin \omega \sin i$$

$R_{32} = \cos\omega \sin i$

$R_{33} = \cos i$

Having determined the elements of the rotation matrix, it only remains to find **r** and **v** in terms of IJK components. Thus

$$\begin{bmatrix} r_I \\ r_J \\ r_K \end{bmatrix} = \tilde{\mathbf{R}} \begin{bmatrix} r_P \\ r_Q \\ r_W \end{bmatrix} \quad \text{and} \quad \begin{bmatrix} v_I \\ v_J \\ v_K \end{bmatrix} = \tilde{\mathbf{R}} \begin{bmatrix} v_P \\ v_Q \\ v_W \end{bmatrix}.$$

This method is not recommended unless other transformations are not practical. Often it is possible to go to the IJK frame using equations (2.5-1) and (2.5-4) with **P** and **Q** known in IJK coordinates.

Special precautions must be taken when the orbit is equatorial or circular or both. In this case either Ω or ω or both are undefined. In the case of the circular orbit ν is also undefined so it is necessary to measure true anomaly from some arbitrary reference such as the ascending node or (if the orbit is also equatorial) from the unit vector **I**. Because of these and other difficulties the method of updating **r** and **v** via the classical orbital elements leaves much to be desired. Other more general methods of updating **r** and **v** that do not suffer from these defects will be presented in Chapter 4.

2.7 ORBIT DETERMINATION FROM A SINGLE RADAR OBSERVATION

A radar installation located on the surface of the earth can measure the position and velocity of a satellite *relative to the radar site*. But the radar site is not located at the center of the earth so the position vector measured is not the **r** we need. Also the earth is rotating so the velocity of the satellite relative to the radar site is not the same as the velocity, **v**, relative to the center of the IJK frame which we need to compute the orbital elements.

Before showing you how we can obtain **r** and **v** *relative to the center of the earth* from radar tracking data we must digress long enough to describe the coordinate system in which the radar site makes and expresses its measurements.

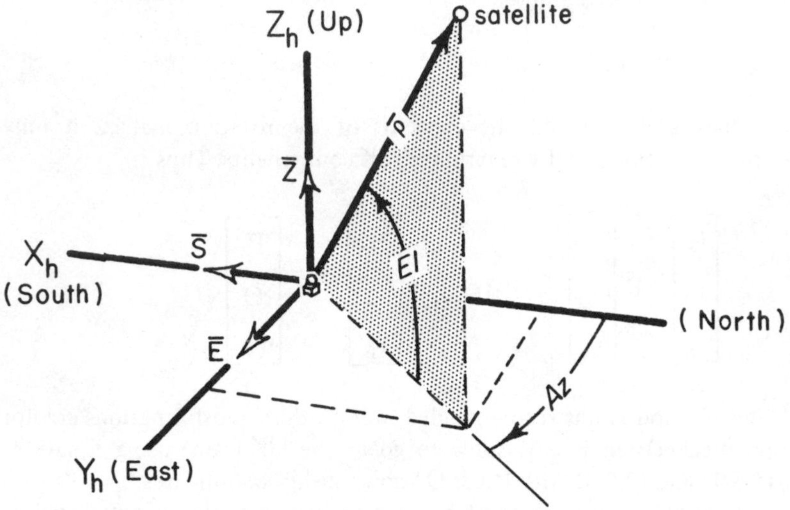

Figure 2.7-1 Topocentric-horizon coordinate system

2.7.1 The Topocentric-Horizon Coordinate System. The origin of the topocentric-horizon system is the point on the surface of the earth (called the "topos") where the radar is located. The fundamental plane is the horizon and the X_h-axis points south. The Y_h-axis is east and the Z_h-axis is up. It seems hardly necessary to point out that the topocentric-horizon system is not an inertial reference frame since it rotates with the earth. The unit vectors **S**, **E** and **Z** shown in Figure 2.7-1 will aid us in expressing vectors in this system.

2.7.2 Expressing Position and Velocity Relative to the Topocentric-Horizon Reference Frame. The radar site measures the range and direction to the satellite. The range is simply the magnitude of the vector $\boldsymbol{\rho}$ (rho) shown in Figure 2.7-1. The direction to the satellite is determined by two angles which can be picked off the gimbal axes on which the radar antenna is mounted. The azimuth angle, Az, is measured clockwise from north; the elevation angle, El, (not to be confused with L!) is measured from the horizontal to the radar line-of-sight. If the radar is capable of detecting a shift in frequency in the returning echo (Doppler effect), the rate at which range is changing, $\dot{\rho}$, can also be measured. The sensors on the gimbal axes are capable of measuring the rate-of-change of the azimuth and elevation angles, $\dot{Az}$

and $\dot{\text{El}}$, as the radar antenna follows the satellite across the sky. Thus, we have the raw material for expressing the position and velocity of the satellite relative to the radar in the six measurements: ρ, Az, El, $\dot{\rho}$, $\dot{\text{Az}}$, $\dot{\text{El}}$.

We will express the position vector as

$$\boldsymbol{\rho} = \rho_S \mathbf{S} + \rho_E \mathbf{E} + \rho_Z \mathbf{Z} \tag{2.7-1}$$

where, from the geometry of Figure 2.7-1,

$$\boxed{\begin{aligned} \rho_S &= -\rho \cos \text{El} \cos \text{Az} \\ \rho_E &= \rho \cos \text{El} \sin \text{Az} \\ \rho_Z &= \rho \sin \text{El}. \end{aligned}} \tag{2.7-2}$$

The velocity relative to the radar site is just $\dot{\bar{\rho}}$

$$\dot{\boldsymbol{\rho}} = \dot{\rho}_S \mathbf{S} + \dot{\rho}_E \mathbf{E} + \dot{\rho}_Z \mathbf{Z}. \tag{2.7-3}$$

Differentiating equations (2.7-2) we get the three components of $\dot{\boldsymbol{\rho}}$:

$$\boxed{\begin{aligned} \dot{\rho}_S &= -\dot{\rho} \cos \text{El} \cos \text{Az} + \rho \sin \text{El}(\dot{\text{El}}) \cos \text{Az} + \\ &\quad \rho \cos \text{El} \sin \text{Az}(\dot{\text{Az}}) \\ \dot{\rho}_E &= \dot{\rho} \cos \text{El} \sin \text{Az} - \rho \sin \text{El}(\dot{\text{El}}) \sin \text{Az} + \\ &\quad \rho \cos \text{El} \cos \text{Az}(\dot{\text{Az}}) \\ \dot{\rho}_Z &= \dot{\rho} \sin \text{El} + \rho \cos \text{El}(\dot{\text{El}}). \end{aligned}} \tag{2.7-4}$$

2.7.3 Position and Velocity Relative to the Geocentric Frame. There is an extremely simple relationship between the topocentric position vector $\boldsymbol{\rho}$ and the geocentric position vector $\mathbf{r}$. From Figure 2.7-2 we see that

$$\boxed{\mathbf{r} = \mathbf{R} + \boldsymbol{\rho}} \tag{2.7-5}$$

where $\mathbf{R}$ is the vector from the center of the earth to the origin of the

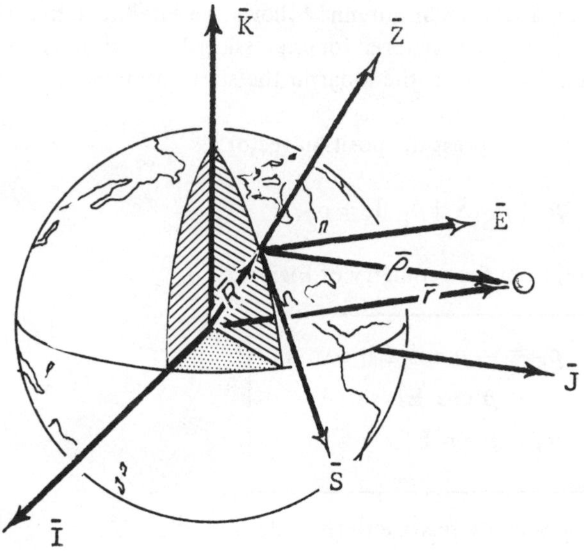

Figure 2.7-2 $\mathbf{r} = \mathbf{R} + \boldsymbol{\rho}$

topocentric frame. If the earth were perfectly spherical then the Z_h axis which defines the local vertical at the radar site would pass through the center of the earth if extended downward and the vector **R** would be

$$\mathbf{R} = r_{\oplus}\mathbf{Z} \tag{2.7-6}$$

where $r_{\oplus}$ is the radius of the earth.

Unfortunately, things are not that simple and to avoid errors of several miles in the position of the radar site relative to the geocenter it is necessary to use a more accurate model for the shape of the earth. A complete discussion of determining **R** on an oblate earth is included later in this chapter. For the moment we will assume that equation (2.7-6) is valid and proceed to the determination of **v**.

The general method of determining velocity relative to a "fixed" frame (hereafter referred to as the "true" velocity) when you are given the velocity relative to a moving frame may be stated in words as follows:

$$\begin{pmatrix}\text{Vel of object}\\ \text{rel to}\\ \text{fixed frame}\end{pmatrix} = \begin{pmatrix}\text{Vel of object}\\ \text{rel to}\\ \text{moving frame}\end{pmatrix} + \begin{pmatrix}\text{True vel of pt in}\\ \text{moving frame where}\\ \text{object is located}\end{pmatrix}$$

The sketches shown in Figure 2.7-3 illustrate this general principle for both a translating and rotating frame.

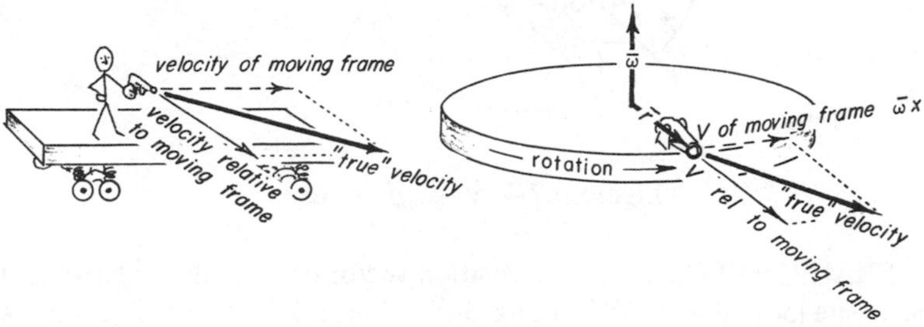

Figure 2.7-3 Relationship between relative and true velocity for translating and rotating reference frames

If you visualize the three-dimensional volume of space (Figure 2.7-4) defined by the topocentric-horizon reference frame, you will see that every point in this frame moves with a different velocity relative to the center of the earth. If $\mathbf{r}$ is the position vector from the center of the earth to the satellite, the velocity of that point in the topocentric frame where the satellite is located is simply $\boldsymbol{\omega}_\oplus \times \mathbf{r}$, where $\boldsymbol{\omega}_\oplus$ is the angular velocity of the earth (hence, the angular velocity of the **SEZ** frame). It is this velocity that must be added vectorially to $\dot{\boldsymbol{\rho}}$ to obtain the "true" velocity $\mathbf{v}$. Therefore,

$$\boxed{\mathbf{v} = \dot{\boldsymbol{\rho}} + \boldsymbol{\omega}_\oplus \times \mathbf{r}.} \tag{2.7-7}$$

You may recognize this expression as a simple application of the Coriolis theorem which will be derived in the next section as the general problem of derivatives in moving coordinate systems is discussed.

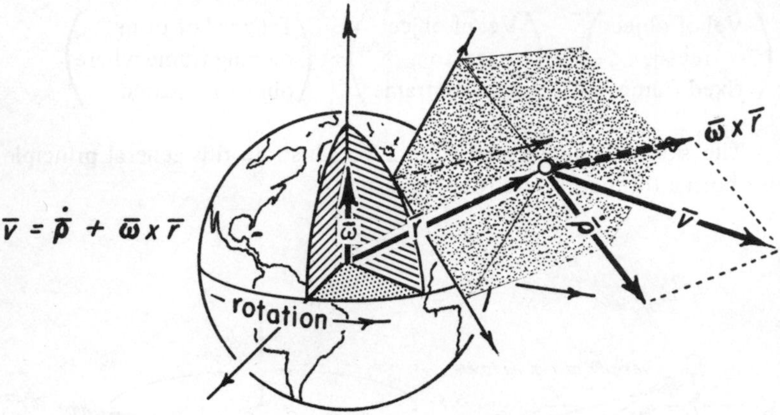

Figure 2.7-4 $\mathbf{v} = \dot{\boldsymbol{\rho}} + \boldsymbol{\omega} \times \mathbf{r}$

EXAMPLE PROBLEM. The position vector of a satellite relative to a radar site located at 169° W Long, 30° N Lat, is $2\mathbf{S} - \mathbf{E} + .5\mathbf{Z}$ (units are DU). The angle to Greenwich is 304°. Find the position vector of the satellite relative to fixed geocentric IJK coordinates. Assume the site is at sea level on a spherical earth.

$$\mathbf{r} = \mathbf{R} + \boldsymbol{\rho} = 2\mathbf{S} - \mathbf{E} + 1.5\mathbf{Z}$$

Convert $\mathbf{r}$ to IJK coordinates using (2.6-12)

$$L = 30^\circ \qquad \text{(Given)}$$

$$\theta = \theta_g + \lambda_E = 304^\circ + (-169^\circ) = 135^\circ$$

$$\widetilde{\mathbf{D}}^{1} = \begin{bmatrix} -.3535 & -.707 & -.612 \\ .3535 & -.707 & .612 \\ -.866 & 0 & .5 \end{bmatrix}$$

$$\begin{bmatrix} r_I \\ r_J \\ r_K \end{bmatrix} = \widetilde{\mathbf{D}}^{-1} \begin{bmatrix} r_S \\ r_E \\ r_Z \end{bmatrix} = \widetilde{\mathbf{D}}^{-1} \begin{bmatrix} 2 \\ -1 \\ 1.5 \end{bmatrix} = \begin{bmatrix} -.918 \\ 2.332 \\ -.982 \end{bmatrix}$$

$$\underline{\underline{\mathbf{r} = -.918\mathbf{I} + 2.332\mathbf{J} - .982\mathbf{K}}}$$

2.7.4 Derivatives in a Moving Reference Frame. Let any vector **A** be defined as a function of time in a "fixed" coordinate frame (x, y, z). Also let there be another coordinate system (u, v, w), rotating *with respect to* the (x, y, z) system (Figure 2.7-5). At this point you should note that this analysis applies to the *relative* motion between *any* two coordinate systems. It is not necessary that one system be fixed in inertial space!

The vector **A** may be *expressed* in either of these coordinate systems as

$$\underbrace{\mathbf{A} = A_I\mathbf{I} + A_J\mathbf{J} + A_K\mathbf{K}}_{\text{FIXED}} = \underbrace{A_U\mathbf{U} + A_V\mathbf{V} + A_W\mathbf{W}}_{\text{ROTATING}}. \tag{2.7-8}$$

Differentiating equation (2.7-8) with respect to time leads to:

$$\begin{aligned} \frac{d\mathbf{A}}{dt} &= \dot{A}_I\mathbf{I} + \dot{A}_J\mathbf{J} + \dot{A}_K\mathbf{K} + A_I\dot{\mathbf{I}} + A_J\dot{\mathbf{J}} + A_K\dot{\mathbf{K}} \\ &= \dot{A}_U\mathbf{U} + \dot{A}_V\mathbf{V} + \dot{A}_W\mathbf{W} + A_U\dot{\mathbf{U}} + A_V\dot{\mathbf{V}} + A_W\dot{\mathbf{W}}. \end{aligned} \tag{2.7-9}$$

Now the unit vectors **I, J, K** may be moving with time if the "fixed" system is not inertial, however, if we specify the time derivative *with respect to the "fixed" coordinate system*, then

$$\dot{\mathbf{I}} = \dot{\mathbf{J}} = \dot{\mathbf{K}} \equiv 0.$$

The unit vectors **U, V** and **W** will move relative to the fixed system so that in general their derivatives are not zero.

Assume that we know the angular velocity of the moving reference frame relative to the fixed frame. We will denote this quantity by ω. Then using the results of the unit vector analysis we know that the time derivative of a unit vector is perpendicular to the unit vector and has magnitude equal to $\boldsymbol{\omega}$. You should prove to yourself that

$$\dot{\mathbf{U}} = \boldsymbol{\omega} \times \mathbf{U},\ \dot{\mathbf{V}} = \boldsymbol{\omega} \times \mathbf{V} \text{ and } \dot{\mathbf{W}} = \boldsymbol{\omega} \times \mathbf{W}.$$

Hence, equation (2.7-9) reduces to

$$\begin{aligned} \frac{d\mathbf{A}}{dt} &= \dot{A}_I\mathbf{I} + \dot{A}_J\mathbf{J} + \dot{A}_K\mathbf{K} \\ &= \dot{A}_U\mathbf{U} + \dot{A}_V\mathbf{V} + \dot{A}_W\mathbf{W} \\ &\quad + A_U(\boldsymbol{\omega}\times\mathbf{U}) + A_V(\boldsymbol{\omega}\times\mathbf{V}) + A_W(\boldsymbol{\omega}\times\mathbf{W}). \end{aligned} \tag{2.7-10}$$

But

$$\dot{A}_U\mathbf{U} + \dot{A}_V\mathbf{V} + \dot{A}_W\mathbf{W} = \left.\frac{d\mathbf{A}}{dt}\right|_R$$

where $\left.\frac{d\mathbf{A}}{dt}\right|_R$ means the time derivative of $\mathbf{A}$ with respect to the rotating reference frame. Similarly,

$$\dot{A}_I\mathbf{I} + \dot{A}_J\mathbf{J} + \dot{A}_K\mathbf{K} = \left.\frac{d\mathbf{A}}{dt}\right|_F$$

where $\left.\frac{d\mathbf{A}}{dt}\right|_F$ means the time derivative of $\mathbf{A}$ with respect to the fixed reference frame. The remaining terms of equation (2.7-10) may be rewritten as

$$\boldsymbol{\omega}\times(A_U\mathbf{U}) + \boldsymbol{\omega}\times(A_V\mathbf{V}) + \boldsymbol{\omega}\times(A_W\mathbf{W}) = \boldsymbol{\omega}\times\mathbf{A}.$$

Hence, equation (2.7-10) reduces further to

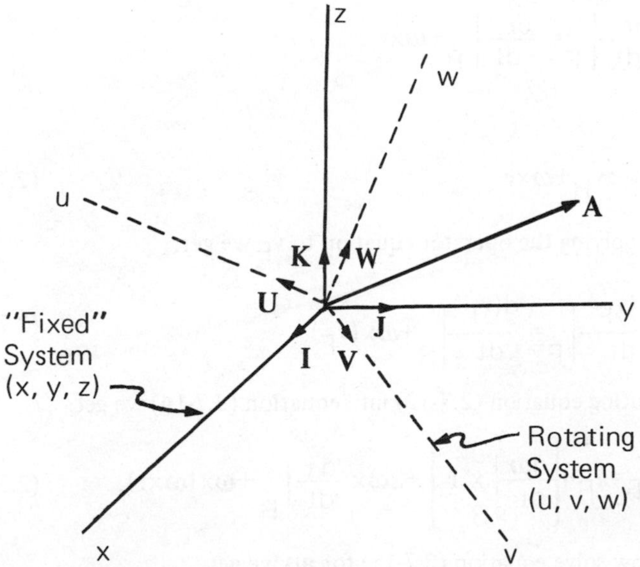

Figure 2.7-5 Fixed and rotating coordinate system

$$\left.\frac{d\mathbf{A}}{dt}\right|_F = \left.\frac{d\mathbf{A}}{dt}\right|_R + \boldsymbol{\omega}\times\mathbf{A}. \tag{2.7-11}$$

We now have a very general equation in which **A** is *any* vector. Equation (2.7-11) may be considered as a true operator

$$\left.\frac{d(\ )}{dt}\right|_F = \left.\frac{d(\ )}{dt}\right|_R + \boldsymbol{\omega}\times(\ ) \tag{2.7-12}$$

where $\boldsymbol{\omega}$ is the instantaneous angular velocity of the rotating frame with respect to the fixed reference frame. Equation (2.7-12) is referred to as the Coriolis theorem.

We will now apply the operator equation (2.7-12) to the position vector $\mathbf{r}$,

$$\left.\frac{d\mathbf{r}}{dt}\right|_F = \left.\frac{d\mathbf{r}}{dt}\right|_R + \boldsymbol{\omega} \times \mathbf{r}$$

or

$$\mathbf{v}_F = \mathbf{v}_R + \boldsymbol{\omega} \times \mathbf{r} \tag{2.7-13}$$

Again applying the operator equation to $\mathbf{v}_F$ we get.

$$\left.\frac{d(\mathbf{v}_F)}{dt}\right|_F = \left.\frac{d(\mathbf{v}_F)}{dt}\right|_R + \boldsymbol{\omega} \times (\mathbf{v}_F). \tag{2.7-14}$$

Substituting equation (2.7-13) into equation (2.7-14) we get

$$\mathbf{a}_F = \mathbf{a}_R + \left[\left.\frac{d\boldsymbol{\omega}}{dt}\right|_R \times \mathbf{r}\right] + 2\boldsymbol{\omega} \times \left.\frac{d\mathbf{r}}{dt}\right|_R + \boldsymbol{\omega} \times (\boldsymbol{\omega} \times \mathbf{r}). \tag{2.7-15}$$

If we now solve equation (2.7-15) for $\mathbf{a}_R$ we get

$$\mathbf{a}_R = \mathbf{a}_F - \left[\left.\frac{d\boldsymbol{\omega}}{dt}\right|_R \times \mathbf{r}\right] - 2\boldsymbol{\omega} \times \left.\frac{d\mathbf{r}}{dt}\right|_R - \boldsymbol{\omega} \times (\boldsymbol{\omega} \times \mathbf{r}). \tag{2.7-16}$$

Let us now apply this result to a problem of practical interest. Suppose we are standing on the surface of the earth. The angular velocity of our rotating platform is a constant $\boldsymbol{\omega}_\oplus$. Then equation (2.7-16) reduces to

$$\mathbf{a}_R = \mathbf{a}_F - 2(\boldsymbol{\omega}_\oplus \times \mathbf{v}_R) - \boldsymbol{\omega}_\oplus \times (\boldsymbol{\omega}_\oplus \times \mathbf{r}).$$

Inspection of this result indicates that the rotating observer sees the acceleration in the fixed system *plus* two others:

$$-2\boldsymbol{\omega}_\oplus \times \mathbf{v}_R = \text{coriolis acceleration}$$

and

$$-\boldsymbol{\omega}_{\oplus} \times (\boldsymbol{\omega}_{\oplus} \times \mathbf{r}) = \text{centrifugal acceleration.}$$

The first is present only when there is relative motion between the object and the rotating frame. The second depends only on the position of the object from the axis of rotation.

2.8 SEZ TO IJK TRANSFORMATION USING AN ELLIPSOID EARTH MODEL

To fully complete the orbit determination problem of the previous section we need to convert the vectors we have expressed in SEZ components to IJK components. We could simply use a spherical earth as a model and write the transformation matrix as discussed earlier. To be more accurate we will first discuss a nonspherical earth model.

"Station coordinates" is the term used to denote the position of a tracking or launch site on the surface of the earth. If the earth were perfectly spherical, latitude and longitude could be considered as spherical coordinates with the radius being just the earth's radius plus the elevation above sea level. It is known, however, that the earth is not a perfect geometric sphere. Therefore, in order to increase the accuracy of our calculations, a model for the geometric shape of the earth must be adopted. We will take as our approximate model an oblate spheroid. The first Vanguard satellite showed the earth to be slightly pear-shaped but this distortion is so small that the oblate spheroid is still an excellent representation.

We will discover that latitude can no longer be interpreted as a spherical coordinate and that the earth's radius is a function of latitude. We will find it most convenient to express the station coordinates of a point in terms of two rectangular coordinates and the longitude. (The interpretation of longitude is the same on an oblate earth as it is on a spherical earth.)

2.8.1 The Reference Ellipsoid. In the model which has been adopted, a cross section of the earth along a meridian is an ellipse whose semi-major axis, a_e, is just the equatorial radius and whose semi-minor axis, b_e, is just the polar radius of the earth. Sections parallel to the equator are, of course, circles. Recent determinations of these radii and the consequent eccentricity of the elliptical cross section are as follows:

Equatorial radius $(a_e) = 6378.145$ km

Polar radius $(b_e) = 6356.785$ km

Eccentricity $(e) = 0.08182$

The reference ellipsoid is a good approximation to that hypothetical surface commonly referred to as "mean sea level." The actual mean sea level surface is called the "geoid" and it deviates from the reference ellipsoid slightly because of the uneven distribution of mass in the earth's interior. The geoid is a true equipotential surface and a plumb bob would hang perpendicular to the surface of the geoid at every point.

2.8.2 The Measurement of Latitude. While an oblate earth introduces no unique problems in the definition or measurement of terrestrial longitude, it does complicate the concept of latitude. Consider Figure 2.8-1. It illustrates the two most commonly used definitions of latitude.

The angle L is called "geocentric latitude" and is defined as the angle between the equatorial plane and the radius from the geocenter.

The angle L is called "geodetic latitude" and is defined as the angle between the equatorial plane and the normal to the surface of the ellipsoid. The word "latitude" usually means geodetic latitude. This is the basis for most of the maps and charts we use. The normal to the surface is the direction that a plumb bob would hang were it not for local anomalies in the earth's gravitational field.

The angle between the equatorial plane and the actual "plumb bob vertical" uncorrected for these gravitational anomalies is called L_a the "astronomical latitude." Since the difference between the true geoid and our reference ellipsoid is slight, the difference between L and L_a is usually negligible.

When you are given the latitude of a place, it is safe to assume that it is the geodetic latitude L unless otherwise stated.

2.8.3 Station Coordinates. What we need now is a method of calculating the station coordinates of a point on the surface of our reference ellipsoid when we know the geodetic latitude and longitude of the point and its height above mean sea level (which we will take to be the height above the reference ellipsoid).

Consider an ellipse comprising a section of our adopted earth model

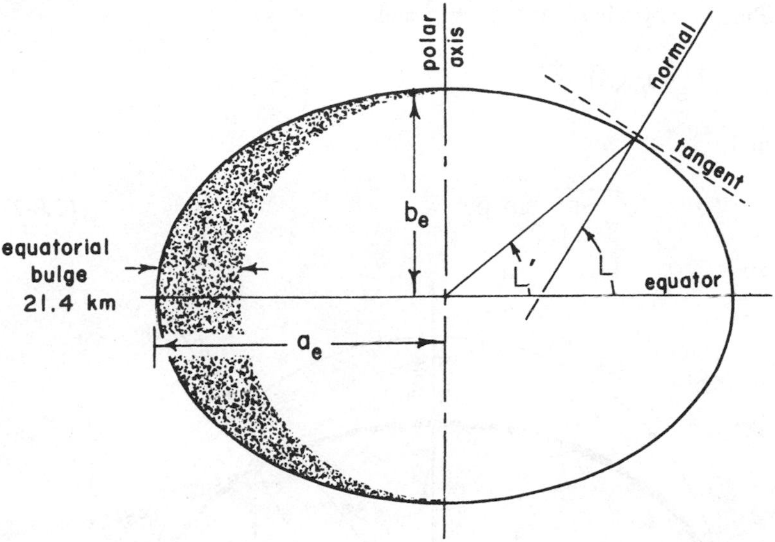

Figure 2.8-1 Geocentric and geodetic latitude

and a rectangular coordinate system as shown in Figure 2.8-2.

We will first determine the x and z coordinates of a point on the ellipse assuming that we know the geodetic latitude, L. It will then be a simple matter to adjust these coordinates for a point which is a known elevation above the surface of the ellipsoid in the direction of the normal.

It is convenient to introduce the angle β, the "reduced latitude," which is illustrated in Figure 2.8-2. The x and z coordinates can immediately be written in terms of β if we note that the ratio of the z ordinate of a point on the ellipse to the corresponding z ordinate of a point on the circumscribed circle is just b_e / a_e. Thus,

$$x = a_e \cos\beta \tag{2.8-1}$$

$$z = \frac{b_e}{a_e} a_e \sin\beta$$

But, for any ellipse, $a^2 = b^2 + c^2$ and $e = c/a$, so

$$b_e = a_e \sqrt{1-e^2}$$

and

$$z = a_e \sqrt{1-e^2} \sin \beta. \tag{2.8-2}$$

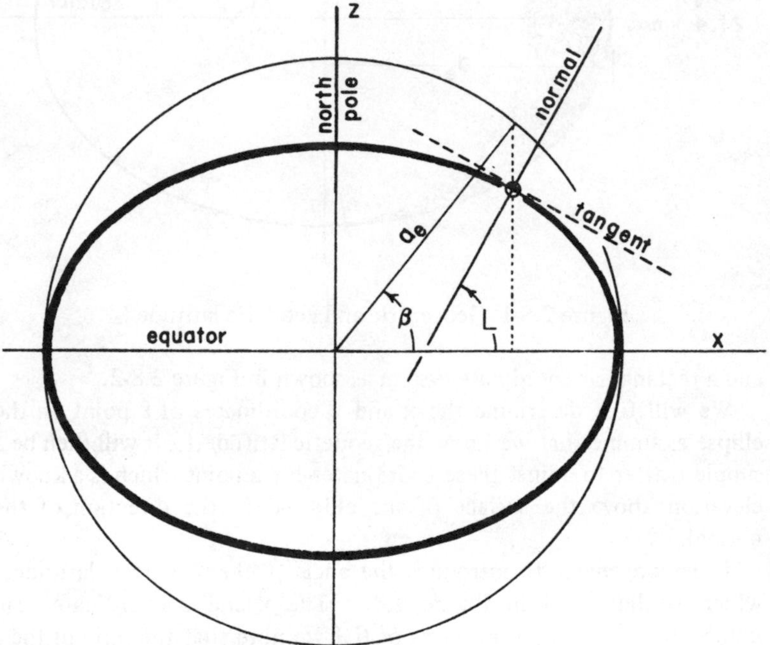

Figure 2.8-2 Station coordinates

We must now express sin β in terms of the geodetic latitude L and the constants a_e and b_e. From elementary calculus we know that the slope of the tangent to the ellipse is just dz/dx and the slope of the normal is $-dx/dz$.

Since the slope of the normal is just tan L, we can write

$$\tan L = -\frac{dx}{dz}.$$

The differentials dx and dz can be obtained by differentiating the expressions for x and z above. Thus,

$$dx = -a_e \sin\beta\, d\beta$$

$$dz = a_e\sqrt{1-e^2}\cos\beta\, d\beta$$

and

$$\tan L = \frac{\tan\beta}{\sqrt{1-e^2}}$$

or

$$\tan\beta = \sqrt{1-e^2}\tan L = \frac{\sqrt{1-e^2}\sin L}{\cos L}. \tag{2.8-3}$$

Suppose we consider this last expression as the quotient

$$\tan\beta = \frac{A}{B}$$

where $A = \sqrt{1-e^2}\sin L$ and $B = \cos L$.

$$\sin\beta = \frac{A}{\sqrt{A^2+B^2}} = \frac{\sqrt{1-e^2}\sin L}{\sqrt{1-e^2\sin^2 L}}$$

$$\cos\beta = \frac{B}{\sqrt{A^2+B^2}} = \frac{\cos L}{\sqrt{1-e^2\sin^2 L}} \tag{2.8-4}$$

We can now write the x and z coordinates for a point on the ellipse.

$$x = \frac{a_e\cos L}{\sqrt{1-e^2\sin^2 L}} \tag{2.8-5}$$

$$z = \frac{a_e(1-e^2)\sin L}{\sqrt{1-e^2\sin^2 L}} \tag{2.8-5}$$

For a point which is a height H above the ellipsoid (which we take to be mean sea level), it is easy to show that the x and z components of the elevation or height (H) normal to the adopted ellipsoid are

$$\begin{aligned} \Delta x &= H\cos L \\ \Delta z &= H\sin L. \end{aligned} \tag{2.8-6}$$

Adding these quantities to the relations for x and z we get the following expressions for the two rectangular station coordinates of a point in terms of geodetic latitude, elevation above mean sea level, and earth equatorial radius and eccentricity:

$$\begin{aligned} x &= \left|\frac{a_e}{\sqrt{1-e^2\sin^2 L}} + H\right|\cos L \\ z &= \left|\frac{a_e(1-e^2)}{\sqrt{1-e^2\sin^2 L}} + H\right|\sin L. \end{aligned} \tag{2.8-7}$$

The third station coordinate is simply the east longitude of the point. If the Greenwich sidereal time, θ_g, is known, it can be combined with east longitude to find local sidereal time, θ. The x and z coordinates plus the angle θ completely locate the observer or launch site in the geocentric-equatorial frame as shown below.

From Figure 2.8-3 it is obvious that the vector **R** from the geocenter to the site on an oblate earth is simply

$$\mathbf{R} = x\cos\theta\,\mathbf{I} + x\sin\theta\,\mathbf{J} + z\,\mathbf{K}. \tag{2.8-8}$$

2.8.4 Transforming a Vector From SEZ To IJK Components. The only remaining problem is how to convert the vectors which we have expressed in SEZ components into the IJK components of the geocentric frame.

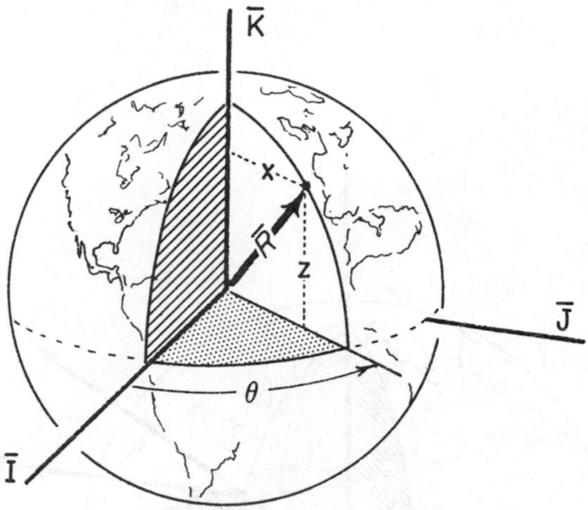

Figure 2.8.3 Vector from geocenter to site

Recall that the geodetic latitude of the radar site, L, is the angle between the equatorial plane and the extension of the local vertical at the radar site (on a spherical or oblate earth). Although Figure 2.8-4 shows a spherical earth, the transformation matrix derived as follows is equally valid for an ellipsoid earth model where L is the geodetic latitude of the site.

The angle between the unit vector **I** (vernal equinox direction) and the Greenwich meridian is called θ_g—the "greenwich sidereal time." If we let λ_E be the geographic longitude of the radar site measured eastward from Greenwich, then

$$\boxed{\theta = \theta_g + \lambda_E} \tag{2.8-9}$$

where θ is called "local sidereal time."

The angles L and θ completely determine the relationship between the IJK frame and the SEZ frame. Obviously we need a method of determining θ at some general time t.

If we knew θ_{go} at some particular time t_o (say 0^h Universal Time on 1 Jan) we could determine θ at time t from

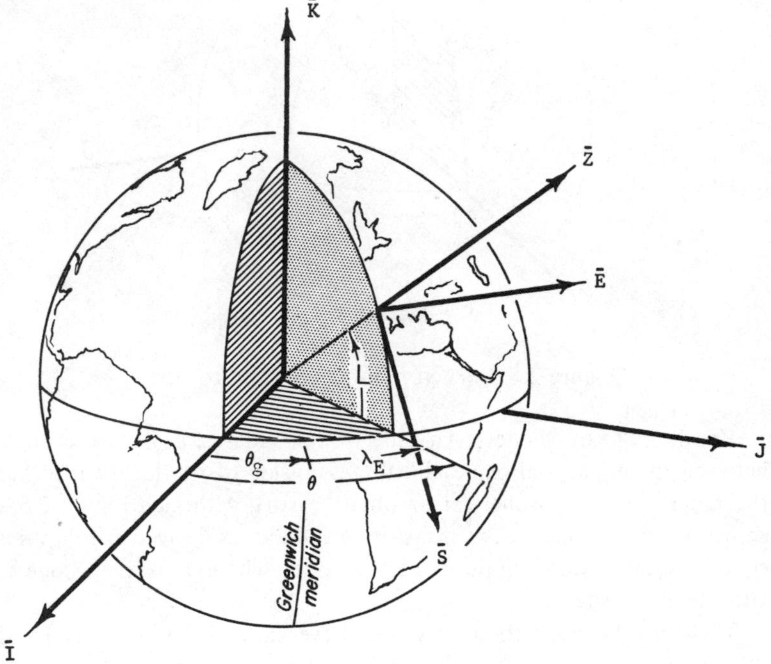

Figure 2.8-4 Angular relationship between frames

$$\theta = \theta_{go} + \omega_{\oplus}(t - t_o) + \lambda_E \tag{2.8-10}$$

where $\omega_{\oplus}$ is the angular velocity of the earth.

The American Ephemeris and Nautical Almanac lists the value of θ_g at 0^h UT for every day of the year. For a more complete discussion of sidereal time see the next section.

We now have all we need to determine the rotation matrix $\widetilde{\mathbf{D}}^{-1}$ that transforms a vector from SEZ to IJK components. From equation (2.6-12)

$$\widetilde{\mathbf{D}}^{-1} = \begin{bmatrix} \sin L \cos\theta & -\sin\theta & \cos L \cos\theta \\ \sin L \sin\theta & \cos\theta & \cos L \sin\theta \\ -\cos L & 0 & \sin L \end{bmatrix} . \qquad (2.8\text{-}11)$$

If a_I, a_J, a_K and a_S, a_E, a_Z are the components of a vector **a** in each of the two systems then

$$\begin{bmatrix} a_I \\ a_J \\ a_K \end{bmatrix} = \widetilde{\mathbf{D}}^{-1} \begin{bmatrix} a_S \\ a_E \\ a_Z \end{bmatrix} \qquad (2.6\text{-}12)$$

2.9 THE MEASUREMENT OF TIME

Time is used as a fundamental dimension in almost every branch of science. When a scientist or layman uses the terms "hours, minutes or seconds" he is understood to mean units of *mean solar time*. This is the time kept by ordinary clocks. Since we will need to talk about another kind of time called "sidereal" time, it will help to understand exactly how each is defined.

2.9.1 Solar and Sidereal Time. It is the sun more than any other heavenly body that governs our daily activity cycle; so, it is no wonder that ordinary time is reckoned by the sun. The time between two successive upper transits of the sun across our local meridian is called an *apparent solar day*. The earth has to turn through slightly more than one complete rotation on its axis relative to the "fixed" stars during this interval. The reason is that the earth travels about 1/365th of the way around its orbit in one day. This should be made clear by Figure 2.9-1.

A *sidereal day* consisting of 24 sidereal hours is defined as the time required for the earth to rotate once on its axis relative to the stars. This occurs in about $23^h\ 56^m\ 4^s$ of ordinary solar time and leads to the following relationships:

1 day of mean solar time = 1.0027379093 days of mean sidereal time
= $24^h 03^m 56\overset{s}{.}55536$ of sidereal time
= 86636.55536 mean sidereal seconds

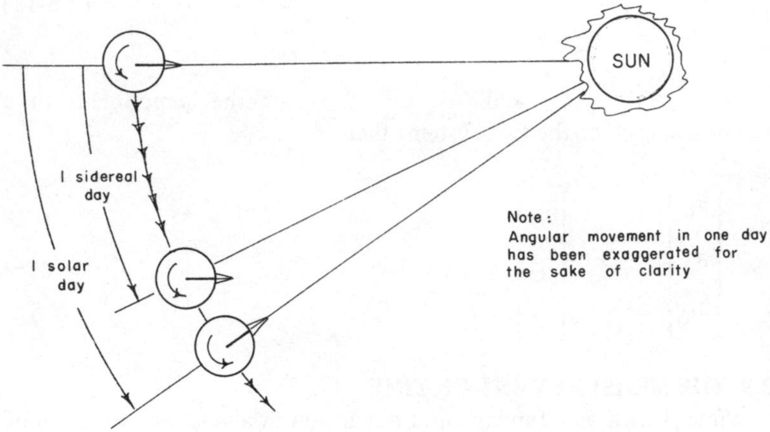

Figure 2.9-1 Solar and sidereal day

1 day of mean sidereal time = .9972695664 days of solar time
$= 23^{h}56^{m}04^{s}.09054$ of mean solar time
= 86164.09054 mean solar seconds.

So far, we have really defined only sidereal time and "apparent" solar time. Based on the definition of an apparent solar day illustrated in the figure, no two solar days would be exactly the same length because the earth's axis is not perpendicular to the plane of its orbit and because the earth's orbit is slightly elliptical. In early January, when the earth is near perihelion, it moves farther around its orbit in 1 day than it does in early July when it is near aphelion. In order to avoid this irregularity in the length of a solar day, a mean solar day is defined based on the assumption that the earth is in a circular orbit whose period matches the actual period of the earth and that the axis of rotation is perpendicular to the orbital plane. An ordinary clock which ticks off 24 hours in one mean solar day would show the sun arriving at our local meridian a little early at certain times of the year and a little late at other times of the year.

2.9.2 Local Mean Solar Time and Universal Time. The earth is divided into 24 time zones approximately 15° of longitude apart. The local mean solar time in each zone differs from the neighboring zones by 1 hour. (A few countries have adopted time zones which differ by only 1/2 hour from the adjacent zones.)

The local mean solar time on the Greenwich meridian is called Greenwich Mean Time (GMT), Universal Time (UT), or Zulu (Z) time.

A time given in terms of a particular time zone can be converted to Universal Time by simply adding or subtracting the correct number of hours. For example, if it is 1800 Eastern Standard Time (EST) the Universal Time is 2300. The conversion for time zones in the United States is as follows:

$$\begin{aligned} &\text{EST} + 5 \text{ hrs} = \text{UT} \\ &\text{CST} + 6 \text{ hrs} = \text{UT} \\ &\text{MST} + 7 \text{ hrs} = \text{UT} \\ &\text{PST} + 8 \text{ hrs} = \text{UT} \end{aligned}$$

2.9.3 Finding the Greenwich Sidereal Time when Universal Time is Known. Often it is desired to relate observations made in the topocentric-horizon system to the IJK unit vectors of the geocentric-equatorial system and vice versa. The geometrical relationship between these two systems depends on the latitude and longitude of the topos and the Greenwich sidereal time, θ_g, (expressed as an angle) at the date and time of the observation. This relationship is illustrated on page 99.

What we need is a convenient way to calculate the angle θ_g for any date and time of day. If we knew what θ_g was on a particular day and time we could calculate θ_g for any future time since we know that in one day the earth turns through 1.0027379093 complete rotations on its axis.

Suppose we take the value of θ_g at 0^h UT on 1 January of a particular year and call it θ_{go}. Also, let us number the days of the year consecutively beginning with 1 January as day 0. Thus,

31 January	= 030	31 July	= 211(212)
28 February	= 058	31 August	= 242(243)
31 March	= 089(090)*	30 September	= 272(273)
30 April	= 119(120)	31 October	= 303(304)
31 May	= 150(151)	30 November	= 333(334)
30 June	= 180(181)	31 December	= 364(365)

*figures in parentheses refer to leap year

If, in addition, we express time in decimal fractions of a day, we can convert a particular date and time into a single number which indicates the number of days which have elapsed since our "time zero." If we call this number D, then

$$\theta_g = \theta_{g_0} + 1.0027379093 \times 360^\circ \times D \text{ [degrees]}$$

or

$$\theta_g = \theta_{g_0} + 1.0027379093 \times 2\pi \times D \text{ [radians]}.$$

Values for θ_{g_0} taken from page 10 of the *American Ephemeris and Nautical Almanac* are given below for several years.

0^h UT

1 Jan	θ_{g_0} [hr,min,sec]	θ_{g_0} [deg]	θ_{g_0} [rad]
1968	$6^h38^m53^s.090$	$99^\circ.721208$	1.74046342
1969	$6^h41^m52^s.353$	$100^\circ.468137$	1.75349981
1970	$6^h40^m55^s.061$	$100.^\circ229421$	1.74933340
1971	$6^h39^m57^s.769$	$99^\circ.990704$	1.74516701

2.9.4 Precession of the Equinoxes. To understand what the values of θ_{g_0} in the table preceeding represent, it is necessary to discuss the slow shifting of the vernal equinox direction known as precession.

The direction of the equinox is determined by the line-of-intersection of the ecliptic plane (the plane of the earth's orbit) and the equatorial plane.

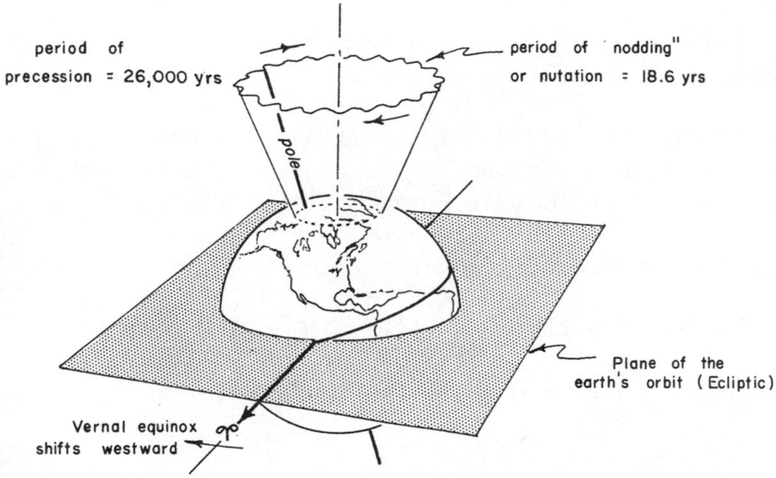

Figure 2.9-2 Precession of the equinoxes

While the plane of the ecliptic is fixed relative to the stars, the equatorial plane is not. Due to the asphericity of the earth, the sun produces a torque on the earth which results in a wobbling or precessional motion similar to that of a simple top. Because the earth's equator is tilted $23\frac{1}{2}^{\circ}$ to the plane of the ecliptic, the polar axis sweeps out a cone-shaped surface in space with a semi-vertex angle of $23\frac{1}{2}^{\circ}$. As the earth's axis precesses, the line-of-intersection of the equator and the ecliptic swings westward slowly. The period of the precession is about 26,000 years, so the equinox direction shifts westward about 50 arc-seconds per year.

The moon also produces a torque on the earth's equatorial bulge. However, the moon's orbital plane precesses due to solar perturbation with a period of about 18.6 years, so the lunar-caused precession has this same period. The effect of the moon is to superimpose a slight nodding motion called "nutation," with a period of 18.6 years, on the slow westward precession caused by the sun.

The *mean equinox* is the position of the equinox when solar precession alone is taken into account. The *apparent equinox* is the actual equinox direction when both precession and nutation are included.

The values of θ_{g_0} in the table preceding are referred to the mean equinox and equator of the dates shown.

EXAMPLE PROBLEM. What is the inertial position vector of a point 6.378 km above mean sea level on the equator, 57.296 degrees west longitude, at 0600 GMT, 2 January 1970?

The given information in consistent units is:

UT = 0600 hrs, Day = 1 (1 Jan 1970 = 0)

Long = $\lambda = -57.296°$ = -1 radian (east longitude is positive)

H = 6.378km = 0.001 $DU_{\oplus}$, Lat = L = 0°

From section 2.8-2 θ_{g_0} for 1970 is 1.74933340

D = 1.25

$$\theta_g = \theta_{g_0} + 1.0027379093 \times 2\pi \times D = 9.6245 \text{ radians}$$

$$\theta = \theta_g + \lambda = 9.6245 - 1.0 = 8.6245 \text{ radians}$$

From equation (2.8-7)

$$x = \left[\frac{a_e}{\sqrt{1 - e^2 \sin^2 L}} + H\right] \cos L = 1.001$$

$$z = \left[\frac{a_e}{\sqrt{1 - e^2 \sin^2 L}} + H\right] \sin L = 0$$

From equation (2.8-8)

$$\begin{aligned}\mathbf{R} &= x \cos\theta\ \mathbf{I} + x \sin\theta\ \mathbf{J} + Z\mathbf{K}\\ &= 1.001\ \cos(8.6245)\mathbf{I} + 1.001(8.6245)\ \mathbf{J} + 0\mathbf{K}\\ &= -.697\mathbf{I} + .718\mathbf{J} + 0\mathbf{K}\end{aligned}$$

EXAMPLE PROBLEM. At 0600 GST a BMEWS tracking station (Lat 60° N, Long 150° W) detected a space object and obtained the following data:

Slant Range (ρ) = 0.4 $DU_\oplus$

Azimuth (Az) = 90°

Elevation (E1) = 30°

Range Rate ($\dot{\rho}$) = 0

Azimuth Rate ($\dot{Az}$) = 10 rad/$TU_\oplus$

Elevation Rate ($\dot{E1}$) = 5 rad/$TU_\oplus$

What were the velocity and position vectors of the space object at the time of observation?

From equation (2.7-2)

$$\rho_S = -(0.4)\ (\cos 30^\circ)\ (\cos 90^\circ) = 0 DU_\oplus$$

$$\rho_E = (0.4)\ (\cos 30^\circ)\ (\sin\ 90^\circ) = 0.346 DU_\oplus$$

$$\rho_Z = (0.4)\ \sin 30^\circ = 0.2 DU_\oplus$$

Hence

$$\boldsymbol{\rho} = 0.346\mathbf{E} + 0.2\mathbf{Z}\ (DU_\oplus)$$

From equation (2.7-4)

$$\dot{\rho}_S = -\ (0)\ (\cos 30)\ (\cos 90) + (0.4)\ (\sin 30)\ (5) \cos 90 + (0.4)\ (\cos 30)\ (\sin 90)\ (10) = 3.46 \text{ DU/TU}$$

$$\dot{\rho}_E = (0)\ (\cos 30)\ (\sin 90) - (0.4)\ (\sin 30)\ (5)\ (\sin 90) + (0.4)\ (\cos 30)\ (\cos 90)\ (10) = -1.0 \text{DU/TU}$$

$$\dot{\rho}_Z = (0)\ (\sin 30) + (0.4)\ (\cos 30)\ (5) = 1.73 \text{DU/TU}$$

Hence

$$\dot{\boldsymbol{\rho}} = 3.46\mathbf{S} - 1.0\mathbf{E} + 1.73\mathbf{Z} \text{ DU/TU}$$

From equation (2.7-5) $\mathbf{r} = 0.346\mathbf{E} + 1.2\mathbf{Z}$ (DU)

From equation (2.8-10)

$$\theta = (6)\ (15^{\circ}/\text{hour}) - 150^{\circ} = -60^{\circ} \text{ (LST)}$$

The rotation matrix (2.8-11) becomes

$$\widetilde{\mathbf{D}}^{-1} = \begin{bmatrix} 0.433 & 0.866 & 0.25 \\ -0.75 & 0.5 & -0.433 \\ -0.5 & 0 & 0.866 \end{bmatrix} .$$

and

$$\mathbf{r} = \widetilde{\mathbf{D}}^{1} \begin{bmatrix} 0 \\ 0.346 \\ 1.2 \end{bmatrix} = 0.6\mathbf{I} - 0.346\mathbf{J} + 1.04\mathbf{K} \text{ (DU}_{\oplus})$$

Similarly,

$$\dot{\boldsymbol{\rho}} = \widetilde{\mathbf{D}}^{1} \begin{bmatrix} 3.46 \\ -1.0 \\ 1.73 \end{bmatrix} = 1.06\mathbf{I} - 3.84\mathbf{J} - 0.232\mathbf{K} \text{ (DU/TU)}$$

From equation (2.7-7)

$$\mathbf{v} = \dot{\boldsymbol{\rho}} + (0.0588\mathbf{K}) \times \mathbf{r} \text{ DU/TU}$$

$$\mathbf{v} = 1.08\mathbf{I} - 3.8\mathbf{J} - 0.232\mathbf{K} \text{ DU/TU}$$

Thus the position and velocity vectors of the object at one epoch have been found, and the orbit is uniquely determined.

2.10 ORBIT DETERMINATION FROM THREE POSITION VECTORS[16]

In the preceding section we saw how to obtain **r** and **v** from a single radar measurement of ρ, $\dot{\rho}$, El, $\dot{\text{El}}$, Az, $\dot{\text{Az}}$. It may happen that a particular radar site is not equipped to measure Doppler phase shifts and so the rate information may be lacking. In this section we will examine a method for determining an orbit from three position vectors $\mathbf{r}_1$, $\mathbf{r}_2$ and $\mathbf{r}_3$ (assumed to be coplanar). These three vectors may be obtained from successive measurements of ρ, El and Az at three times by the methods of the last section or by any other technique.

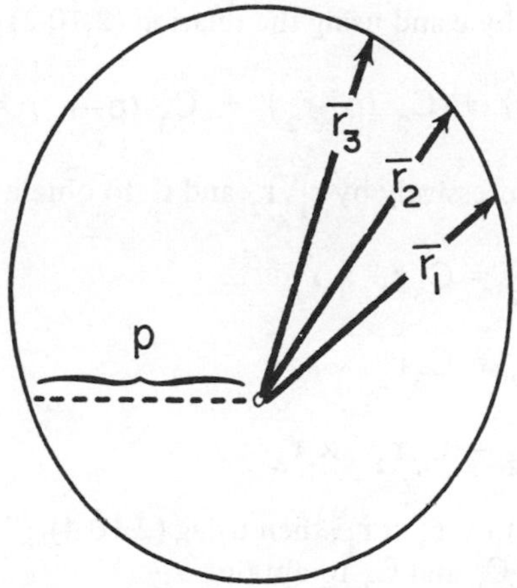

Figure 2.10-1 Orbit through $\mathbf{r}_1$, $\mathbf{r}_2$ and $\mathbf{r}_3$

The scheme to be presented is associated with the name of J. W. Gibbs and has come to be known as the Gibbsian method. As Baker[5] points out, it was developed using pure vector analysis and was historically the first contribution of an American scholar to celestial mechanics. Gibbs is, of course, well known for his contributions to thermodynamics, but the contributions to celestial mechanics of this fine scholar, who in 1863 received our country's first engineering degree, is equally outstanding but less generally remembered.

The Gibbs problem can be stated as follows: Given three nonzero coplanar vectors $\mathbf{r}_1, \mathbf{r}_2$ and $\mathbf{r}_3$ which represent three sequential positions of an orbiting object on one pass, find the parameter p and the eccentricity e of the orbit and the perifocal base vectors **P, Q** and **W**.

Solution: Since the vectors $\mathbf{r}_1$, $\mathbf{r}_2$ and $\mathbf{r}_3$ are coplanar, there must exist scalars C_1, C_2 and C_3 such that

$$C_1\mathbf{r}_1 + C_2\mathbf{r}_2 + C_3\mathbf{r}_3 = 0. \tag{2.10-1}$$

Using the polar equation of a conic section, equation (1.5-4), and the definition of a dot product, we can show that:

$$\mathbf{e} \cdot \mathbf{r} = p - r. \tag{2.10-2}$$

Dotting (2.10-1) by $\mathbf{e}$ and using the relation (2.10-2) gives

$$C_1(p - r_1) + C_2(p - r_2) + C_3(p - r_3) = 0. \tag{2.10-3}$$

Cross (2.10-1) successively by $\mathbf{r}_1$, $\mathbf{r}_2$ and $\mathbf{r}_3$ to obtain

$$C_2\mathbf{r}_1 \times \mathbf{r}_2 = C_3\mathbf{r}_3 \times \mathbf{r}_1 \tag{2.10-4}$$

$$C_1\mathbf{r}_1 \times \mathbf{r}_2 = C_3\mathbf{r}_2 \times \mathbf{r}_3 \tag{2.10-5}$$

$$C_1\mathbf{r}_3 \times \mathbf{r}_1 = C_2\mathbf{r}_2 \times \mathbf{r}_3. \tag{2.10-6}$$

Multiply (2.10-3) by $\mathbf{r}_3 \times \mathbf{r}_1$, then using (2.10-4), (2.10-5) and (2.10-6) we can eliminate C_1 and C_3 to obtain:

$$\begin{aligned} C_2\mathbf{r}_2 \times \mathbf{r}_3(p - r_1) &+ C_2\mathbf{r}_3 \times \mathbf{r}_1(p - r_2) \\ &+ C_2\mathbf{r}_1 \times \mathbf{r}_2(p - r_3) = 0. \end{aligned} \tag{2.10-7}$$

Notice that C_2 may now be divided out. Multiply the factors and collect terms:

$$\begin{aligned} p(\mathbf{r}_1 \times \mathbf{r}_2 + \mathbf{r}_2 \times \mathbf{r}_3 + \mathbf{r}_3 \times \mathbf{r}_1) &= \\ r_3\mathbf{r}_1 \times \mathbf{r}_2 + r_1\mathbf{r}_2 \times \mathbf{r}_3 &+ r_2\mathbf{r}_3 \times \mathbf{r}_1. \end{aligned} \tag{2.10-8}$$

Let the right side of (2.10-8) be defined as a vector $\mathbf{N}$ and the coefficient of p be defined as a vector $\mathbf{D}$, then

$$p\mathbf{D} = \mathbf{N}\,. \tag{2.10-9}$$

Therefore, provided $\mathbf{N} \cdot \mathbf{D} = N\,D$, i.e., $\mathbf{N}$ and $\mathbf{D}$ have the same direction,

$$p = \frac{N}{D}\,. \tag{2.10-10}$$

It can be shown that for any set of vectors suitable for the Gibbsian method as described above, $\mathbf{N}$ and $\mathbf{D}$ do have the same direction and that this is the direction of the angular momentum vector $\mathbf{h}$. This is also the direction of $\mathbf{W}$ in the perifocal coordinate system.

Since $\mathbf{P}$, $\mathbf{Q}$ and $\mathbf{W}$ are orthogonal unit vectors

$$\mathbf{Q} = \mathbf{W} \times \mathbf{P}\,. \tag{2.10-11}$$

Since $\mathbf{W}$ is a unit vector in the direction of $\mathbf{N}$ and $\mathbf{P}$ is a unit vector in the direction of $\mathbf{e}$, equation (2.10-11) can be written

$$\mathbf{Q} = \frac{1}{Ne}\,(\mathbf{N} \times \mathbf{e})\,, \tag{2.10-12}$$

Now substitute for $\mathbf{N}$ from its definition in (2.10-8)

$$Ne\mathbf{Q} = r_3\,(\mathbf{r}_1 \times \mathbf{r}_2) \times \mathbf{e} + r_1\,(\mathbf{r}_2 \times \mathbf{r}_3) \times \mathbf{e} + r_2\,(\mathbf{r}_3 \times \mathbf{r}_1) \times \mathbf{e}\,. \tag{2.10-13}$$

Now use the general relationship for a vector triple product

$$(\mathbf{a} \times \mathbf{b}) \times \mathbf{c} = (\mathbf{a} \cdot \mathbf{c})\,\mathbf{b} - (\mathbf{b} \cdot \mathbf{c})\,\mathbf{a} \tag{2.10-14}$$

To rewrite (2.10-13):

$$\begin{aligned} Ne\mathbf{Q} = {} & r_3\,(\mathbf{r}_1 \cdot \mathbf{e})\,\mathbf{r}_2 - r_3\,(\mathbf{r}_2 \cdot \mathbf{e})\,\mathbf{r}_1 \\ & + r_1\,(\mathbf{r}_2 \cdot \mathbf{e})\,\mathbf{r}_3 - r_1\,(\mathbf{r}_3 \cdot \mathbf{e})\,\mathbf{r}_2 \\ & + r_2\,(\mathbf{r}_3 \cdot \mathbf{e})\,\mathbf{r}_1 - r_2\,(\mathbf{r}_1 \cdot \mathbf{e})\,\mathbf{r}_3\,. \end{aligned} \tag{2.10-15}$$

Again using the relationship (2.10-2) and factoring p from the right side we have:

$$\mathrm{N}e\mathbf{Q}=\mathrm{p}\,[\,(\mathrm{r}_2-\mathrm{r}_3)\,\mathbf{r}_1+(\mathrm{r}_3-\mathrm{r}_1)\,\mathbf{r}_2+(\mathrm{r}_1-\mathrm{r}_2)\,\mathbf{r}_3\,]\equiv \mathrm{p}\mathbf{S} \tag{2.10-16}$$

Where **S** is defined by the bracketed quantity in (2.10-16).

Therefore, since $\mathrm{N}e\mathbf{Q} = \mathrm{p}\mathbf{S}$, $\mathrm{N} = \mathrm{pD}$ and **Q** and **S** have the same direction

$$e = \frac{\mathrm{S}}{\mathrm{D}} \tag{2.10-17}$$

and

$$\mathbf{Q} = \frac{\mathbf{S}}{\mathrm{S}} \tag{2.10-18}$$

$$\mathbf{W} = \frac{\mathbf{N}}{\mathrm{N}}\,. \tag{2.10-19}$$

Since **P**, **Q** and **W** are orthogonal,

$$\mathbf{P} = \mathbf{Q} \times \mathbf{W}, \tag{2.10-20}$$

Thus, to solve this problem use the given **r** vectors to form the **N**, **D** and **S** vectors. Before solving the problem, check $\mathbf{N} \neq 0$ and $\mathbf{D} \cdot \mathbf{N} > 0$. Then use (2.10-10) to find p; (2.10-17) to find e; (2.10-18) to find **Q**; (2.10-19) to find **W**; and (2.10-20) to find **P**. The information thus obtained may be used in formula (2.5-4) to obtain the velocity vector corresponding to any of the given position vectors. However, it is possible to develop an expression that gives the **v** vector directly in terms of the **D**, **N** and **S** vectors.

From equation (1.5-2) we can write

$$\dot{\mathbf{r}} \times \mathbf{h} = \mu\left(\frac{\mathbf{r}}{\mathrm{r}} + \mathbf{e}\right). \tag{2.10-21}$$

Cross $\mathbf{h}$ into (2.10-21) to obtain

$$\mathbf{h} \times (\dot{\mathbf{r}} \times \mathbf{h}) = \mu \left(\frac{\mathbf{h} \times \mathbf{r}}{r} + \mathbf{h} \times \mathbf{e}\right)$$

Using the identity; $\mathbf{a} \times (\mathbf{b} \times \mathbf{c}) = (\mathbf{a} \cdot \mathbf{c})\,\mathbf{b} - (\mathbf{a} \cdot \mathbf{b})\,\mathbf{c}$
the left side becomes $(\mathbf{h} \cdot \mathbf{h})\,\mathbf{v} - (\mathbf{h} \cdot \mathbf{v})\,\mathbf{h}$ and $\mathbf{h} \cdot \mathbf{v} = 0$.
Thus

$$h^2 \mathbf{v} = \mu \left(\frac{\mathbf{h} \times \mathbf{r}}{r} + \mathbf{h} \times \mathbf{e}\right).$$

We can write $\mathbf{h} = h\mathbf{W}$ and $\mathbf{e} = e\mathbf{P}$, so

$$\mathbf{v} = \frac{\mu}{h} \left(\frac{\mathbf{W} \times \mathbf{r}}{r} + e\mathbf{W} \times \mathbf{P}\right)$$

$$\mathbf{v} = \frac{\mu}{h} \left(\frac{\mathbf{W} \times \mathbf{r}}{r} + e\mathbf{Q}\right) \tag{2.10-22}$$

Using the fact that $h = \sqrt{N\mu/D}$,

$$e = \frac{S}{D},\ \mathbf{Q} = \frac{\mathbf{S}}{S} \text{ and } \mathbf{W} = \frac{\mathbf{D}}{D}$$

we have $\mathbf{v} = \frac{1}{r}\sqrt{\frac{\mu}{ND}}\,\mathbf{D} \times \mathbf{r} + \sqrt{\frac{\mu}{ND}}\,\mathbf{S}$ (2.10-23)

To streamline the calculations let us define a scalar and a vector

$$\mathbf{B} \triangleq \mathbf{D} \times \mathbf{r} \tag{2.10-24}$$

$$L \triangleq \sqrt{\frac{\mu}{DN}} \tag{2.10-25}$$

Then $\mathbf{v} = \frac{L}{r}\mathbf{B} + L\mathbf{S}$ (2.10-26)

Thus, to find any of the three velocities directly, proceed as follows: For example, find $\mathbf{v}_2$:

1. Test: $\mathbf{r}_1 \cdot \mathbf{r}_2 \times \mathbf{r}_3 = 0$ for coplanar vectors.

2. Form the **D**, **N** and **S** vectors.

3. Test: $D \neq 0$, $N \neq 0$, $\mathbf{D} \cdot \mathbf{N} > 0$ to assure that the vectors describe a possible two body orbit.

4. Form $\mathbf{B} = \mathbf{D} \times \mathbf{r}_2$.

5. Form $L = \sqrt{\frac{\mu}{DN}}$.

6. Finally, $\mathbf{v}_2 = \frac{L}{r_2}\mathbf{B} + L\mathbf{S}$.

There are a number of other derivations of the Gibbsian method of orbit determination, but all of them have some problems like quadrant resolution which makes computer implementation difficult. This method, using the stated tests, appears to be foolproof in that there are no known special cases.

There are several general features of the Gibbsian method worth noting that set it apart from other methods of orbit determination. Earlier in this chapter we noted that six independent quantities called "orbital elements" are needed to completely specify the size, shape and orientation of an orbit and the position of the satellite in that orbit. By specifying three position vectors we appear to have nine independent quantities–three components for each of the three vectors–from which to determine the six orbital elements. This is not exactly true. The fact that the three vectors must lie in the same plane means that they are not independent.

Another interesting feature of the Gibbsian method is that it is purely geometrical and vectorial and makes use of the theorem that "one and only one conic section can be drawn through three coplanar position vectors such that the focus lies at the origin of the three position vectors." Thus, the only feature of orbital motion that is exploited is the fact that the path is a conic section whose focus is the center of the earth. The time-of-flight between the three positions is not used in the calculations. If we make use of the dynamical equation of motion of the satellite it is possible to obtain the orbit from only two position vectors, $\mathbf{r}_1$ and $\mathbf{r}_2$, and the time-of-flight between these positions. This is such an important problem that we will devote Chapter 5 entirely to its solution.

EXAMPLE PROBLEM. Radar observations of an earth satellite during a single pass yield in chronological order the following positions (canonical units).

$$\mathbf{r}_1 = 1.000\mathbf{K}$$

$$\mathbf{r}_2 = -0.700\mathbf{J} - 0.8000\mathbf{K}$$

$$\mathbf{r}_3 = 0.9000\mathbf{J} + 0.5000\mathbf{K}$$

Find: **P**, **Q** and **W** (the perifocal basis vectors expressed in the **I J K** system), the semi-latus rectum, eccentricity, period and the velocity vector at position two.

Form the **D**, **N**, and **S** vectors:

$$\mathbf{D} = 1.970\mathbf{I}$$

$$\mathbf{N} = 2.047\mathbf{I}\ \text{DU}$$

$$\mathbf{S} = -0.0774\mathbf{J} - 0.0217\mathbf{K}$$

Test: $\mathbf{r}_1 \cdot \mathbf{r}_2 \times \mathbf{r}_3 = 0$ to verify that the observed vectors are coplanar.

Since $D \neq 0, N \neq 0$ and $\mathbf{D} \cdot \mathbf{N} > 0$ we know that the given data present a solvable problem.

$$p = \frac{N}{D} = \frac{2.047}{1.97} = 1.039 \text{ DU (3578 nautical miles)}$$

$$e = \frac{S}{D} = \frac{.0804}{1.97} = 0.04081$$

$$a = 1.041 \text{ DU (3585 nautical miles)}$$

$$\mathbb{P} = 2\pi\sqrt{\frac{a^3}{\mu}} = 6.67 \text{ TU (89.7 minutes)}$$

$$\mathbf{Q} = \frac{\mathbf{S}}{S} = -0.963\mathbf{J} - 0.270\mathbf{K}$$

$$\mathbf{W} = \frac{\mathbf{N}}{N} = 1.000\mathbf{I}$$

$$\mathbf{P} = \mathbf{Q} \times \mathbf{W} = -0.270\mathbf{J} + 0.963\mathbf{K}$$

Now form the **B** vector:

$$\mathbf{B} = \mathbf{D} \times \mathbf{r}_2 = 1.576\mathbf{J} - 1.379\mathbf{K}$$

Form a scalar:

$$L = 1/\sqrt{DN} = .4979,$$

then $$\mathbf{v}_2 = \frac{L}{r_2}\mathbf{B} + L\mathbf{S} = 0.700\mathbf{J} - 0.657\mathbf{K} \text{ DU/TU}$$

$$v_2 = 0.960 \text{ DU/TU (4.10 nm/sec)}$$

The same equations used above are very efficient for use in a computer solution to problems of this type. Another approach is to immediately solve for $\mathbf{v}_2$ and then use $\mathbf{r}_2$, $\mathbf{v}_2$ and the method of section 2.4 to solve for the elements.

2.11 ORBIT DETERMINATION FROM OPTICAL SIGHTINGS

The modern orbit determination problem is made much simpler by the availability of radar range and range-rate information. However the angular pointing accuracy and resolution of radar sensors is far below that of optical sensors such as the Baker-Nunn camera. As a result, optical methods still yield the most accurate preliminary orbits and some method of orbit determination proceeding from angular data only (e.g. topocentric right ascension and declination) is required.

Six independent quantities suffice to completely specify a satellite's orbit. These may be the six classical orbital elements or they may be the six components of the vectors **r** and **v** at some epoch. In either case, an optical observation yields only two independent quantities such as El and Az or right ascension and declination, so a minimum of three observations is required at three different times to determine the orbit.

Since astronomers had to determine the orbits of comets and minor planets (asteroids) using angular data only, the method presented below has been in long use and was first suggested by Laplace in 1780.[6]

2.11.1 Determining the Line of Sight Unit Vectors. Let us assume that we have the topocentric right ascension and declination of a satellite at three separate times, $\alpha_1, \delta_1, \alpha_2, \delta_2, \alpha_3, \delta_3$. These could easily be obtained from a photograph of the satellite against the star background. If we let $\mathbf{L}_1$, $\mathbf{L}_2$ and $\mathbf{L}_3$ be unit vectors along the line-of-sight to the satellite at the three observation times, then

$$\mathbf{L}_i = \begin{bmatrix} L_I \\ L_J \\ L_K \end{bmatrix}_i = \begin{bmatrix} \cos\delta_i \cos\alpha_i \\ \cos\delta_i \sin\alpha_i \\ \sin\delta_i \end{bmatrix}_i, \quad i = 1, 2, 3, \qquad (2.11\text{-}1)$$

Now since $\mathbf{L}_i$ are unit vectors directed along the slant range vector $\bar{\rho}$ from the observation site to the satellite, we may write

$$\mathbf{r} = \rho\mathbf{L} + \mathbf{R} \qquad (2.11\text{-}2)$$

where subscripts have been omitted for simplicity and where ρ is the slant range to the satellite, **r** is the vector from the center of the earth to the satellite, and **R** is the vector from the center of the earth to the observation site (see Figure 2.7-2).

We may differentiate equation (2.11-2) twice to obtain

$$\dot{\mathbf{r}} = \dot{\rho}\mathbf{L} + \rho\dot{\mathbf{L}} + \dot{\mathbf{R}} \tag{2.11-3}$$

$$\ddot{\mathbf{r}} = 2\dot{\rho}\dot{\mathbf{L}} + \ddot{\rho}\mathbf{L} + \rho\ddot{\mathbf{L}} + \ddot{\mathbf{R}}\,. \tag{2.11-4}$$

From the equation of motion we have the dynamical relationship

$$\ddot{\mathbf{r}} = -\mu\frac{\mathbf{r}}{r^3}\,.$$

Substituting this into equation (2.11-4) and simplifying yields

$$\boxed{\mathbf{L}\,\ddot{\rho} + 2\dot{\mathbf{L}}\,\dot{\rho} + \left(\ddot{\mathbf{L}} + \frac{\mu}{r^3}\mathbf{L}\right)\rho = -\left(\ddot{\mathbf{R}} + \mu\frac{\mathbf{R}}{r^3}\right).} \tag{2.11-5}$$

At a specified time, say the middle observation, the above vector equation represents three component equations in 10 unknowns. The vectors $\mathbf{L}$, $\mathbf{R}$ and $\ddot{\mathbf{R}}$ are known at time t_2; $\dot{\mathbf{L}}$, $\ddot{\mathbf{L}}$, ρ, $\dot{\rho}$, $\ddot{\rho}$ and r, however, are not known.

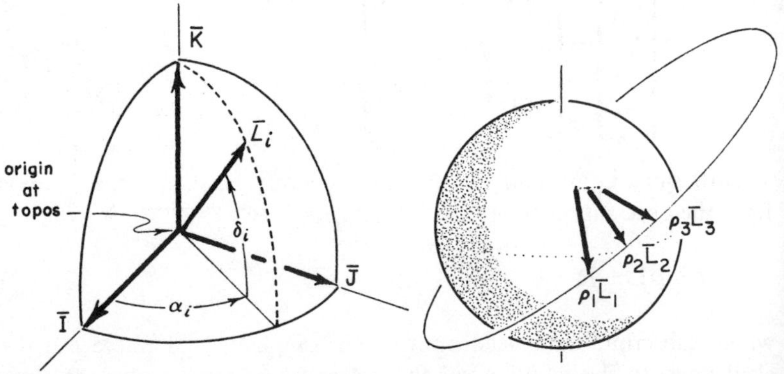

Figure 2.11-1 Line-of-sight vectors

2.11.2 Derivatives of the Line-of-Sight Vector. Since we have the value of $\mathbf{L}$ at three times, t_1, t_2 and t_3, we can numerically differentiate to obtain $\dot{\mathbf{L}}$ and $\ddot{\mathbf{L}}$ at the central date, t_2, provided the three observations are not too far apart in time. We may use the Lagrange interpolation formula to write a general analytical expression for $\mathbf{L}$ as a function of time:

$$\mathbf{L}(t) = \frac{(t-t_2)(t-t_3)}{(t_1-t_2)(t_1-t_3)}\mathbf{L}_1 + \frac{(t-t_1)(t-t_3)}{(t_2-t_1)(t_2-t_3)}\mathbf{L}_2 + \frac{(t-t_1)(t-t_2)}{(t_3-t_1)(t_3-t_2)}\mathbf{L}_3$$

Note that this second order polynomial in t reduces to $\mathbf{L}_1$ when $t = t_1$, $\mathbf{L}_2$ when $t = t_2$ and $\mathbf{L}_3$ when $t = t_3$. Differentiating this equation twice yields $\dot{\mathbf{L}}$ and $\ddot{\mathbf{L}}$, thus

$$\dot{\mathbf{L}}(t) = \frac{2t-t_2-t_3}{(t_1-t_2)(t_1-t_3)}\mathbf{L}_1 + \frac{2t-t_1-t_3}{(t_2-t_1)(t_2-t_3)}\mathbf{L}_2 + \frac{2t-t_1-t_2}{(t_3-t_1)(t_3-t_2)}\mathbf{L}_3 \tag{2.11-6}$$

$$\ddot{\mathbf{L}}(t) = \frac{2}{(t_1-t_2)(t_1-t_3)}\mathbf{L}_1 + \frac{2}{(t_2-t_1)(t_2-t_3)}\mathbf{L}_2 + \frac{2}{(t_3-t_1)(t_3-t_2)}\mathbf{L}_3 . \tag{2.11-7}$$

By setting $t = t_2$ in equations (2.11-6) and (2.11-7) we can obtain numerical values for $\dot{\mathbf{L}}$ and $\ddot{\mathbf{L}}$ at the central date. It should be noted, however, that if more than three observations are available, a more accurate value of $\dot{\mathbf{L}}$ and $\ddot{\mathbf{L}}$ at the central date may be obtained by fitting higher order polynomials with the Lagrange interpolation formula to the observations or, better yet, making a least squares polynomial fit to the observations. This, in fact, must be done if $\dddot{\mathbf{L}}$ and higher order derivatives are not negligible.[4]

Equation (2.11-5) written for the central date now represents three component equations in four unknowns, ρ, $\dot{\rho}$, $\ddot{\rho}$ and r.

2.11-3 Solving for the Vector r. For the time being, let us assume that we know r and solve equation (2.11-5) for ρ using Cramer's rule. The determinant of the coefficients is clearly

$$D = \begin{vmatrix} L_I & 2\dot{L}_I & \ddot{L}_I + \mu L_I/r^3 \\ L_J & 2\dot{L}_J & \ddot{L}_J + \mu L_J/r^3 \\ L_K & 2\dot{L}_K & \ddot{L}_K + \mu L_K/r^3 \end{vmatrix}.$$

Since the value of the determinant is not changed if we subtract μ / r^3 times the first column from the third column, D reduces to

$$D = 2 \begin{vmatrix} L_I & \dot{L}_I & \ddot{L}_I \\ L_J & \dot{L}_J & \ddot{L}_J \\ L_K & \dot{L}_K & \ddot{L}_K \end{vmatrix}. \tag{2.11-8}$$

Applying Cramer's rule to equation (2.11-5), it is evident that

$$D\rho = -\begin{vmatrix} L_I & 2\dot{L}_I & \ddot{R}_I + \mu R_I/r^3 \\ L_J & 2\dot{L}_J & \ddot{R}_J + \mu R_J/r^3 \\ L_K & 2\dot{L}_K & \ddot{R}_K + \mu R_K/r^3 \end{vmatrix}.$$

This determinant can be conveniently split to produce

$$\mathsf{D}\rho = -2\begin{vmatrix} \mathsf{L}_\mathsf{I} & \dot{\mathsf{L}}_\mathsf{I} & \ddot{\mathsf{R}}_\mathsf{I} \\ \mathsf{L}_\mathsf{J} & \dot{\mathsf{L}}_\mathsf{J} & \ddot{\mathsf{R}}_\mathsf{J} \\ \mathsf{L}_\mathsf{K} & \dot{\mathsf{L}}_\mathsf{K} & \ddot{\mathsf{R}}_\mathsf{K} \end{vmatrix} -2\frac{\mu}{r^3}\begin{vmatrix} \mathsf{L}_\mathsf{I} & \dot{\mathsf{L}}_\mathsf{I} & \mathsf{R}_\mathsf{I} \\ \mathsf{L}_\mathsf{J} & \dot{\mathsf{L}}_\mathsf{J} & \mathsf{R}_\mathsf{J} \\ \mathsf{L}_\mathsf{K} & \dot{\mathsf{L}}_\mathsf{K} & \mathsf{R}_\mathsf{K} \end{vmatrix}. \tag{2.11-9}$$

For convenience let us call the first determinant D_1 and the second D_2. Then

$$\rho = \frac{-2\,\mathsf{D}_1}{\mathsf{D}} - \frac{2\mu\,\mathsf{D}_2}{r^3\mathsf{D}}, \quad \mathsf{D} \neq 0. \tag{2.11-10}$$

Provided that the determinant of the coefficients, D, is not zero we have succeeded in solving for ρ as a function of the still unknown r. The conditions that result in D being zero will be discussed in a later section.

From geometry we know that ρ and r are related by

$$\boxed{\mathbf{r} = \rho\mathbf{L} + \mathbf{R}} \tag{2.11-2}$$

Dotting this equation into itself yields

$$r^2 = \rho^2 + 2\rho\,\mathbf{L}\cdot\mathbf{R} + R^2 \tag{2.11-11}$$

Equations (2.11-10) and (2.11-11) represent two equations in two unknowns, ρ and r. Substituting equation (2.11-10) into (2.11-11) leads to an eighth order equation in r which may be solved by iteration.

Once the value of r at the central date is known equation (2.11-10) may be solved for ρ and the vector $\mathbf{r}$ obtained from equation (2.11-2).

2.11.4 Solving for Velocity. Applying Cramer's rule again to equation (2.11-5), we may solve for $\dot{\rho}$ in a manner exactly analogous to that of the preceding section. If we do this we find that

$$\mathsf{D}\dot{\rho} = -\begin{vmatrix} \mathsf{L}_\mathsf{I} & \ddot{\mathsf{R}}_\mathsf{I} & \ddot{\mathsf{L}}_\mathsf{I} \\ \mathsf{L}_\mathsf{J} & \ddot{\mathsf{R}}_\mathsf{J} & \ddot{\mathsf{L}}_\mathsf{J} \\ \mathsf{L}_\mathsf{K} & \ddot{\mathsf{R}}_\mathsf{K} & \ddot{\mathsf{L}}_\mathsf{K} \end{vmatrix} - \frac{\mu}{r^3}\begin{vmatrix} \mathsf{L}_\mathsf{I} & \mathsf{R}_\mathsf{I} & \ddot{\mathsf{L}}_\mathsf{I} \\ \mathsf{L}_\mathsf{J} & \mathsf{R}_\mathsf{J} & \ddot{\mathsf{L}}_\mathsf{J} \\ \mathsf{L}_\mathsf{K} & \mathsf{R}_\mathsf{K} & \ddot{\mathsf{L}}_\mathsf{K} \end{vmatrix}. \tag{2.11-12}$$

For convenience let us call the first determinant D_3 and the second D_4. Then

$$\dot{\rho} = -\frac{D_3}{D} - \frac{\mu}{r^3}\frac{D_4}{D}, \quad D \neq 0 \tag{2.11-13}$$

Since we already know r we can solve equation (2.11-13) for $\dot{\rho}$. To obtain the velocity vector, **v**, at the central date we only need to differentiate **r** in equation (2.11-2).

$$\boxed{\mathbf{v} = \dot{\mathbf{r}} = \dot{\rho}\mathbf{L} + \rho\dot{\mathbf{L}} + \dot{\mathbf{R}}\,.} \tag{2.11-14}$$

2.11.5 Vanishing of the Determinant, D. In the preceding analysis we have assumed that the determinant D is not zero so that Cramer's rule may be used to solve for ρ and $\dot{\rho}$. Moulton[7] has shown that D will be zero only if the three observations lie along the arc of a great circle as viewed from the observation site at time t_2. This is another way of saying that if the observer lies in the plane of the satellite's orbit at the central date, Laplace's method fails.

A somewhat better method of orbit determination from optical sightings will be presented in Chapter 5.

2.12 IMPROVING A PRELIMINARY ORBIT BY DIFFERENTIAL CORRECTION

The preceding sections dealt with the problem of determining a preliminary orbit from a minimum number of observations. This preliminary orbit may be used to predict the position and velocity of the satellite at some future date. As a matter of fact, one of the first things that is done when a new satellite is detected is to compute an ephemeris for the satellite so that a downrange tracking station can acquire the satellite and thus improve the accuracy of the preliminary orbit by making further observations.

There are two ways in which further observations of a satellite may be used to improve the orbital elements. If the downrange station can get another "six-dimensional fix" on the satellite, such as ρ, $\dot{\rho}$, El, $\dot{El}$, Az, $\dot{Az}$ or three optical sightings of $\alpha_1, \delta_1, \alpha_2, \delta_2, \alpha_3, \delta_3$, then a complete redetermination of the orbital elements can be made and the new or "improved" orbital elements taken as the average of all preceding determinations of the elements.

It may happen, however, that the downrange station cannot obtain the type of six-dimensional fix required to redetermine the orbital elements. For example, a future observation may consist of six closely spaced observations of range-rate only. The question then arises, "can this information be used to improve the accuracy of the preliminary orbit?" The answer is "yes." By using a technique known as "differential correction" any six or more subsequent observations may be used to improve our knowledge of the orbital elements.

2.12.1 Computing Residuals. Differential correction is based upon the concept of residuals. A residual is the difference between an actual observation and what the observation would have been if the satellite traveled exactly along the nominal orbit. Because such a nominal orbit will not be exactly correct due to sensor errors or uncertainty in the original observation station's geographical coordinates, the observational data collected by a downrange station (e.g. Doppler range-rate, $\dot{\rho}$) will differ from the computed data.

Suppose that the six components of $\mathbf{r}_O$ and $\mathbf{v}_O$ are taken as the preliminary orbital elements of a satellite at some epoch t_O. Now assume that by some analytical method based on two-body orbital mechanics or some numerical method based on perturbation theory we can predict that the range-rates relative to some downrange observing site will be at six times, t_1, t_2, t_3, t_4, t_5 and t_6. These predictions are, of course, based on the assumption that the six nominal elements $[r_I, r_J, r_K, v_I, v_J, v_K]$ at $t = t_O$ are correct.

The downrange station now makes its observations of $\dot{\rho}$ at the six prescribed times and forms a set of six residuals based on the difference between the predicted values of $\dot{\rho}$ and the actual observations. The six residuals are $\Delta\dot{\rho}_1$, $\Delta\dot{\rho}_2$, $\Delta\dot{\rho}_3$, $\Delta\dot{\rho}_4$, $\Delta\dot{\rho}_5$, and $\Delta\dot{\rho}_6$.

2.12.2 The Differential Correction equations. Assuming that the residuals are small, we can write the following six first-order equations

$$\Delta\dot{\rho}_1 = \frac{\partial\dot{\rho}_1}{\partial r_I}\Delta r_I + \frac{\partial\dot{\rho}_1}{\partial r_J}\Delta r_J + \frac{\partial\dot{\rho}_1}{\partial r_K}\Delta r_K + \frac{\partial\dot{\rho}_1}{\partial v_I}\Delta v_I + \frac{\partial\dot{\rho}_1}{\partial v_J}\Delta v_J + \frac{\partial\dot{\rho}_1}{\partial v_K}\Delta v_K$$

$$\vdots \qquad \vdots \qquad \vdots \qquad \vdots \qquad \vdots \qquad \vdots \qquad \vdots$$

$$\Delta\dot{\rho}_6 = \frac{\partial\dot{\rho}_6}{\partial r_I}\Delta r_I + \frac{\partial\dot{\rho}_6}{\partial r_J}\Delta r_J + \frac{\partial\dot{\rho}_6}{\partial r_K}\Delta r_K + \frac{\partial\dot{\rho}_6}{\partial v_I}\Delta v_I + \frac{\partial\dot{\rho}_6}{\partial v_J}\Delta v_J + \frac{\partial\dot{\rho}_6}{\partial v_K}\Delta v_K$$

(2.12-1)

Assuming that the partial derivatives can be numerically evaluated, equations (2.12-1) constitute a set of six simultaneous linear equations in six unknowns, $\Delta r_I, \Delta r_J, \Delta r_K, \Delta v_I, \Delta v_J, \Delta v_K$. Using matrix methods these equations can be inverted and solved for the correction terms which will then be added to the preliminary orbital elements yielding a "corrected" or "improved" set of orbital elements

$$[r_I + \Delta r_I, r_J + \Delta r_J, \ldots\ldots, v_K + \Delta v_K].$$

These corrected elements are then used to recompute the predicted range-rates at the six observation times. New residuals are formed and the whole process repeated until the residuals cease to become smaller with further iterations. In essence the differential correction process is just a six-dimensional Newton iteration where we are trying, by trial and error, to find the value of the orbital elements at time t_0 that will correctly predict the observations, i.e., reduce the residuals to zero.

2.12.3 Evaluation of Partial Derivatives. The inversion of equations (2.12-1) requires a knowledge of all 36 partial derivatives such as $\partial \dot{\rho}_1 / \partial r_I$, etc. Usually it is impossible to obtain such derivatives analytically. With the aid of a digital computer, however, it is a simple matter to obtain them numerically. All that is required is to introduce a small variation, such as Δr_I, to each of the original orbital elements in turn and compute the resulting variation in each of the predicted $\dot{\rho}$'s. (A variation of 1 or 2 percent in the original elements is usually sufficient.) Then, for example,

$$\frac{\partial \dot{\rho}_1}{\partial r_I} \approx \frac{\dot{\rho}_1(r_I + \Delta r_I, r_J, r_K, \ldots, v_K) - \dot{\rho}_1(r_I, r_J, \ldots, v_K)}{\Delta r_I}.$$

Although the preceding analysis was based on using range-rate data to differentially correct an orbit, the general method is valid no matter what type of data the residuals are based upon. The only requirement is that at least six independent observations are necessary. The following example problems demonstrate the use of the method.

The following notation will be used in the following examples of differential correction:

n = number of observations

p = number of elements (parameters in the equation, usually 6 for orbit problems)

$\tilde{W}$ is an $n \times n$ diagonal matrix whose elements consist of the square of the confidence placed in the corresponding measurements (e.g. for 90% confidence, use 0.81). This way we can reduce the effect of questionable data without completely disregarding it.

α and β are the elements (two here, since $p = 2$ below).

$\tilde{A}$ is an $n \times p$ matrix of partial derivatives of each quantity with respect to each of the elements measured.

$\tilde{b}$ is the $n \times 1$ matrix of residuals based on the previous estimate of the elements.

$\tilde{\Delta z}$ is the $p \times 1$ matrix of computed corrections to our estimates of the elements

x_i independent variable measurement,

y_i dependent variable measurement corresponding to x_i,

$\bar{y}_i$ computed (predicted) value of dependent variable using previous values for the elements (α and β).

If $p > n$ we do not have enough data to solve the problem.

If $p = n$ the result is a set of simultaneous equations that can be solved explicitly for the exact solution.

If $p < n$ we have more equations than we do unknowns so there is no unique solution. In this case we seek the best solution in a "least-squares" sense. This means we find the curve that causes the sum of the squares of the residuals to be a minimum. Or equivalently we minimize the square root of the arithmetic mean of the squared residuals, called the root mean square (RMS). This can occur even though the curve may not pass *through* any points.

The solution to weighted least squares iterative differential correction is given by the following equation:

$$\tilde{\Delta z} = (\tilde{A}^T \tilde{W} \tilde{A})^{-1} \tilde{A}^T \tilde{W} \tilde{b} \tag{2.12-2}$$

Follow this procedure:

1. Solve equation (2.12-2) for the changes to the elements.
2. Correct the elements ($\alpha_{new} = \alpha_{old} + \Delta\alpha$).
3. Compute new residuals using the same data with new elements.
4. Repeat steps 1 through 3 until the residuals are:
   a. zero, for the exactly determined case ($p = n$) or
   b. minimum as described above.

EXAMPLE PROBLEM.[17] To make this example simple let us first use two observations and two elements. In this case the problem is exactly determined ($n = p = 2$).

Let the elements be α and β ($\widetilde{\Delta\alpha} = \widetilde{\Delta z_1}$ and $\widetilde{\Delta\beta} = \widetilde{\Delta z_2}$) in the relationship

$$\bar{y} = \alpha + \beta x$$

Assume equal confidence in all data. Therefore W is the identity matrix and will not be carried through the calculations. Choose $\alpha = 2$ and $\beta = 3$ for the first estimates.

Given:

Observation	x_i	y_i	$\bar{y}_i$	Residual
1	2	1	8	-7
2	3	2	11	-9

where we have predicted the values of the variable y for the two values of x,

$$x_1 = 2, \quad \bar{y}_1 = 2 + 3x_1 = 8$$

$$x_2 = 3, \quad \bar{y}_2 = 2 + 3x_2 = 11$$

and computed the residuals for the (2 x 1) matrix

$$\tilde{b} = \begin{bmatrix} y_1 - \bar{y}_1 \\ y_2 - \bar{y}_2 \end{bmatrix} = \begin{bmatrix} -7 \\ -9 \end{bmatrix}, \quad \widetilde{W} = \begin{bmatrix} 1 & 0 \\ 0 & 1 \end{bmatrix}.$$

From the fitting equation $y = \alpha + \beta x$ the partial derivatives are:

$$\frac{\partial y_i}{\partial \alpha} = 1 \qquad \frac{\partial y_i}{\partial \beta} = x_i$$

The matrix $\tilde{A}$ (2 x 2) is:

$$\tilde{A} = \begin{bmatrix} \dfrac{\partial y_1}{\partial \alpha} & \dfrac{\partial y_1}{\partial \beta} \\ \dfrac{\partial y_2}{\partial \alpha} & \dfrac{\partial y_2}{\partial \beta} \end{bmatrix} = \begin{bmatrix} 1 & 2 \\ 1 & 3 \end{bmatrix}$$

We will now solve equation (2.12-2) for the elements of the (2 x 1) matrix

$$\tilde{\Delta z} = \begin{bmatrix} \Delta\alpha \\ \Delta\beta \end{bmatrix}$$

$$\tilde{A}^T\tilde{A} = \begin{bmatrix} 2 & 5 \\ 5 & 13 \end{bmatrix}, \qquad (\tilde{A}^T\tilde{A})^{-1} = \begin{bmatrix} 13 & -5 \\ -5 & 2 \end{bmatrix}$$

$$\tilde{A}^T\tilde{b} = \begin{bmatrix} -16 \\ -41 \end{bmatrix}$$

$$\tilde{\Delta z} = \begin{bmatrix} \Delta\alpha \\ \Delta\beta \end{bmatrix} = \begin{bmatrix} 13 & -5 \\ -5 & 2 \end{bmatrix} \begin{bmatrix} -16 \\ -41 \end{bmatrix} = \begin{bmatrix} -3 \\ -2 \end{bmatrix}$$

Thus,

$$\alpha_{new} = \alpha_{old} + \Delta\alpha = -1, \quad \beta_{new} = \beta_{old} + \Delta\beta = 1$$

The fitting equation is now

$$y = -1 + x$$

which yields the following:

Observation	x_i	y_i	$\overline{y}_i$	Residual
1	2	1	1	0
2	3	2	2	0

Notice that we achieved the exact result in one iteration. This is not surprising since our equation is that of a straight line and we used only two points.

The relationships between the orbit elements and components of position and velocity are very nonlinear. In practice, large numbers of measurements are used to determine the orbit. This means that in a realistic problem 100 x 6 or larger matrices (100 observations, 6 orbit elements) would be used. This obviously implies the use of a digital computer for the solution. To illustrate the *method* involved, we will use a simple, yet still nonlinear relationship for our "elements" and a small, yet still overspecified number of measurements.

EXAMPLE PROBLEM. Let the elements be α and β in the relationship:

$$y = \alpha x^{\beta} \qquad (p=2)$$

Given the following measurements of equal confidence, let us assume that our curve passes *through* points 3 and 4. This gives us a first guess of

$$\alpha = 0.474 \text{ and } \beta = 3.360$$

Thus using

$$\overline{y}_i = .474\ x_i^{\ 3.36}$$

we predict values of y_i corresponding to the given x_i and compute residuals.

x_i	y_i	$\bar{y}_i$	$y_i - \bar{y}_i$	$(y_i - \bar{y}_i)^2$
1	2.500	0.474	2.026	4.105
2	8.000	4.865	3.135	9.828
3	19.000	19.007	-0.007	0
4	50.000	49.969	0.031	0.001
				13.934

$$\text{RMS residual} = \sqrt{\frac{\sum_{i=1}^{n} (y_i - \bar{y}_i)^2}{n}} = \sqrt{\frac{13.934}{4}} = 1.866$$

The partial derivatives for matrix $\tilde{A}$ are given by:

$$\frac{\partial y_i}{\partial \alpha} = x_i^{\beta} = x_i^{3.36}$$

$$\frac{\partial y_i}{\partial \beta} = \alpha x_i^{\beta} \log_e x_i = .474 x_i^{3.36} \log_e x_i$$

The residuals for matrix $\tilde{b}$ are given by

$$\tilde{b}_i = y_i - .474 x_i^{3.36}$$

Since the data is of equal confidence

$$\tilde{W} = \begin{bmatrix} 1 & 0 & 0 & 0 \\ 0 & 1 & 0 & 0 \\ 0 & 0 & 1 & 0 \\ 0 & 0 & 0 & 1 \end{bmatrix}$$

After computing the elements of the matrices we have

$$\tilde{A} = \begin{bmatrix} 1.00 & 0.00 \\ 10.27 & 3.37 \\ 40.10 & 20.88 \\ 105.41 & 69.27 \end{bmatrix}, \quad \tilde{b} = \begin{bmatrix} 2.026 \\ 3.133 \\ -0.007 \\ 0.031 \end{bmatrix}$$

$$\tilde{A}^T\tilde{W}\tilde{A} = \begin{bmatrix} 12{,}830 & 8{,}175 \\ 8{,}175 & 5{,}246 \end{bmatrix}$$

$$(\tilde{A}^T\tilde{W}\tilde{A})^{-1} = \begin{bmatrix} 0.011 & -.017 \\ -.017 & 0.027 \end{bmatrix}, \quad \tilde{A}^T\tilde{W}\tilde{b} = \begin{bmatrix} 37.21 \\ 12.58 \end{bmatrix}$$

Now using (2.12-2) we have

$$\tilde{\Delta z} = (\tilde{A}^T\tilde{W}\tilde{A})^{-1}\tilde{A}^T\tilde{W}\tilde{b} = \begin{bmatrix} 0.196 \\ -0.304 \end{bmatrix}$$

Thus

$$\alpha_{new} = .474 + .196 \qquad \beta_{new} = 3.36 - .304$$

$$\alpha_{new} = 0.670 \qquad \beta_{new} = 3.056$$

Computation using the new residuals (based on the new elements and the observed data) yields an RMS residual of 2.360, which is larger than what we started with, so we iterate.

The second iteration yields:

$$\Delta\alpha = .062\,, \qquad \Delta\beta = -.018$$

Thus

$$\alpha_{new} = 0.733, \quad \beta_{new} = 3.039$$

Recalculation of residuals using these newer parameters yields an RMS residual of 1.582.

We will consider our RMS residual to be a minimum if its value on two successive iterations differs by less than 0.001. This is not yet the case so we iterate again. The next values of the parameter are:

$$\alpha = .735 \qquad \beta = 3.038.$$

This time the RMS residual is 1.581, which means that to three significant figures we have converged on the best values of α and β.

This best relationship between the given data and our fitting equation is given by

$$y = 0.735 \ x^{3.038}$$

2.12.4 Unequally Weighted Data. If the data is not weighted equally we can observe several things:

First: If a piece of data that is actually exact should be weighted less than other good data, the algorithm above will still converge to the same best value but it will take more iterations.

Second: Weighting incorrect data less than the other data will cause the final values of the elements to be much closer to the values obtained using correct data than if the bad data had been weighted equally. However, the RMS residual value is larger the lower the weight given the bad data. The number of iterations needed for convergence increases the lower the weight.

2.13 SPACE SURVEILLANCE

In the preceding sections we have seen how, in theory, we can determine the orbital elements of a satellite from only a few observations. In practice, however, a handful of observations on new orbiting objects can't secure the degree of precision needed for orbital surveillance and prediction. Typical requirements are for 100-200 observations per object per day during the first few days of orbit, 20-50 observations per object per day to update already established orbits, and finally, during orbital decay, 200-300 observations to confirm and locate reentry.[8]

In 1975 there were nearly 3,500 detected objects in orbit around the earth. By 1980 this number is expected to grow to about 5,000.

2.13.1 The Spacetrack System. The task of keeping track of this growing space population belongs to the 14th Aerospace Force of the Aerospace Defense Command. The data needed to identify and catalogue orbiting objects comes from a network of electronic and optical sensors scattered around the world and known as the "496L Spacetrack System." Spacetrack is a synthesis of many systems: it receives inputs from the Ballistic Missile Early Warning System (BMEWS), the Electronic Intelligence System (ELINT), the Navy's Space Surveillance System (SPASUR), and over-the-horizon radars (OTH). In addition, other sensors are available on an on-call basis: observations are received from the Eastern and Western Test Ranges, the White Sands Missile Range, the Smithsonian Astrophysical Observatory's optical tracking network, and from Air Force Cambridge Research Laboratories' Millstone Hill radar.

In addition to just cataloguing new space objects the mission of Spacetrack has been extended to reconnaissance satellite payload recovery, antisatellite targeting, manned spacecraft/debris collision avoidance, spacecraft failure diagnosis, and midcourse ICBM interception. At present, the greatest effort is being expended on just keeping track of the existing space traffic—a job made more difficult by the Soviet's accidental or deliberate explosion in orbit of satellites and boosters, forming "clouds" of space debris.

2.14 TYPE AND LOCATION OF SENSORS

Since space surveillance is an outgrowth of ballistic trajectory monitoring, it is not surprising that all of our radar sensors are located in the Northern Hemisphere. Satellite tracking cameras deployed around the world by the Smithsonian Astrophysical Observatory (SAO) in support of the recent International Geophysical Year (IGY) provide the data from the Southern Hemisphere.

2.14.1 Radar Sensors. Radar sensors can be broadly categorized into two types: detection fans and trackers. The detection fans—most of which are part of the BMEWS system—consist of two horizontal fan-shaped beams, about 1° in width and $3\frac{1}{2}^{\circ}$ apart in elevation sent out from football-field-size antennas. The horizontal sweep rate is fast enough that a missile or satellite cannot pass through the fans undetected. These detection radars with a range of 2,500-3,500 miles make about 12,000 observations per day—mostly of already catalogued objects. If an "unknown" ballistic object is detected, the precomputed impact area is determined by a "table look-up" procedure at the site based on where the object crossed the two fans and the elapsed time interval between fan crossings.

At present there are two FPS-17 detection radars at Diyarbakir, Turkey, and three more at Shemya in the Aleutians. Four of the larger FPS-50s are deployed at Thule, Greenland, while three are located at Clear, Alaska. One of the earlier detection radars, the FPS-43, at the Trinidad site of the Eastern Test Range, is now on active Spacetrack alert. The only other detection fan which supplies occasional data to Spacetrack on request is located at Kwajalein Island.

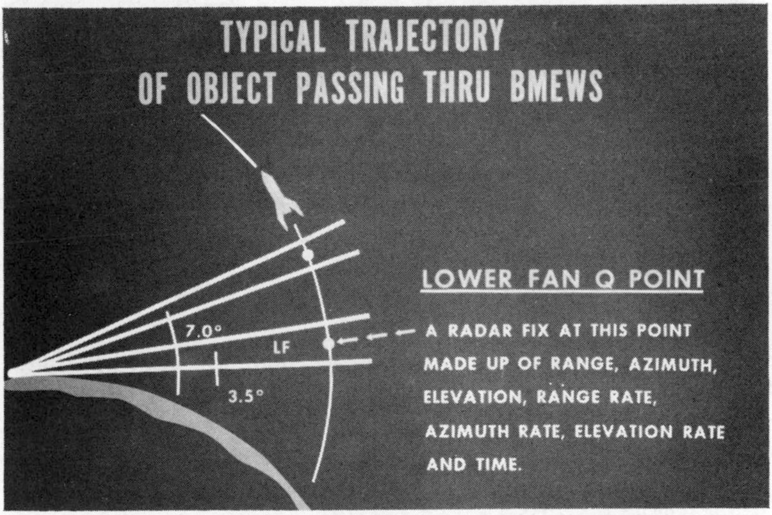

Figure 2.14-1 BMEWS diagram

The best orbit determination data on new satellites comes from the tracking radars scattered around the Spacetrack net. There is usually one tracker associated with each detection radar that can quickly acquire a new target from a simple extrapolation of its track through the detection fans. A typical tracker such as the FPS-49 has an 85-foot mechanically-steered dish antenna weighing 106 tons and is capable of scan rates up to 10° per second.[8] The prototype is located at Moorestown, New Jersey, and is on active spacetrack alert. One FPS-49

Figure 2.14-2. *This BMEWS station at Clear, Alaska, uses a combination of one RCA FPS-92 tracker, under the 140-ft radome at the left, plus three GE FPS-50 detection radars which stand 165 ft high and 400 ft long. BMEWS which has other sites at Thule and Fylingdales Moor supplies data on orbiting satellites to Spacetrack regularly.*

is at Thule and three are at Fylingdales Moor in Yorkshire, United Kingdom. An advanced version of this tracker (the FPS-92) featuring more elaborate receiver circuits and hydrostatic bearings is operating at Clear, Alaska.

In addition, there is an FPS-79 at Diyarbakir and an FPS-80 at Shemya. The one at Diyarbakir has a unique feature which enhances its spacetrack usefulness. A variable-focus feed horn provides a wide beam for detection and a narrow beamwidth for tracking. Pulse compression is used to improve both the gain and resolution of the 35-foot dish antenna.

An interesting new development in tracking radars is the FPS-85 with a fixed "phased-array" antenna and an electronically-steered beam. The prototype located at Eglin AFB, Florida, gives radar coverage of the Caribbean area. It is capable of tracking several targets simultaneously.

One other radar sensor that contributes to the Spacetrack System is

Figure 2.14-3. *The Bendix FPS-85 phased-array tracking radar at Eglin AFB. Two antenna arrays are used for electronic beam steering; the square array is the transmitter while the octagon serves as the receiver. High sensor beam speed permits tracking of multiple targets but only at the expense of accuracy.*

over-the-horizon radar with transmitters located in the Far East and receivers scattered in Western Europe. OTH radar operates on the principle of detecting launches and identifying the signature of a particular booster by the disturbance it causes in the ionosphere.

2.14.2 Radio Interferometers. Another class of sensors which provide accurate directional information on a satellite is based on the principle of radio interferometry. The original system using this technique was Minitrack—used to track Vanguard. It was a passive system requiring radio transmitters aboard the satellite. The Navy's SPASUR net is an active system of three transmitters and six receiving antennas stretching across the country along 33° N latitude from California to Georgia. The transmitters send out a continuous carrier wave at 108 Mc in a thin vertical fan. When a satellite passes through this "fence" a satellite reflected signal is received at the ground. The zenith angle of arrival of the signal is measured precisely by a pair of antennas spaced along the ground at the receiving site. When two or

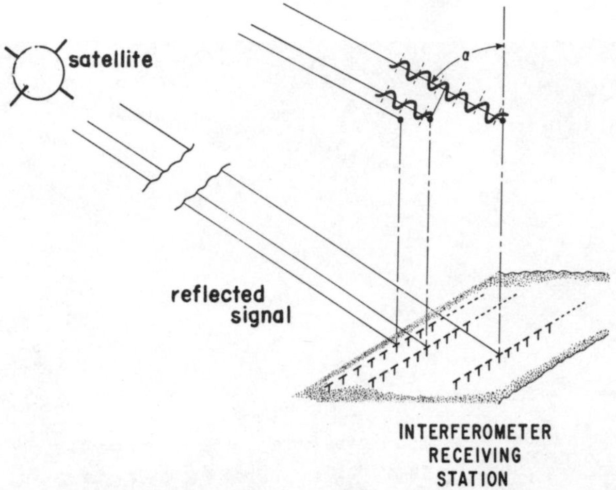

Figure 2.14-4 Determining direction to satellite from phase difference in signal

more receiving sites are used, the position of the satellite passing through the fence is determined by triangulation. To obtain a preliminary orbit from the first pass through the fence the rate of change of phase between the most widely spaced antenna pairs in the East-West and North-South directions is used to determine the velocity vector.[9] An orbit obtained in this way is very crude, but is useful in predicting the next pass through the system. After the second pass a refinement can be made as the period, and therefore, the semi-major axis, is well established. These observations give information from only one part of the satellite orbit, but after 12 hours the earth rotates the system under the "backside" of the orbit allowing further improvement of the orbital elements. Considering the type of observational data received, SPASUR is best utilized in the role of updating already established orbits by differential correction techniques.

2.14.3 Optical Sensors. SAO operates more than a dozen optical tracking stations around the world, each equipped with a Baker-Nunn telescopic camera. In addition to these, two Baker-Nunns are operated by 14th Aerospace Force at Edwards AFB and Sand Island in the Pacific, and the RCAF operates one at Cold Lake, Alberta, Canada. The Baker-Nunn instrument is an F/1 Schmidt camera of 20-inch focal

Figure 2.14-5. *A Baker-Nunn satellite camera operated by the RCAF at Cold Lake, Alberta. Optical tracking with Baker-Nunn cameras plays a gap-filler and calibration role for Spacetrack.*

length with a field of view 5° by 30°. The camera alternately tracks the satellite and then the star background. A separate optical system superimposes, on the same strip of Cinemascope film, the image of a crystal-controlled clock which is periodically illuminated by strobe lights to establish a time reference. From the photograph the position (topocentric right ascension and declination) of the satellite can be accurately determined by comparison with the well-known positions of the background stars.[10]

Under favorable conditions, the instrument can photograph a 16th magnitude object; it recorded the 6-inch diameter Vanguard I at a distance of 2,400 miles.

Despite the high accuracy and other desirable features, the Baker-Nunn data has certain inherent disadvantages. For a good photograph the weather must be favorable, seeing conditions must be good, and the lighting correct. The latter condition means the site must be in darkness and the satellite target in sunlight. As a result, it is usually impossible to get more than a few observations of the orbit at a desired point for any particular spacecraft. Further, precise data reduction cannot be done in the field and, in any case, takes time.

One possibility of reducing the data processing time is with an image orthicon detector coupled to the Schmidt telescope by fiber optics. This would theoretically allow real-time analysis of the tracking data and is under development at the Cloudcroft, New Mexico site.

In any case, the Baker-Nunn cameras provide one of the few sources of data from the Southern Hemisphere and are extensively used for calibration of the radar sensors in the Spacetrack net.

2.14.4 Typical Sensor Errors. With all the radar trackers located in the Northern Hemisphere, it is not surprising that the predicted position of a new satellite after one revolution can be in error by as much as 110 km.[8] An in-track error of this amount would make the satellite nearly 15 seconds early or late in passing through a detection fan or the SPASUR fence. Several factors combine to make these first-pass residuals large. (A residual is the difference between some orbital coordinate predicted on the basis of the preliminary orbital elements and the measured value of that coordinate.) Sensor errors themselves contribute to the residuals. For detection radars, satellite position uncertainties can be as high as 5,000 meters, while for tracking radars the uncertainty can vary from 100 to 500 meters depending on whether they use pulse compression. Doppler range-rate information, on the other hand, is relatively accurate. Radial velocity uncertainties may be as low as 1/6 meter/sec.[11] Most of the Spacetrack radars can achieve pointing accuracies of 36 arc-seconds.[11] The 120-foot dish of the Millstone Hill tracker is good to 18 arc-seconds.[12] Unfortunately radar sensors need almost constant recalibration to maintain these accuracies.

The radio interferometer technique (Minitrack, SPASUR) yields directional information accurate to 20-40 seconds of arc and time of passage through the radio fence accurate to 2-4 milliseconds.[11]

The most accurate angular fix is obtained from Baker-Nunn camera data. On-site film reduction is accurate to only 30 seconds of arc but films sent to Cambridge, Massachusetts, for laboratory analysis yield satellite positions accurate to 3 arc-seconds.[13]

Another source of sensor inaccuracy is the uncertainty in the geodetic latitude and longitude of the tracking site. These uncertainties contribute 30-300 meters of satellite prediction error.

Figure 2.14-6. *The RCA TRADEX instrumentation radar on Kwajalein. Both TRADEX and Millstone Hill are called upon for Spacetrack sightings although they are assigned to other projects.*

Even if all sensor errors could be eliminated, persistant residuals of about 5 km in position or 0.7 seconds in time would remain. The persistant residue levels are due to departures from two-body orbital motion caused by the earth's equatorial bulge, nonuniform gravitational fields, lunar attraction, solar radiation pressure and atmospheric drag. Although general and special perturbation techniques are used to account for these effects, more accurate models for the earth and its atmosphere are needed to reduce the residuals still further.

2.14.5 Future Developments. Some of the most sophisticated instrumentation radar in the world is now operating or is scheduled for the radar complex at Kwajalein atoll in the Pacific.

The TRADEX (Target Resolution and Discrimination Experiments) system at Roi Namur is a highly sensitive 84-foot tracker which operates in three radar bands, providing test data on multiple targets at extreme range. Not only can this radar detect and track a number of targets simultaneously, but it also records target tracks for later playback.

The ALTAIR (ARPA Long-Range Tracking and Instrumentation Radar) system is an advanced version of TRADEX operating on two bands and featuring a larger (120-foot) antenna and higher average power for extreme sensitivity.

An improved system planned for the early 1970s is ALCOR (ARPA-Lincoln C-Band Observables Radar). Operating in the SHF band, it will provide extremely fine-grain resolution data in both range and range-rate for the same type and complex of targets as TRADEX and ALTAIR.[14]

The trend in future tracking radars seems to be toward some form of electronically-steered "agile beam" radar. This may take the form of a hybrid between the mechanically-steered dish antenna and the fixed phased-array configuration. It is certain that solid-state circuitry will play an increasing role in surveillance and tracking systems.

One of the most promising innovations in tracking is the use of laser technology. Theoretically, the laser is capable of providing real-time tracking data more accurately than any other method.

Until recently most laser tracking systems required an optical reflector on the target satellite. Now, a continuous-wave, Doppler-type laser radar that does not require a cooperative target is under development at MIT's Lincoln Laboratory. Many improvements, however, are required before a workable laser radar system can be realized.

In the system under development, the target-reflected laser beam, which is shifted in frequency by the moving target, is combined with a sample of the laser's output producing a signal at the difference frequency of the two mixed beams that is proportional to the range-rate, $\dot{\rho}$, of the target.[15]

A serious disadvantage of laser systems is that they only work in good weather.

2.15 GROUND TRACK OF A SATELLITE

While knowing the orbital elements of a satellite enables you to visualize the orbit and its orientation in the IJK inertial reference frame, it is often important to know what the ground track of a

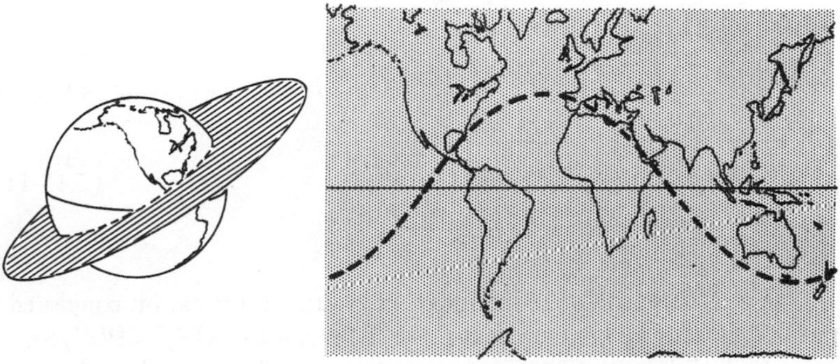

Figure 2.15-1 Ground trace

satellite is. One of the most valuable characteristics of an artificial earth satellite is its ability to pass over large portions of the earth's surface in a relatively short time. As a result it has tremendous potential as an instrument for scientific or military surveillance.

2.15.1 Ground Track on a Nonrotating Earth. The orbit of an earth satellite always lies in a plane passing through the center of the earth. The track of this plane on the surface of a nonrotating spherical earth is a great circle. If the earth did not rotate the satellite would retrace the same ground track over and over. Figure 2.15-1 shows what the ground track would look like on a Mercator projection.

Notice that the maximum latitude north or south of the equator that the satellite passes over is just equal to the inclination, i, of the orbit. For a retrograde orbit the most northerly or southerly latitude on the ground track is $180^\circ - i$.

2.15.2 Effect of Launch Site Latitude and Launch Azimuth on Orbit Inclination. We can determine the effect of launch site latitude and launch azimuth on orbit inclination by studying Figure 2.15-2. Suppose a satellite is launched from point C on the earth whose latitude and longitude are L_o and λ_o, respectively with a launch azimuth, β_o. The ground track of the resulting orbit crosses the equator at a point A at an angle equal to the orbital inclination. The arc CB which forms the third side of a spherical triangle is formed by the meridian passing through the launch site and subtends the angle L_o at the center of the earth. Since we know two angles and the included side of this triangle we can solve for the third angle, i:

$$\cos i = -\overset{0}{\cancel{\cos 90^\circ}}\ \cos \beta_o + \overset{1}{\cancel{\sin 90^\circ}}\ \sin \beta_o\ \cos L_o$$

$$\boxed{\cos i = \sin \beta_o\ \cos L_o .} \tag{2.15-1}$$

There is a tremendous amount of interesting information concealed in this innocent-looking equation. For a direct orbit $(0 \leqslant i < 90^\circ)$ cos i must be positive. Since L_o can range between 0° and 90° for launch sites in the northern hemisphere and between 0° and -90° for launch sites in the southern hemisphere, cos L_o must always be positive. A direct orbit requires, therefore, that the launch azimuth, β_o, be easterly, i.e., between 0° and 180°.

Suppose we now ask "what is the minimum orbital inclination we can achieve from a launch site at latitude, L_o?" If i is to be minimized, cos i must be maximized which implies that launch azimuth, β_o, should be 90°. For a due east launch equation (2.15-1) tells us that the orbital inclination will be the minimum possible from a launch site at latitude, L_o and i will be precisely equal to L_o!

Among other things, this tells us that a satellite cannot be put directly into an equatorial orbit $(i = 0^\circ)$ from a launch site which is not on the equator. The Soviet Union is at a particular disadvantage in this regard because none of its launch sites is closer than 45° to the equator so it cannot launch a satellite whose inclination is less than 45°. If the Soviets wish to establish an equatorial orbit it requires a plane change of at least 45° after the satellite is established in its initial orbit. This is an expensive maneuver as we shall see in the next chapter.

2.15.3 Effect of Earth Rotation on the Ground Track. The orbital plane of a satellite remains fixed in space while the earth turns under the orbit. The net effect of earth rotation is to displace the ground track westward on each successive revolution of the satellite by the number of degrees the earth turns during one orbital period. The result is illustrated in Figure 2.15-3.

Instead of retracing the same ground track over and over a satellite

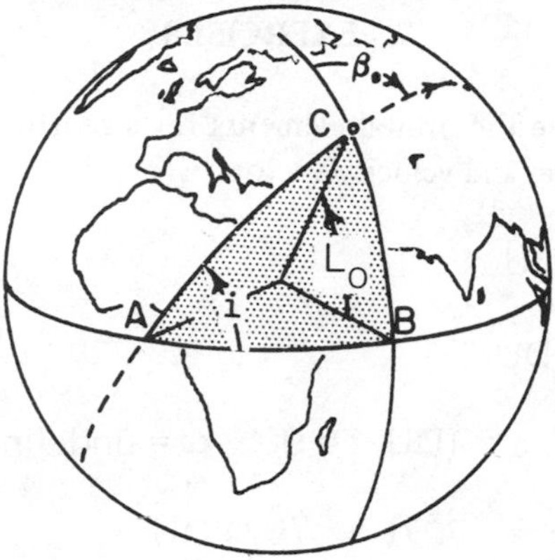

Figure 2.15-2 Effect of launch azimuth and latitude on inclination

eventually covers a swath around the earth between latitudes north and south of the equator equal to the inclination. A global surveillance satellite would have to be in a polar orbit to overfly all of the earth's surface.

If the time required for one complete rotation of the earth on its axis (23 hrs 56 min) is an exact multiple of the satellite's period then eventually the satellite will retrace exactly the same path over the earth as it did on its initial revolution. This is a desirable property for a reconnaissance satellite where you wish to have it overfly a specific target once each day. It is also desirable in manned spaceflights to overfly the primary astronaut recovery areas at least once each day.

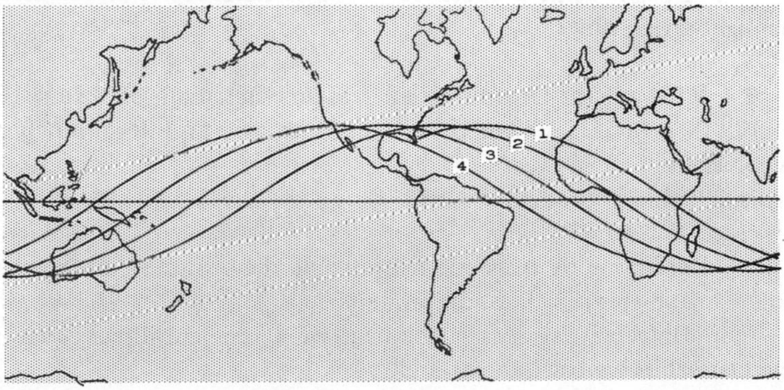

Figure 2.15-3 Westward displacement of ground trace due to earth rotation

EXERCISES

2.1 Determine the orbital elements for an earth orbit which has the following position and velocity vectors:

$$\mathbf{r} = 1\mathbf{K} \text{ DU}$$

$$\mathbf{v} = 1\mathbf{I} \text{ DU/TU}$$

(Partial answers: $a = 1\text{DU}$, $i = 90^\circ$, $\omega =$ undefined)

2.2 Given $\mathbf{r} = -.707\mathbf{I} + .707\mathbf{J}$ DU

$\mathbf{v} = 1/2\mathbf{J}$ DU/TU)

Determine the orbital elements and sketch the orbit.
(Partial answers: $p = 1/8\text{DU}$, $e = .885$, $\nu_0 = 173^\circ$)

2.3 Answer the following:

a. Which takes longer, a solar day or sidereal day?

b. What causes an apparent solar day to be different from a mean solar day?

c. What was the local sidereal time (radians) of Greenwich, England, at 0448 hours (local) on 3 January 1970?

d. What was the local sidereal time (radians) of the US Air Force Academy, Colorado (104.89° W Long) at that same time?

e. Does it make a difference whether Colorado is on standard or daylight saving time?

2.4 What was the Greenwich sidereal time in radians on 3 June 1970 at $17^h\ 00^m\ 00^s$ UT? What is the remainder over an integer number of revolutions?
(Ans. 970.171487 radians = GST)

2.5 Determine by inspection if possible the orbital elements for the following objects (earth canonical units):

a. Object A is crossing the negative **J** axis in a direct equatorial circular orbit at an altitude of 1 DU.

b. Object B departs from a point $\mathbf{r} = -1\mathbf{K}$ DU with local escape speed in the $-\mathbf{I}$ direction.

c. Object C departs from a point $\mathbf{r} = 1\mathbf{K}$ DU $\mathbf{v} = \sqrt{2}\,\mathbf{I} + \sqrt{2}\,\mathbf{J}$ DU/TU.

OBJECT	p	e	i	Ω	ω	ν_o
A	____	____	____	____	____	____
B	____	____	____	____	____	____
C	____	____	____	____	____	____

2.6 Radar readings determine that an object is located at $1.2\mathbf{K}$ DU with a velocity of $(0.4\mathbf{I} - 0.3\mathbf{K})$ DU/TU. Determine $p, e, i, u_o, \Omega, \omega, \nu_o, \ell_o$ and the latitude of impact.
(Answer [partial] $e = .82075$, $\ell_o = 270^\circ$)

2.7 A radar site reduces a set of observed quantities such that:

$$\mathbf{r}_o = -\mathbf{I} - \mathbf{J} - \mathbf{K} \quad (\mathrm{DU}_\oplus)$$

$$\mathbf{v}_o = \frac{1}{3}(\mathbf{I} - \mathbf{J} + \mathbf{K}) \quad (\mathrm{DU}_\oplus/\mathrm{TU}_\oplus)$$

in the geocentric equatorial coordinate system. Determine the orbital elements.
(Partial answers: $p = 8/9$, $\omega = 104^\circ 51'$)

2.8 For the following orbital elements:

$p = .23\ \mathrm{DU}_\oplus$	$\Omega = 180^\circ$
$e = .82$	$\omega = 260^\circ$
$i = 90^\circ$	$\nu_o = 190^\circ$
$u_o = 90^\circ$	$\ell_o = 270^\circ$

a. Express the $\mathbf{r}$ and $\mathbf{v}$ vectors for the satellite in the perifocal system along the unit vectors **P, Q, W.**

b. By a suitable coordinate transformation technique express the **r** and **v** vectors in the geocentric equatorial system in the IJK system. (Answer: **r** = 1.195**K**)

2.9 A radar station at Sunnyvale, California, makes an observation on an object at 2048 hours, PST, 10 January 1970. Site longitude is 121.5° West. What is the local sidereal time?
(Ans. LST = 63.8927170 radians)

2.10 A basis IJK requires 3 rotations before it can be lined up with another basis UVW; the 1st rotation is 30° about the first axis; the 2nd rotation is 60° about the second axis; the final rotation is 90° about the third axis.

a. Find the matrix required to transform a vector from the IJK basis to the UVW basis.

b. Transform **r** = 2**I** - **J** + 4**K** to the UVW basis.

2.11 The values for θ_{go} for 1968 through 1971 given in the text cluster near a value of approximately 100°. Explain why you would expect this to be so.

2.12 Determine the orbital elements by inspection for an object crossing the positive Y axis in a retrograde, equatorial, circular orbit at an altitude of 1 DU.

2.13 Determine the orbital elements of the following objects using the Gibbs method. Use of a computer is suggested, but not necessary. Be sure to make all the tests since all the orbits may not be possible. Units are earth canonical units.

		I	**J**	**K**
a.	$\mathbf{r}_1$	1.41422511	0	1.414202
	$\mathbf{r}_2$	1.81065659	1.06066883	0.3106515
	$\mathbf{r}_3$	1.35353995	1.4142251l	−0.6464495

(Partial answer: e = .171, p = 1.76)

b.	$\mathbf{r}_1$	0.70711255	0	0.70710101
	$\mathbf{r}_2$	−0.89497879	0.56568081	−0.09496418
	$\mathbf{r}_3$	−0.09497879	−0.56568081	−0.89497724
c.	$\mathbf{r}_1$	1.0	0	0
	$\mathbf{r}_2$	−0.8	0.6	.0
	$\mathbf{r}_3$	0.8	−0.6	0
d.	$\mathbf{r}_1$	0.20709623	3.53552813	1.2071255
	$\mathbf{r}_2$	0.91420062	4.9497417	1.91423467
	$\mathbf{r}_3$	1.62130501	6.36395526	2.62134384

(Partial answer: Straight line orbit - hyperbola with infinite eccentricity)

e.	$\mathbf{r}_1$	1.0	0	0
	$\mathbf{r}_2$	0	1.0	0
	$\mathbf{r}_3$	−1.0	0	0
f.	$\mathbf{r}_1$	7.0	2.0	0
	$\mathbf{r}_2$	1.0	1.0	0
	$\mathbf{r}_3$	2.0	7.0	0
g.	$\mathbf{r}_1$	0	2.7	0
	$\mathbf{r}_2$	2.97	0	0
	$\mathbf{r}_3$	−2.97	0	0

***2.14** A BMEWS site determines that an object is located at $1.2\mathbf{K}$ DU with a velocity of $.4\mathbf{I} - .3\mathbf{K}$ DU/TU. Determine the orbital elements of the orbit and the latitude of impact of the object.

***2.15** Given the orbital elements for objects A, B, C and D fill in the blank to correctly complete the following statements:

OBJECT	i	Ω	Π	ℓ_0
A	0°	undefined	210°	30°
B	4°	180°	260°	90°
C	110°	90°	110°	140°
D	23°	60°	260°	160°

a. Object ____ is in retrograde motion.

b. Object ____ has a true anomaly at epoch of 180°.

c. Object ____ has its perigee south of the equatorial plane.

d. Object ______ has a line of nodes which coincides with the vernal equinox direction.

e. Object ______ has an argument of perigee of 200°.

***2.16** A radar site located in Greenland observes an object which has components of the position and velocity vectors only in the **K** direction of the geocentric-equatorial coordinate system. Draw a sketch of the orbit and discuss the orbit type.

***2.17** A radar tracking site located at 30° N, 97.5° W obtains the following data at 0930 GST for a satellite passing directly overhead:

$$\rho = 0.1 \text{ DU} \qquad \dot{\rho} = 0$$

$$Az = 30^\circ \qquad \dot{Az} = 0$$

$$El = 90^\circ \qquad \dot{El} = 10 \text{ RAD/TU}$$

a. Determine the rectangular coordinates of the object in the topocentric-horizon system.

b. What is the velocity of the satellite relative to the radar site in terms of south, east and zenith (up) components?

c. Express the vector **r** in terms of topocentric-horizon coordinates.

d. Transform the vector **r** into geocentric-equatorial coordinates.

e. Determine the velocity **v** in terms of geocentric-equatorial coordinates.
(Ans. $\mathbf{v} = 0.6201\,\mathbf{I} - 0.0078\,\mathbf{J} - 0.75\,\mathbf{K}$ DU/TU)

LIST OF REFERENCES

1. Gauss, Carl Friedrich. *Theory of the Motion of the Heavenly Bodies Moving About the Sun in Conic Sections.* A translation of *Theoria Motus (1857)* by Charles H. Davis. New York, Dover Publications, Inc, 1963.

2. Armitage, Angus. *Edmond Halley*. London, Thomas Nelson and Sons Ltd, 1966.

3. Newman, James R. Comentary on "An Ingenious Army Captain and on a Generous and Many-sided Man," in *The World of Mathematics*. Vol 3. New York, NY, Simon and Schuster, 1956.

4. Escobal, Pedro Ramon. *Methods of Orbit Determination*. New York, NY, John Wiley & Sons, Inc, 1965.

5. Baker, Robert M. L. Jr. *Astrodynamics: Applications and Advanced Topics*. New York and London, Academic Press, 1967.

6. Laplace, Pierre Simon. *Memoires de l' Academie Royale des Sciences de Paris*, 1780.

7. Moulton, Forest Ray. *An Introduction to Celestial Mechanics*. Second Revised Edition, New York, NY, The Macmillan Company, 1914.

8. Thomas, Paul G. "Space Traffic Surveillance," *Space/Aeronautics*. Vol 48, No 6, November 1967.

9. Cleeton, C. E. "The U. S. Navy Space Surveillance System," *Proceedings of the IAS Symposium on Tracking and Command of Aerospace Vehicles*. New York, NY, Institute of the Aerospace Sciences, 1962.

10. Henize, K. G. "The Baker-Nunn Satellite Tracking Camera," *Sky and Telescope*. Vol 16, pp 108-111, Cambridge, Mass, Sky Publishing Corp, 1957.

11. Davies, R. S. "Aspects of Future Satellite Tracking and Control," *Proceedings of the IAS Symposium on Tracking and Command of Aerospace Vehicles.* New York, NY, Institute of the Aerospace Sciences, 1962.

12. *Deep Space and Missile Tracking Antennas.* Proceedings of a Symposium held Nov 27–Dec 1, 1966. New York, NY, American Society of Mechanical Engineers, 1966.

13. Briskman, Robert D. "NASA Ground Support Instrumentation," *Proceedings of the IAS Symposium on Tracking and Command of Aerospace Vehicles.* New York, NY, Institute of the Aerospace Sciences, 1962.

14. Cheetham, R. P. "Future Radar Developments," *The Journal of the Astronautical Sciences.* Vol 14, No 5, Sep–Oct 1967.

15. *Electronic.* Vol 40, No 20, p 39. New York, NY, McGraw-Hill, Inc, Oct 2, 1967.

16. Bate, Roger R. and Eller, Thomas J. "An Improved Approach to the Gibbsian Method," Department of Astronautics and Computer Science Report A-70-2, USAF Academy, Colorado, 1970.

17. Carson, Gerald C. "Computerized Satellite Orbit Determination," Satellite Control Facility, Air Force Systems Command, Sunnyvale, California, 1966.

CHAPTER 3

BASIC ORBITAL MANEUVERS

> But if we now imagine bodies to be projected in the directions of lines parallel to the horizon from greater heights, as of 5, 10, 100, 1,000 or more miles, or rather as many semidiameters of the earth, those bodies, according to their different velocity, and the different force of gravity in different heights, will describe arcs either concentric with the earth, or variously eccentric, and go on revolving through the heavens in those orbits just as the planets do in their orbits.
>
> –Isaac Newton[1]

The concept of artificial satellites circling the earth was introduced to scientific literature by Sir Isaac Newton in 1686. After that, for the next 250 years, the idea seems to have been forgotten. The great pioneers of rocketry–Ziolkovsky, Goddard, and Oberth–were the first to predict that high performance rockets together with the principle of "staging" would make such artificial satellites possible. It is a curious fact that these pioneers foresaw and predicted manned satellites but none of them could see a use for unmanned earth satellites. Ley[2] suggests that the absence of reliable telemetering techniques in the early 1930s would explain this oversight.

Who was the first to think of an unmanned artificial satellite after the concept of a manned space station had been introduced by Hermann Oberth in 1923 still remains to be established. It very likely may have been Wernher von Braun or one of the others at Peenemünde–the first place on earth where a group of people with space-travel inclinations was paid to devote all their time, energy and imagination to rocket research.[2] Walter Dornberger in his book *V-2* mentions that the discussions about future developments at Peenemünde included the rather macabre suggestion that the space-travel pioneers be honored by placing their embalmed bodies into glass spheres which would be put into orbit around the earth.[3]

The earliest plans to actually fire a satellite into earth orbit were proposed by von Braun at a meeting of the Space-Flight Committee of the American Rocket Society in the Spring of 1954. His plan was to use a Redstone rocket with successive clusters of a small solid propellant rocket called "Loki" on top. The scheme was endorsed by the Office of Naval Research and dubbed "Project Orbiter." Launch date was tentatively set for midsummer of 1957.

However, on 29 July 1955, the White House announced that the United States would orbit an artificial earth satellite called "Vanguard" as part of its International Geophysical Year Program.

Project Orbiter, since it contemplated the use of military hardware, was not considered appropriate for launching a "strictly scientific" satellite and the plan was shelved.

On 4 October 1957, the Soviet Union successfully orbited Sputnik I. After several Vanguard failures the von Braun team was at last given its chance and put up Explorer I on 1 February 1958. Vanguard I was finally launched successfully 1½ months later.

In this chapter we will describe the methods used to establish earth satellites in both low and high altitude orbits and the techniques for maneuvering them from one orbit to another.

3.1 LOW ALTITUDE EARTH ORBITS

Manned spaceflight, still in its infancy, has largely been confined to regions of space very near the surface of the earth. The reason for this is neither timidity nor lack of large booster rockets. Rather, the environment of near-earth space conspires to limit the altitude of an artificial satellite, particularly if it is manned, to a very narrow region just above the earth's sensible atmosphere.

Altitudes below 100 nm are not possible because of atmospheric drag and the Van Allen radiation belts limit manned flights to altitudes below about 300 nm.

3.1.1 Effect of Orbital Altitude on Satellite Lifetimes. The exact relationship between orbital altitude and satellite lifetime depends on several factors. For circular orbits of a manned satellite about the size

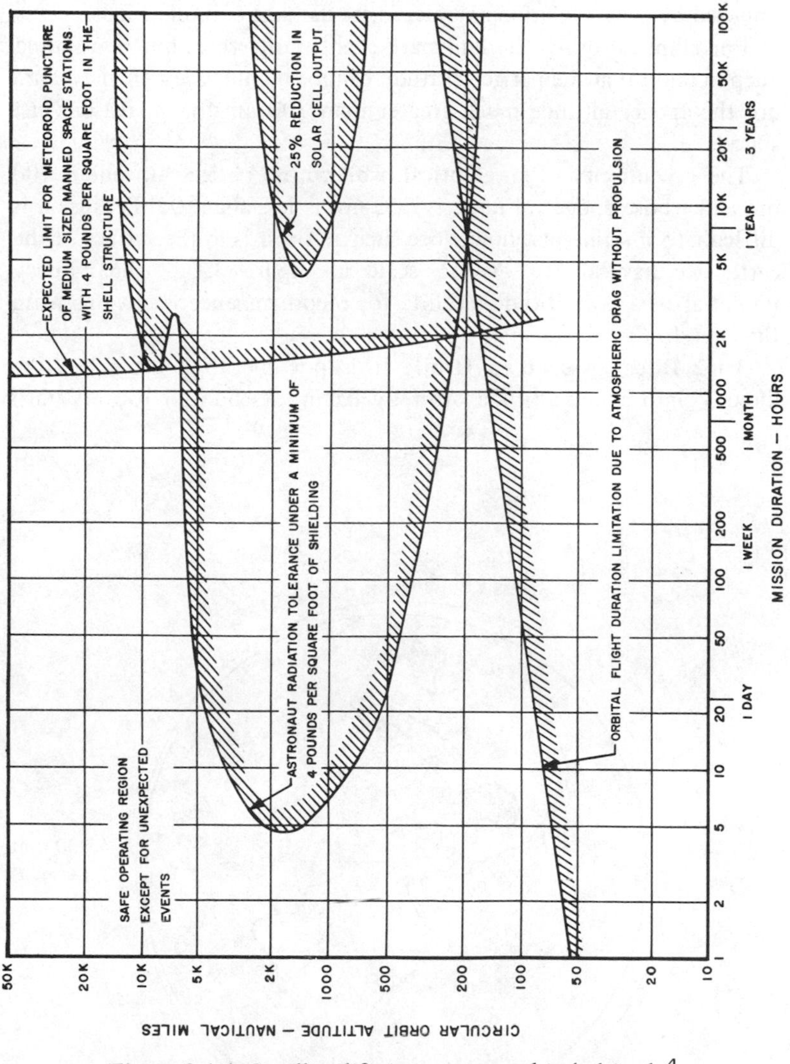

Figure 3.1-1 Satellite lifetime versus orbital altitude[4]

of the Gemini or Apollo spacecraft, Figure 3.1-1 shows the limits imposed by drag, radiation and meteorite damage considerations.

For elliptical orbits the limits are slightly different, but you should keep in mind that the perigee altitude cannot be much less than 100 nm nor the apogee altitude much greater than 300 nm for the reasons just stated.

The eccentricity of an elliptical orbit whose perigee altitude is 100 nm and whose apogee altitude is 300 nm is less than .03! Because it is difficult to imagine just how close such an orbit is to the surface of the earth, we have drawn one to scale in Figure 3.1-2. The military potential of a low altitude satellite for reconnaissance is obvious from this sketch.

3.1.2 Direct Ascent to Orbit. It is possible to inject a satellite directly into a low altitude orbit by having its booster rockets burn

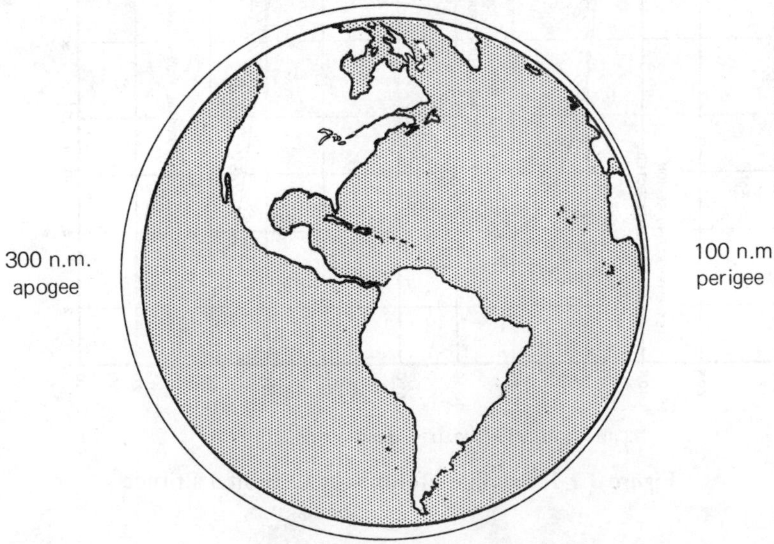

Figure 3.1-2 Typical low altitude earth orbit

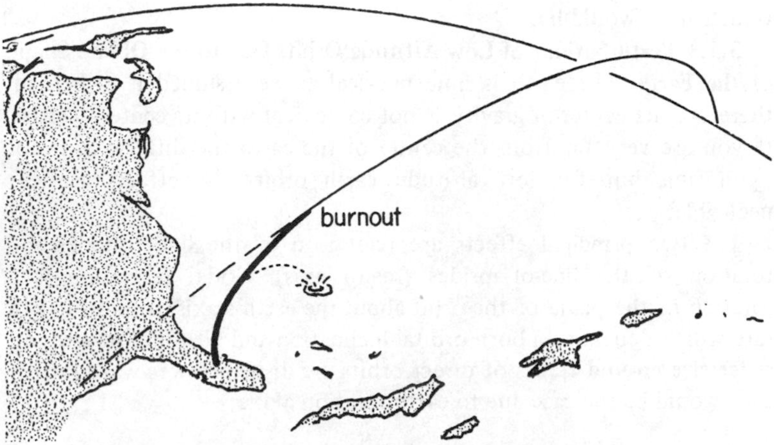

Figure 3.1-3 Direct ascent into a low altitude orbit

continuously from lift-off to a burnout point somewhere on the desired orbit. The injection or burnout point is usually planned to occur at perigee with a flight-path angle at burnout of 0°. Any deviation from the correct burnout speed or flight-path angle could be catastrophic for, as you can see from Figure 3.1-2, there is very little clearance between the orbit and the surface of the earth.

It normally takes at least a two-stage booster to inject a two- or three-man vehicle into low earth orbit. The vehicle is not allowed to coast between first stage booster separation and second stage ignition. The powered flight trajectory looks something like what is shown in Figure 3.1-3. The vehicle rises vertically from the launch pad, immediately beginning a roll to the correct azimuth. The pitch program–a slow tilting of the vehicle to the desired flight-path angle–normally begins about 15 seconds after lift-off and continues until the vehicle is traveling horizontally at the desired burnout altitude. The final burnout point is usually about 300 nm downrange of the launch point.

The first stage booster falls to the earth several hundred miles downrange but the final stage booster, since it has essentially the same speed and direction at burnout as the satellite itself, may orbit the earth for several revolutions before atmospheric drag causes its orbit to decay and it re-enters. The lower ballistic coefficient (weight to area ratio) of the empty booster causes it to be affected more by drag than the vehicle itself would be.

3.1.3 Perturbations of Low Altitude Orbits Due to the Oblate Shape of the Earth. The earth is not spherical as we assumed it to be and, therefore, its center-of-gravity is not coincident with its center-of-mass. If you are very far from the center of the earth the difference is not significant, but for low altitude earth orbits the effects are not negligible.

The two principal effects are regression of the line-of-nodes and rotation of the line-of-apsides (major axis). Nodal regression is a rotation of the plane of the orbit about the earth's axis of rotation at a rate which depends on both orbital inclination and altitude. As a result, successive ground traces of direct orbits are displaced westward farther than would be the case due to earth rotation alone.

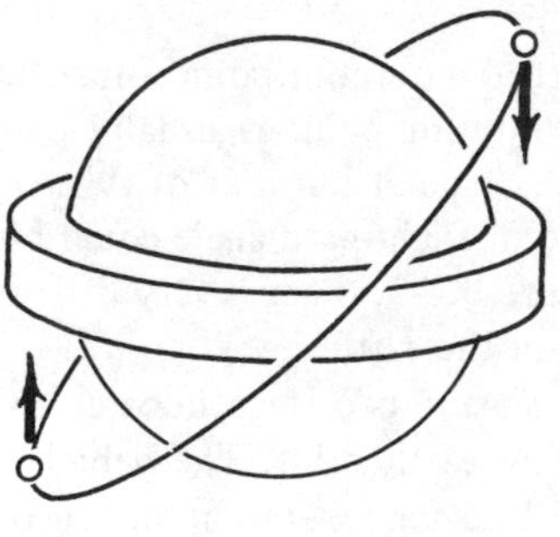

Figure 3.1-4 Perturbative torque caused by earth's equatorial bulge

The gravitational effect of an oblate earth can more easily be visualized by picturing a spherical earth surrounded by a belt of excess matter representing the equatorial bulge. When a satellite is in the positions shown in Figure 3.1-4 the net effect of the bulges is to produce a slight torque on the satellite about the center of the earth. This torque will cause the plane of the orbit to precess just as a gyroscope would under a similar torque. The result is that the nodes move westward for direct orbits and eastward for retrograde orbits.

The nodal regression rate is shown in Figure 3.1-5. Note that for low altitude orbits of low inclination the rate approaches 9° per day. You

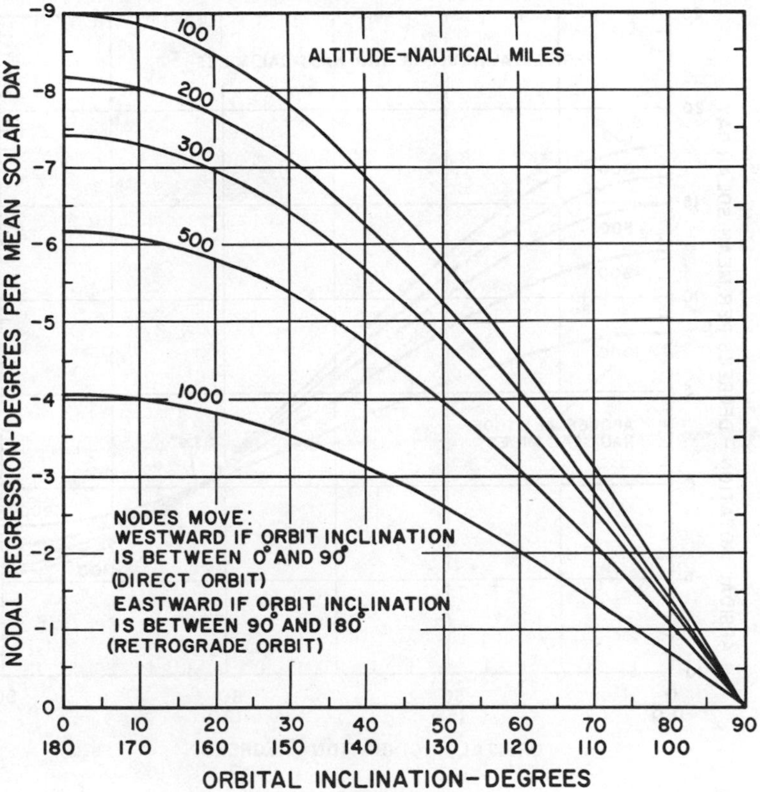

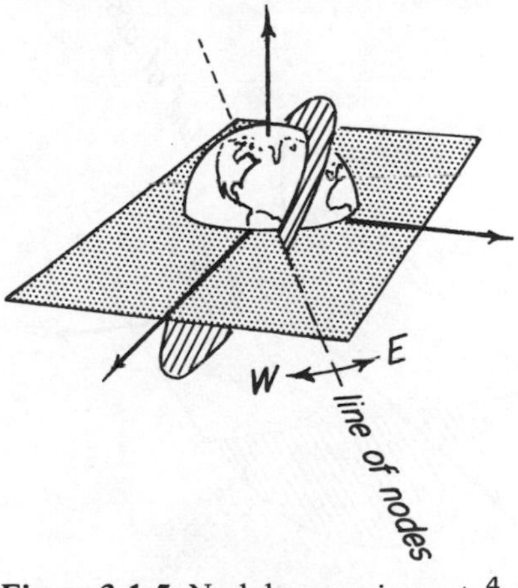

Figure 3.1-5 Nodal regression rate[4]

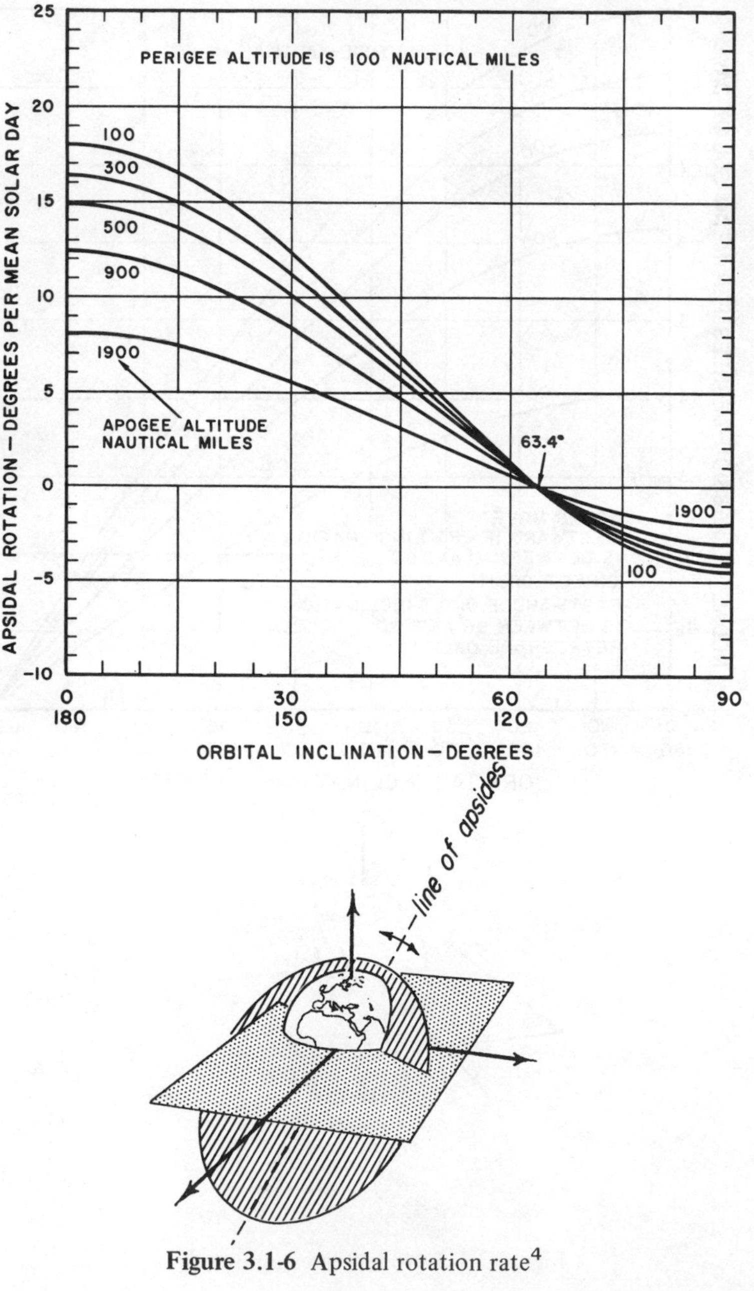

Figure 3.1-6 Apsidal rotation rate[4]

should be able to look at Figure 3.1-4 and see why the nodal regression is greatest for orbits whose inclination is near 0° or 180° and goes to zero for polar orbits.

The rotation of the line of apsides is only applicable to eccentric orbits. With this perturbation, which also is due to oblateness, the major axis of an elliptical trajectory will rotate in the direction of motion of the satellite if the orbital inclination is less than 63.4^o or greater than 116.6^o, and opposite to the direction of motion for inclinations between 63.4^o and 116.6^o. The rate at which the major axis rotates is a function of both orbit altitude and inclination angle. Figure 3.1-6 shows the apsidal rotation rate versus inclination angle for a perigee altitude of 100 nm and various apogee altitudes.

EXAMPLE PROBLEM. A satellite is orbiting the earth in a 500 nm circular orbit. The ascending node moves to the west, completing one revolution every 90 days.

a. What is the inclination of the orbit?

b. It is desired that the ascending node make only one revolution every 135 days. Calculate the new orbital inclination required if the satellite remains at the same altitude.

1) The given information is:

$$h = 500 \text{ n mi}$$

$$\Delta\Omega = 360^o \text{ per 90 days}$$

$$\therefore \text{Nodal regression per day} = \frac{360}{90} = 4^o/\text{day}$$

From Figure 3.1-5 $i = \underline{\underline{50^o}}$

2) It is desired to have $\Delta\Omega = 360^o$ per 135 days. Therefore

$$\text{Nodal regression per day} = \frac{360}{135} = 2.67^o/\text{day}$$

From Figure 3.1-5 $i \approx \underline{\underline{64^o}}$

3.2 HIGH ALTITUDE EARTH ORBITS

Figure 3.1-1 shows that to be safe from radiation a high altitude manned satellite would have to be above 5,000 nm altitude. On the scale of Figure 3.1-2 this would be 2½ inches from the surface. Of course, unmanned satellites can operate safely anywhere above 100 nm altitude.

The reasons for wanting a higher orbit could be to get away from atmospheric drag entirely or to be able to see a large part of the earth's surface at one time. For communications satellites this latter consideration can be all important. As you go to higher orbits the period of the satellite increases. If you go high enough the period can be made exactly equal to the time it takes the earth to rotate once on its axis (23 hr 56 min). This is the so-called "stationary" or synchronous satellite.

3.2.1 The Synchronous Satellite. If the period of a *circular direct equatorial* orbit is exactly 23 hr 56 min it will appear to hover motionless over a point on the equator. This, of course, is an illusion since the satellite is not at rest in the inertial IJK frame but only in the noninertial topocentric-horizon frame.

The correct altitude for a synchronous circular orbit is 19,300 nm or about 5.6 earth radii above the surface. Such a satellite would be well above the dangerous Van Allen radiation and could be manned.

The great utility of such a satellite for communications is obvious. Its usefulness as a reconnaissance vehicle is debatable. It might be able to detect missile launchings by their infrared "signatures." High resolution photography of the earth's surface would be difficult from that altitude, however.

There is much misunderstanding even among otherwise well-informed persons concerning the ground trace of a synchronous satellite. Many think it possible to "hang" a synchronous satellite over any point on the earth–Red China, for example. This is, unfortunately, not the case. The satellite can only appear to be motionless over a point on the equator. If the synchronous circular orbit is inclined to the equator its ground trace will be a "figure-eight" curve which carries it north and south approximately along a meridian. Figure 3.2-1 shows why.

3.2.2 Launching a High Altitude Satellite–The Ascent Ellipse. Launching a high altitude satellite is a two-step operation requiring two burnouts separated by a coasting phase. The first stage booster climbs

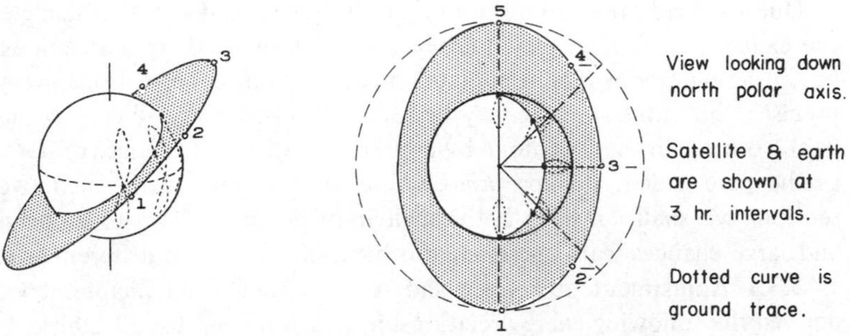

Figure 3.2-1 Ground trace of an inclined synchronous satellite

almost vertically, reaching burnout with a flight-path angle of 45° or more. This first burn places the second stage in an elliptical orbit (called an "ascent ellipse") which has its apogee at the altitude of the desired orbit. At apogee of the ascent ellipse, the second stage booster engines are fired to increase the speed of the satellite and establish it in its final orbit.

This burn-coast-burn technique is normally used any time the altitude of the orbit injection point exceeds 150 nm.

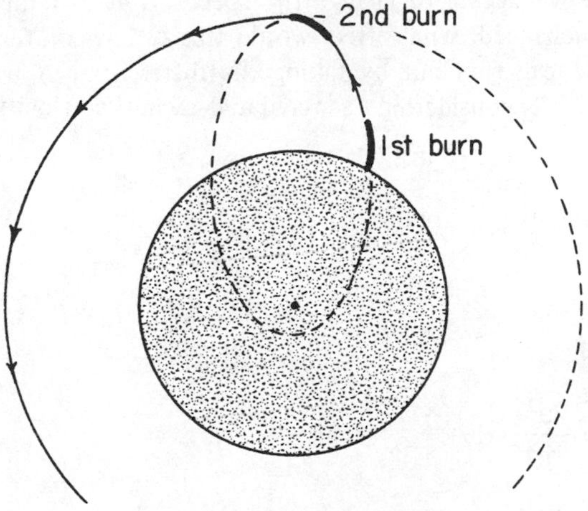

Figure 3.2-2 Ascent ellipse

3.3 IN-PLANE ORBIT CHANGES

Due to small errors in burnout altitude, speed, and flight-path angle, the exact orbit desired may not be achieved. Usually this is not serious, but, if a rendezvous is contemplated or if, for some other reason, a very precise orbit is required, it may be necessary to make small corrections in the orbit. This may be done by applying small speed changes or Δv's, as they are called, at appropriate points in the orbit. In the next two sections we shall consider both small in-plane corrections to an orbit and large changes from one circular orbit to a new one of different size.

3.3.1 Adjustment of Perigee and Apogee Height. In Chapter 1 we derived the following energy relationship which is valid for all orbits:

$$\mathcal{E} = \frac{v^2}{2} - \frac{\mu}{r} = -\frac{\mu}{2a}. \tag{3.3-1}$$

If we solve for v^2 we get

$$v^2 = \mu\left(\frac{2}{r} - \frac{1}{a}\right). \tag{3.3-2}$$

Suppose we decide to change the speed, v, at a point in an orbit leaving r unchanged. What effect would this Δv have on the semi-major axis, a? We can find out by taking the differential of both sides of equation (3.3-2) considering r as fixed and Δv in the velocity direction.

$$2v\,dv = \frac{\mu}{a^2}\,da \tag{3.3-3}$$

or

$$da = \frac{2a^2}{\mu}\,v\,dv\,. \tag{3.3-4}$$

For an infinitesimally small change in speed, dv, we get a change in semi-major axis, da, of the orbit given by equation (3.3-4). Since the major axis is $2a$, the length of the orbit changes by twice this amount or $2da$.

But suppose we make the speed change at perigee? The resulting change in major axis will actually be a change in the height of apogee. Similarly, a Δv applied at apogee will result in a change in perigee height. We can specialize the general relationship shown by equation (3.3-4) for small but finite Δv's applied at perigee and apogee as

$$\boxed{\begin{aligned} \Delta h_a &\approx \frac{4a^2}{\mu} v_p \, \Delta v_p \\ \Delta h_p &\approx \frac{4a^2}{\mu} v_a \, \Delta v_a . \end{aligned}} \tag{3.3-5}$$

This method of evaluating a small change in one of the orbital elements as a result of a small change in some other variable is illustrative of one of the techniques used in perturbation theory. In technical terms what we have just done is called "variation of parameters."

3.3.2 The Hohmann Transfer. Transfer between two circular coplanar orbits is one of the most useful maneuvers we have. It represents an alternate method of establishing a satellite in a high altitude orbit. For example, we could first make a direct ascent to a low altitude "parking orbit" and then transfer to a higher circular orbit by means of an elliptical transfer orbit which is just tangent to both of the circular orbits.

The least speed change (Δv) required for a transfer between two circular orbits is achieved by using such a doubly-tangent transfer ellipse. The first recognition of this principle was by Hohmann in 1925 and such orbits are, therefore, called *Hohmann transfer orbits.*[5] Consider the two circular orbits shown in Figure 3.3-1. Suppose we want to travel from the small orbit, whose radius is r_1, to the large orbit, whose radius is r_2, along the transfer ellipse.

We call the speed at point 1 on the transfer ellipse v_1. Since we know r_1, we could compute v_1 if we knew the energy of

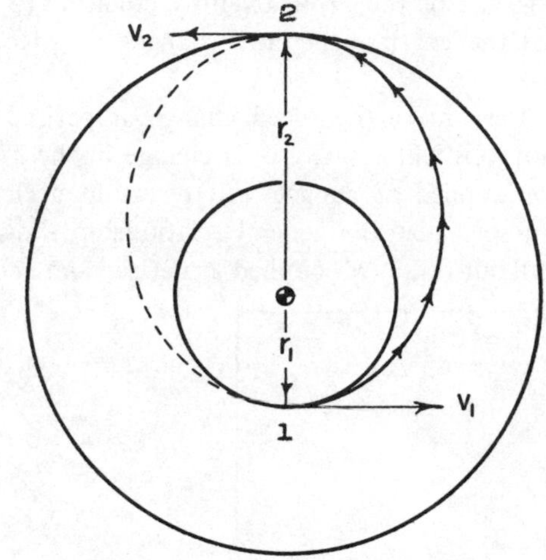

Figure 3.3-1 Hohmann transfer

the transfer orbit, $\mathcal{E}_t$. But from the geometry of Figure 3.3-1

$$2a_t = r_1 + r_2 \tag{3.3-6}$$

and, since $\mathcal{E} = -\mu/2a$,

$$\boxed{\mathcal{E}_t = -\mu/(r_1 + r_2).} \tag{3.3-7}$$

We can now write the energy equation for point 1 of the elliptical orbit and solve it for v_1:

$$\boxed{v_1 = \sqrt{2\left[\frac{\mu}{r_1} + \mathcal{E}_t\right]}.} \tag{3.3-8}$$

Since our satellite already has circular speed at point 1 of the small orbit, its speed is

$$v_{cs1} = \sqrt{\frac{\mu}{r_1}} . \tag{3.3-9}$$

To make our satellite go from the small circular orbit to the transfer ellipse we need to increase its speed from v_{cs} to v_1. So,

$$\Delta v_1 = v_1 - v_{cs_1} . \tag{3.3-10}$$

The speed change required to transfer from the ellipse to the large circle at point 2 can be computed in a similar fashion.

Although, in our example, we went from a smaller orbit to a larger one, the same principles may be applied to a transfer in the opposite direction. The only difference would be that two speed decreases would be required instead of two speed increases.

The time-of-flight for a Hohmann transfer is obviously just half the period of the transfer orbit. Since $\mathbb{P}_t = 2\pi\sqrt{a_t^3/\mu}$ and we know a_t,

$$\text{TOF} = \pi\sqrt{\frac{a_t^3}{\mu}} . \tag{3.3-11}$$

While the Hohmann transfer is the *most economical* from the standpoint of Δv required, it also *takes longer* than any other possible transfer orbit between the same two circular orbits. The other possible transfer orbits between coplanar circular orbits are discussed in the next section.

EXAMPLE PROBLEM. A communications satellite is in a circular orbit of radius 2 DU. Find the minimum Δv required to double the altitude of the satellite.

Minimum Δv implies a Hohmann transfer. For the transfer trajectory, $r_p = 2\,\text{DU}$, $r_a = 3\,\text{DU}$.

$$\mathcal{E}_t = -\frac{\mu}{2a} = -\frac{1}{5}\ \text{DU}^2/\text{TU}^2$$

$$v_{cs_1} = \sqrt{\frac{\mu}{r_1}} = .707 \text{ DU/TU}$$

$$v_{t_1} = \sqrt{2(\mathcal{E}_t + \frac{\mu}{r_1})} = .775 \text{ DU/TU}$$

$$\Delta v_1 = .068 \text{ DU/TU}$$

$$v_{cs_2} = \sqrt{\frac{\mu}{r_2}} = .577 \text{ DU/TU}$$

$$v_{t_2} = \sqrt{2(\mathcal{E}_t + \frac{\mu}{r_2})} = .516 \text{ DU/TU}$$

$$\Delta v_2 = .061 \text{ DU/TU}$$

$$\Delta v_{TOT} = .129 \text{ DU/TU} = 3{,}346 \text{ ft/sec}$$

3.3.3 General Coplanar Transfer Between Circular Orbits. Transfer between circular coplanar orbits merely requires that the transfer orbit intersect or at least be tangent to both of the circular orbits. Figure 3.3-2 shows transfer orbits which are both possible and impossible.

It is obvious from Figure 3.3-2 that the periapsis radius of the transfer orbit must be equal to or less than the radius of the inner orbit *and* the apoapsis radius must be equal to or exceed the radius of the outer orbit if the transfer orbit is to touch both circular orbits. We can express this condition mathematically as

$$r_p = \frac{p}{1+e} \leqslant r_1 \tag{3.3-12}$$

$$r_a = \frac{p}{1-e} \geqslant r_2 \tag{3.3-13}$$

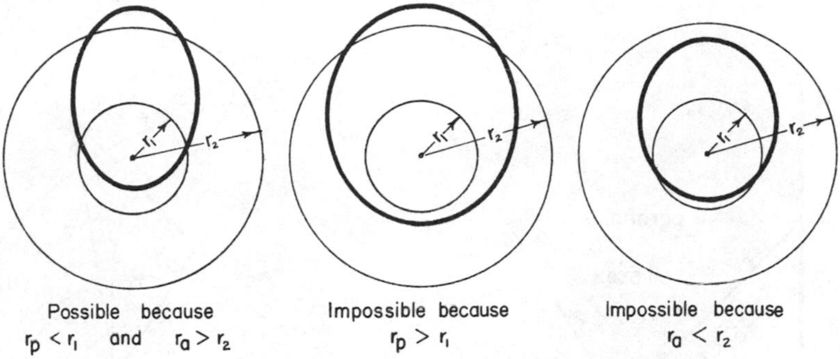

Figure 3.3-2 Transfer orbit must intersect both circular orbits

where p and e are the parameter and eccentricity of the transfer orbit and where r_1 and r_2 are the radii of the inner and outer circular orbits, respectively.

Orbits that satisfy both of these equations *will* intersect or at least be tangent to both circular orbits. We can plot these two equations (See Figure 3.3-3) and interpret them graphically. To satisfy both conditions, p and e of the transfer orbit must specify a point which lies in the shaded area. Values of (p, e) which fall on the limit lines correspond to orbits that are just tangent to one or the other of the circular orbits.

Suppose we have picked values of p and e for our transfer orbit which satisfy the conditions above. Knowing p and e, we can compute the energy, $\mathcal{E}_t$, and the angular momentum, h_t, in the transfer orbit. Since $p = a(1 - e^2)$ and $\mathcal{E} = -\mu / 2a$,

$$\boxed{\mathcal{E}_t = -\mu(1 - e^2)/2p.} \tag{3.3-14}$$

Since $p = h^2/\mu$

$$\boxed{h_t = \sqrt{\mu p}.} \tag{3.3-15}$$

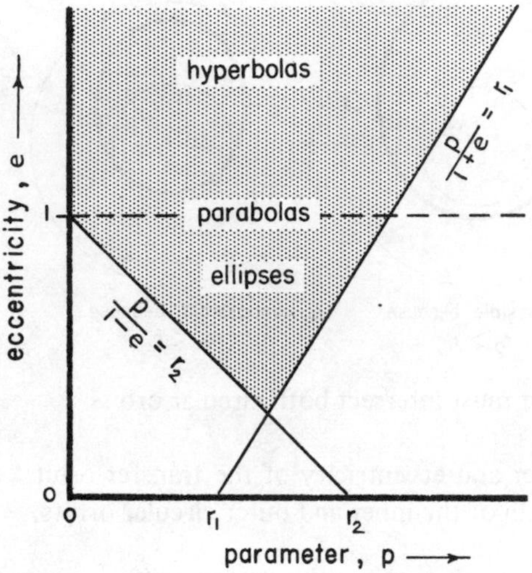

Figure 3.3-3 p versus e

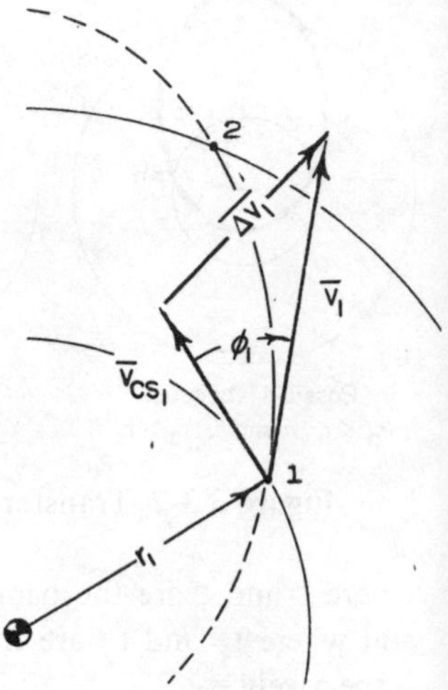

Figure 3.3-4 Δv required at point 1

We now can proceed just as we did in the case of a Hohmann transfer. Solving the energy equation for the speed at point 1 in the transfer orbit, we get

$$v_1 = \sqrt{2\left(\frac{\mu}{r_1} + \mathcal{E}_t\right)} \, . \tag{3.3-8}$$

Since our satellite already has circular speed at point 1 of the small orbit, its speed is

$$v_{cs_1} = \sqrt{\frac{\mu}{r_1}} \, . \tag{3.3-9}$$

The angle between $\mathbf{v}_1$ and $\mathbf{v}_{cs_1}$ is just the flight-path angle, ϕ_1. Since $h = rv\cos\phi$,

$$\cos \phi_1 = h_t / r_1 v_1 . \tag{3.3-16}$$

Now, since we know two sides and the included angle of the vector triangle shown in Figure 3.3-4, we can use the law of cosines to solve for the third side, Δv.

$$\boxed{\Delta v_1^2 = v_1^2 + v_{cs_1}^2 - 2 v_1 v_{cs1} \cos \phi_1 .} \tag{3.3-17}$$

The speed change required at point 2 can be computed in a similar fashion.

Since the Hohmann transfer is just a special case of the problem illustrated above, it is not surprising that equation (3.3-17) reduces to equation (3.3-10) when the flight-path angle, ϕ_1, is zero.

3.4 OUT-OF-PLANE ORBIT CHANGES

A velocity change, $\Delta \mathbf{v}$, which lies in the plane of the orbit can change its size or shape, or it can rotate the line of apsides. To change the orientation of the orbital plane in space requires a $\Delta \mathbf{v}$ component perpendicular to the plane of the orbit.

3.4.1 Simple Plane Change. If, after applying a finite $\Delta \mathbf{v}$, the speed and flight-path angle of the satellite are unchanged, then only the plane of the orbit has been altered. This is called a simple plane change.

An example of a simple plane change would be changing an inclined orbit to an equatorial orbit as shown in Figure 3.4-1. The plane of the orbit has been changed through an angle, θ. The initial velocity and final velocity are identical in magnitude and, together with the $\Delta \mathbf{v}$ required, form an isosceles vector triangle. We can solve for the magnitude of $\Delta \mathbf{v}$ using the law of cosines, assuming that we know v and θ. Or, more simply, we can divide the isosceles triangle into two right triangles, as shown at the right of Figure 3.4-1, and obtain directly

$$\boxed{\Delta v = 2v \sin \frac{\theta}{2}} \quad \text{for circular orbits.} \tag{3.4-1}$$

If the object of the plane change is to "equatorialize" an orbit, the $\Delta \mathbf{v}$ must be applied at one of the nodes (where the

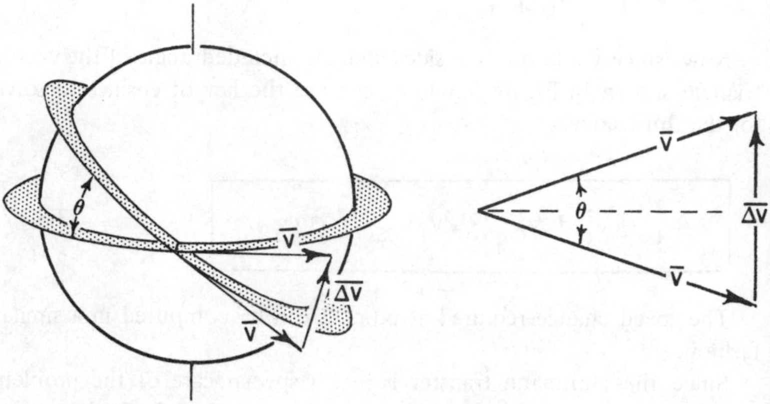

Figure 3.4-1 Simple plane change through an angle θ

satellite passes through the equatorial plane).

Large plane changes are prohibitively expensive in terms of the velocity change required. A plane change of 60°, for example, makes Δv equal to v! The Soviet Union must pay this high price if it wants equatorial satellites. Since it has no launch sites south of latitude 45° N, it cannot launch a satellite whose inclination is less than 45°. Therefore, a turn of at least 45° must be made at the equator if the satellite is to be equatorialized.

EXAMPLE PROBLEM. It is desired to transfer supplies from a 0.1 DU circular parking orbit around the earth to a space station in a 4 DU coplanar circular orbit. The transfer will be accomplished via an elliptical orbit tangent to the lower orbit and crossing the high orbit at the end of the minor axis of the transfer orbit.

a. Determine the total Δv required to accomplish the mission.

b. If the space station orbit is inclined at an angle of 10° to the low parking orbit, calculate the additional Δv required for the simple plane change necessary to accomplish the transfer. Assume the plane change is performed *after* you have established the shuttle vehicle in a 4 DU circular orbit.

a. The given information is:

$$h_1 = 0.1 \text{ DU}$$

$$h_2 = 4 \text{ DU}$$

Therefore $r_1 = r_\oplus + h_1 = 1.1 \text{DU}$

$$r_2 = r_\oplus + h_2 = 5 \text{ DU}$$

Since the transfer ellipse is tangent to the low orbit

$$\Delta v_1 = v_1 - v_{cs_1}$$

We need to know $\mathcal{E}_t$ for the transfer orbit. It is given that the transfer orbit intersects the high orbit at the end of its (transfer orbit) minor axis. Therefore $a_t = r_2$.

Hence, $\mathcal{E}_t = -\dfrac{\mu}{2a_t} = -\dfrac{\mu}{2r_2} = -\dfrac{1}{2(5)} = -0.1 \dfrac{\text{DU}^2}{\text{TU}_2}$

$$\therefore \; v_1 = \sqrt{2(\mathcal{E}_t + \frac{\mu}{r_1})} = \sqrt{2(-.1 + \frac{1}{1.1})}$$

$$= \sqrt{1.618} = 1.27 \text{ DU/TU}$$

$$v_{cs_1} = \sqrt{\frac{\mu}{r_1}} = 0.953 \text{ DU/TU}$$

$$\therefore \; \Delta v_1 = 1.27\text{-}0.953 \text{ DU/TU} = 8211 \text{ ft/sec}$$

or $\Delta v_1 = 8211$ ft/sec

Since v_2 and v_{cs_2} are not tangent

$$|\Delta \mathbf{v}_2| = |\mathbf{v}_{cs_2} - \mathbf{v}_2|$$

and we must use equation (3.3-17) to determine Δv_2.

$$h_t = r_1 v_1 = (1.1)(1.27) = 1.4 \text{ DU}^2/\text{TU}$$

$$v_2 = \sqrt{2(\mathcal{E}_t + \frac{\mu}{r_2})} = 0.447 \text{ DU/TU}$$

$$\therefore \cos\phi_2 = \frac{h_t}{r_2 v_2} = \frac{1.4}{(5)(0.447)} = 0.626$$

$$v_{cs_2} = \sqrt{\frac{\mu}{r_2}} = 0.447 \text{ DU/TU}$$

For the Student

Why does $v_2 = v_{cs_2}$?

From equation (3.3-17)

$$\therefore \Delta v_2^2 = v_2^2 + v_{cs_2}^2 - 2v_2 v_{cs_2} \cos\phi_2$$

$$= 0.2 + 0.2 - 2(.447)(.447)(.626)$$

$$= 0.4 - (0.4)(.626) = 0.15 \text{ DU/TU}$$

$$\Delta v_2 = \sqrt{.15} = 0.387 \text{ DU/TU} = 10{,}042 \text{ ft/sec}$$

Therefore

$$\Delta v_{Total} = \Delta v_1 + \Delta v_2$$

$$= .317 + .387 = \underline{\underline{.704 \text{ DU/TU}}} = 18{,}253 \text{ ft/sec}$$

b. For a simple plane change use equation (3.4-1)

$$\Delta v = 2v \sin \frac{\theta}{2}$$

$$= 2\, v_{cs_2} \sin 5^{\circ} = 2(.447)(.0873)$$

$$= \underline{\underline{.078 \text{ DU/TU}}} = 2023 \text{ ft/sec}$$

EXERCISES

3.1 What is the inclination of a circular orbit of period 100 minutes designed such that the trace of the orbit moves eastward at the rate of 3° per day?

3.2 Calculate the total Δv required to transfer between two coplanar circular orbits of radii $r_1 = 2$ DU and $r_2 = 5$ DU respectively using a transfer ellipse having parameters $p = 2.11$ DU and $e = 0.76$.
(Ans. 0.897 DU/TU)

3.3 Two satellites are orbiting the earth in circular orbits–not at the same altitude or inclination. What sequence of orbit changes and plane changes would most efficiently place the lower satellite in the same orbit as that of the higher one? Assume only one maneuver can be performed at a time, i.e., a plane change *or* an orbit transfer.

3.4 Refer to Figure 3.2-1. What would be the effect on the ground trace of a sychronous satellite if:

a. $e > 0$ but the period remained 23 hours 56 minutes?
b. i changed–all other parameters remained the same.
c. $\mathbb{TP} < 23$ hours 56 minutes.

3.5 You are in a circular earth orbit with a velocity of 1 DU/TU. Your Service Module is in another circular orbit with a velocity of .5 DU/TU. What is the minimum Δv needed to transfer to the Service Module's orbit?
(Ans. 0.449 DU/TU)

3.6 Determine which of the following orbits could be used to transfer between two circular-coplanar orbits with radii 1.2 DU and 4 DU respectively.

a. $r_p = 1$ DU

$e = 0.5$

b. $a = 2.5$ DU

$e = 0.56$

c. $\mathcal{E} = -0.1\ DU^2/TU^2$

$h = 1.34\ DU^2/TU$

d. $p = 1.95$ DU

$e = 0.5$

3.7 Design a satellite orbit which could provide a continuous communications link between Moscow and points in Siberia for at least 1/3 to 1/2 a day.

3.8 Compute the minimum Δv required to transfer between two coplanar elliptical orbits which have their major axes aligned. The parameters for the ellipses are given by:

$r_{p_1} = 1.1$ DU $\qquad r_{p_2} = 5$ DU

$e_1 = 0.290 \qquad e_2 = .412$

Assume both perigees lie on the same side of the earth.
(Ans. 0.311 DU/TU)

3.9 Calculate the total Δv required to transfer from a circular orbit of radius 1 DU to a circular orbit of infinite radius and then back to a circular orbit of 15 DU, using Hohmann transfers. Compare this with the Δv required to make a Hohmann transfer from the 1 DU circular orbit directly to the 15 DU circular orbit. At least 5 digits of accuracy are needed for this calculation.

***3.10** Note that in problem 3.9 it is more economical to use the three impulse transfer mode. This is often referred to as a bielliptical transfer. Find the ratio between circular orbit radii (outer to inner) beyond which it is more economical to use the bielliptical transfer mode.

LIST OF REFERENCES

1. Newton, Sir Isaac. *Principia.* Motte's translation revised by Cajori. Vol 2. Berkeley and Los Angeles, University of California Press, 1962.

2. Ley, Willy. *Rockets, Missiles, and Space Travel.* 2nd revised ed. New York, NY, The Viking Press, 1961.

3. Dornberger, Walter. *V-2.* New York, NY, The Viking Press, 1955.

4. *Space Planners Guide.* Washington, DC, Headquarters, Air Force Systems Command, 1 July 1965.

5. Baker, Robert M. L., and Maud W. Makemson. *An Introduction to Astrodynamics.* New York and London, Academic Press, 1960.

CHAPTER 4

POSITION AND VELOCITY AS A FUNCTION OF TIME

> ... the determination of the true movement of the planets, including the earth ... This was Kepler's first great problem. The second problem lay in the question: What are the mathematical laws controlling these movements? Clearly, the solution of the second problem, if it were possible for the human spirit to accomplish it, presupposed the solution of the first. For one must know an event before one can test a theory related to this event.
>
> –Albert Einstein[1]

4.1 HISTORICAL BACKGROUND

The year Tycho Brahe died (1601) Johannes Kepler, who had worked with Tycho in the 18 months preceding his death, was appointed as Imperial Mathematician to the Court of Emperor Rudolph II. Recognizing the goldmine of information locked up in Tycho's painstaking observations, Kepler packed them up and moved them to Prague with him. In a letter to one of his English admirers he calmly reported:

> "I confess that when Tycho died, I quickly took advantage of the absence, or lack of circumspection, of the heirs, by taking the observations under my care, or perhaps usurping them ... "[2]

Kepler stayed in Prague from 1601 to 1612. It was the most fruitful period of his life and saw the publication of *Astronomia Nova* in 1609 in which he announced his first two laws of planetary motion. The manner in which Kepler arrived at these two laws is fascinating and can be told only because in *New Astronomy* Kepler leads his reader into every blind alley, detour, trap or pitfall that he himself encountered.

> "What matters to me," Kepler points out in his Preface, "is not merely to impart to the reader what I have to say, but above all to convey to him the reasons, subterfuges, and lucky hazards which led me to my discoveries. When Christopher Columbus, Magelhaen and the Portuguese relate how they went astray on their journeys, we not only forgive them, but would regret to miss their narration because without it the whole grand entertainment would be lost."[2]

Kepler selected Tycho's observations of Mars and tried to reconcile them with some simple geometrical theory of motion. He began by making three revolutionary assumptions: (a) that the orbit was a circle with the sun slightly off-center, (b) that the orbital motion took place in a plane which was fixed in space, and (c) that Mars did not necessarily move with uniform velocity along this circle. Thus, Kepler immediately cleared away a vast amount of rubbish that had obstructed progress since Ptolemy.

Kepler's first task was to determine the radius of the circle and the

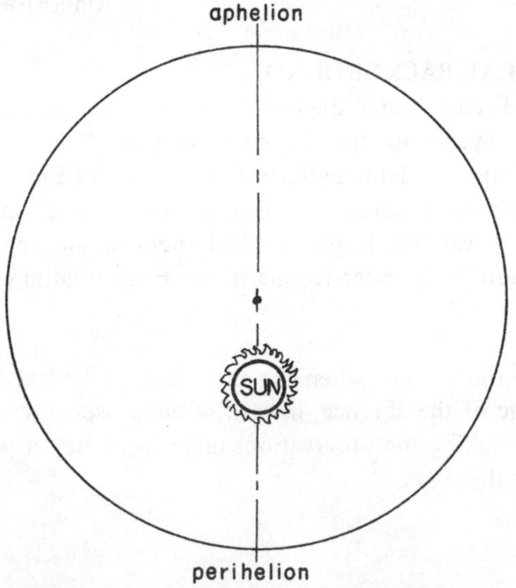

Figure 4.1-1 Kepler first assumed circular orbit with the sun off-center.

direction of the axis connecting perihelion and aphelion. At the very beginning of a whole chapter of excruciating trial-and-error calculations, Kepler absentmindedly put down three erroneous figures for three vital longitudes of Mars, never noticing his error. His results, however, were nearly correct because of several mistakes of simple arithmetic committed later in the chapter which happened very nearly to cancel out his earlier errors.

At the end he seemed to have achieved his goal of representing within 2 arc-minutes the position of Mars at all 10 oppositions recorded by Tycho. But then without a word of transition, in the next two chapters Kepler explains, almost with masochistic delight, how two other observations from Tycho's collection did not fit; there was a discrepancy of 8 minutes of arc. Others might have shrugged off this minor disparity between fact and hypothesis. It is to Kepler's everlasting credit that he made it the basis for a complete reformation of astronomy. He decided that the sacred concept of circular motion had to go.

Before Kepler could determine the true shape of Mars' orbit, without benefit of any preconceived notions, he had to determine precisely the earth's motion around the sun. For this purpose he designed a highly original method and when he had finished his computations he was ecstatic. His results showed that the earth did not move with uniform speed, but faster or slower according to its distance from the sun. Moreover, at the extremums of the orbit (perihelion and aphelion) the earth's velocity proved to be inversely proportional to distance.

At this point Kepler could contain himself no longer and becomes airborne, as it were, with the warning: "Ye physicists, prick your ears, for now we are going to invade your territory." He was convinced that there was "a force in the sun" which moved the planets. What could be more beautifully simple than that the force should vary inversely with distance? He had proved the inverse ratio of speed to distance for only *two points* in the orbit, perihelion and aphelion, yet he made the patently incorrect generalization that this "law" held true for the *entire* orbit. This was the first of the critical mistakes that would cancel itself out "as if by a miracle" and lead him by faulty reasoning to the correct result.

Forgetting his earlier resolve to abandon circular motion he reasoned, again incorrectly, that, since speed was inversely proportional to distance, the line joining the sun (which was off-center in the circle) nd the planet swept out equal areas in the orbit in equal times.

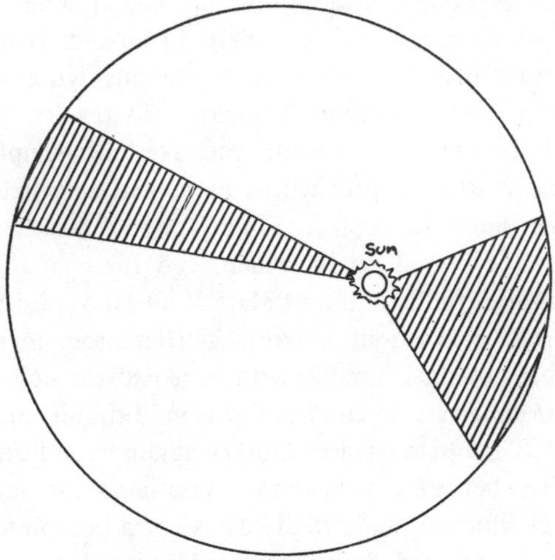

Figure 4.1-2 Kepler's law of equal areas

This was his famous Second Law-discovered before the First–a law of amazing simplicity, arrived at by a series of faulty steps which he himself later recognized with the observation: "But these two errors–it is like a miracle–cancel out in the most precise manner, as I shall prove further down."[2]

The correct result is even more miraculous than Kepler realized since his explanation of why the errors cancelled was also erroneous!

Kepler now turned again to the problem of replacing the circle with another geometric shape; he settled on the oval. But a very special oval: it had the shape of an egg, with the narrow end at the perihelion and the broad end at aphelion. As Koestler[2] observed: "No philosopher had laid such a monstrous egg before."

Finally, a kind of snowblindness seemed to descend upon him: he held the solution in his hands but was unable to see it. On 4 July 1603 he wrote a friend that he was unable to solve the problem of computing the area of his oval; but "if only the shape were a perfect ellipse all the answers could be found in Archimedes' and Apollonius' work."[2] Finally, after struggling with his egg for more than a year he stumbled onto the secret of the Martian orbit. He was able to express the distance from the sun by a simple mathematical formula: but he did not recognize that this formula specifically defined the orbit as an ellipse. Any student of analytical geometry today would have recognized it immediately; but analytical geometry came after Kepler. He had reached his goal, but he did not realize that he had reached it!

He tried to construct the curve represented by his equation, but he did not know how, made a mistake in geometry, and ended up with a "chubby-faced" orbit. The climax to this comedy of errors came when, in a moment of despair, Kepler threw out his equation (which denoted an ellipse) because he wanted to try an entirely new hypothesis: to wit, *an elliptic orbit*! When the orbit fit and he realized what had happened, he frankly confessed:

> "Why should I mince my words? The truth of Nature, which I had rejected and chased away, returned by stealth through the backdoor, disguising itself to be accepted. That is to say, I laid [the original equation] aside and fell back on ellipses, believing that this was a quite different hypothesis, whereas the two . . . are one and the same . . . I thought and searched, until I went nearly mad, for a reason why the planet preferred an elliptical orbit [to mine] . . . Ah, what a foolish bird I have been!"[2]

In the end, Kepler was able to write an empirical expression for the time-of-flight of a planet from one point in its orbit to another–although he still did not know the true reason why it should move in an orbit at all. In this chapter we will, with the benefit of hindsight and concepts introduced by Newton, derive the Kepler time-of-flight equation in much the same way that Kepler did. We will then turn our attention to the solution of what has come to be known as "the Kepler problem"–predicting the future position and velocity of an orbiting object as a function of some known initial position and velocity and the time-of-flight. In doing this we will introduce one of the most recent advances in the field of orbital mechanics–a universal formulation of the time-of-flight relationships valid for all conic orbits.

4.2 TIME-OF-FLIGHT AS A FUNCTION OF ECCENTRIC ANOMALY

Many of the concepts introduced by Kepler, along with the names he used to describe them, have persisted to this day. You are already familiar with the term "true anomaly" used to describe the angle from periapsis to the orbiting object measured in the direction of motion. In this section we will encounter a new term called "eccentric anomaly" which was introduced by Kepler in connection with elliptical orbits. Although he was not aware that parabolic and hyperbolic orbits existed, the concept can be extended to these orbits also as we shall see.

It is possible to derive time-of-flight equations analytically, using only the dynamical equation of motion and integral calculus. We will pursue a lengthier, but more motivated, derivation in which the eccentric anomaly arises quite naturally in the course of the geometrical arguments. This derivation is presented more for its historical value than for actual use. The universal variable approach is strongly recommended as the best method for general use.

4.2.1 Time-of-Flight on the Elliptical Orbit. We have already seen that in one orbital period the radius vector sweeps out an area equal to the total area of an ellipse, i.e., πab. In going part way around an orbit, say from periapsis to some general point, P, where the true anomaly is ν, the radius vector sweeps out the shaded area, A_1, in Figure 4.2-1. Because area is swept out at a constant rate in an orbit (Kepler's Second Law) we can say that

$$\frac{t - T}{A_1} = \frac{\mathbb{P}}{\pi ab} \qquad (4.2\text{-}1)$$

where T is the time of periapsis passage and $\mathbb{P}$ is the period.

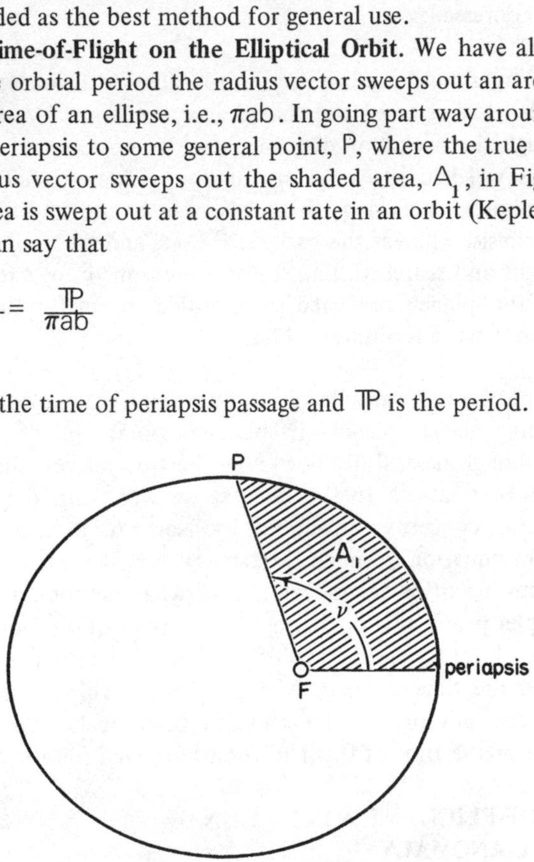

Figure 4.2-1 Area swept out by **r**

The only unknown in equation (4.2-1) is the area A_1. The geometrical construction illustrated in Figure 4.2-2 will enable us to write an expression for A_1.

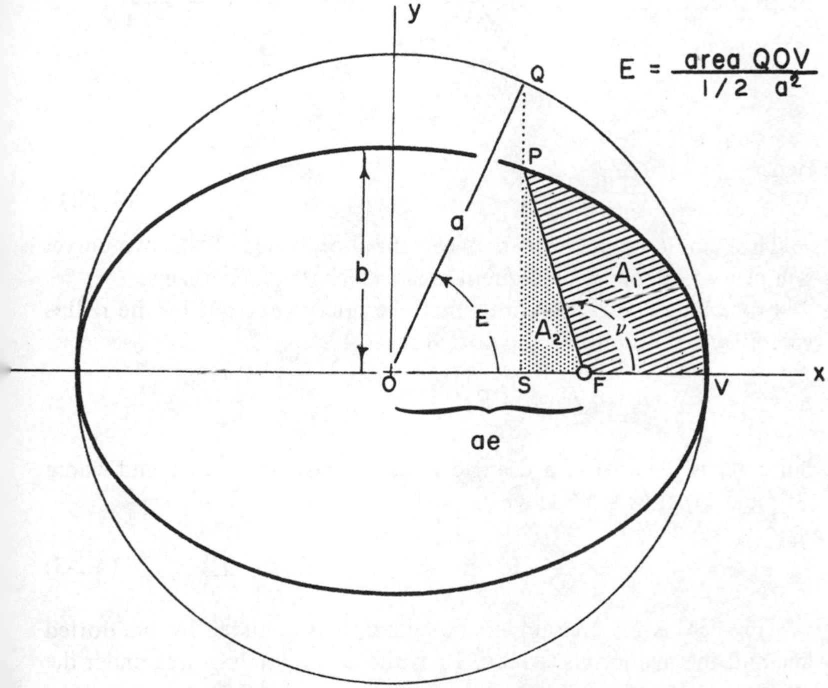

Figure 4.2-2 Eccentric anomaly, E

A circle of radius a has been circumscribed about the ellipse. A dotted line, perpendicular to the major axis, has been extended through P to where it intersects the "auxiliary circle" at Q. The angle E is called the *eccentric anomaly.*

Before proceeding further we must derive a simple relationship between the ellipse and its auxiliary circle. From analytical geometry, the equations of the curves in cartesian coordinates are:

Ellipse: $\dfrac{x^2}{a^2} + \dfrac{y^2}{b^2} = 1$ Circle: $\dfrac{x^2}{a^2} + \dfrac{y^2}{a^2} = 1$

From which

$$y_{ellipse} = \sqrt{\frac{a^2 b^2 - b^2 x^2}{a^2}} = \frac{b}{a}\sqrt{a^2 - x^2}$$

$$y_{circle} = \sqrt{a^2 - x^2}$$

Hence

$$\frac{y_{ellipse}}{y_{circle}} = \frac{b}{a} \tag{4.2-2}$$

This simple relationship between the y-ordinates of the two curves will play a key role in subsequent area and length comparisons.

From Figure 4.2-2 we note that the area swept out by the radius vector is Area PSV minus the dotted area, A_2.

$$A_1 = \text{Area PSV} - A_2 \quad .$$

Since A_2 is the area of a triangle whose base is $ae - a\cos E$ and whose altitude is $b/a\ (a \sin E)$, we can write

$$A_2 = \frac{ab}{2}\ (e \sin E - \cos E \sin E) \ . \tag{4.2-3}$$

Area PSV is the area under the ellipse; it is bounded by the dotted line and the major axis. Area QSV is the corresponding area under the auxiliary circle. It follows directly from equation (4.2-2) that

$$\text{Area PSV} = \frac{b}{a}\ (\text{Area QSV})$$

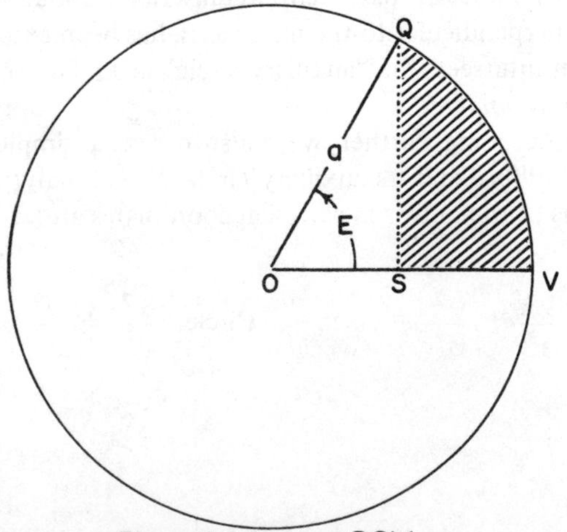

Figure 4.2-3 Area QSV

The Area QSV is just the area of the sector QOV, which is $1/2\ a^2 E$ (where E is in radians), minus the triangle, whose base is (a cos E), and whose altitude is (a sin E). Hence

$$\text{Area PSV} = \frac{ab}{2}(E - \cos E \sin E).$$

Substituting into the expression for area A_1 yields

$$A_1 = \frac{ab}{2}(E - e \sin E) .$$

Finally, substituting into equation (4.2-1) and expressing the period as $2\pi\sqrt{a^3/\mu}$, ,we get

$$\boxed{t - T = \sqrt{\frac{a^3}{\mu}}(E - e \sin E)} \qquad (4.2\text{-}4)$$

Kepler introduced the definition

$$M = E - e \sin E \qquad (4.2\text{-}5)$$

where M is called the "mean anomaly." If we also use the definition

$$n \equiv \sqrt{\mu/a^3}$$

where n is called the "mean motion," then the mean anomaly may be written:

$$M = n(t - T) = E - e \sin E \qquad (4.2\text{-}6)$$

which is often referred to as *Kepler's equation.*

Obviously, in order to use equation (4.2-4), we must be able to relate the eccentric anomaly, E, to its corresponding true anomaly, ν. From Figure 4.2-2,

$$\cos E = \frac{ae + r \cos \nu}{a} . \qquad (4.2\text{-}7)$$

Since $r = \dfrac{a(1 - e^2)}{1 + e \cos \nu}$ equation (4.2-7) reduces to

$$\cos E = \frac{e + \cos \nu}{1 + e \cos \nu} . \qquad (4.2\text{-}8)$$

Eccentric anomaly may be determined from equation (4.2-8). The correct quadrant for E is obtained by noting that ν and E are always in the same half-plane; when ν is between 0 and π, so is E.

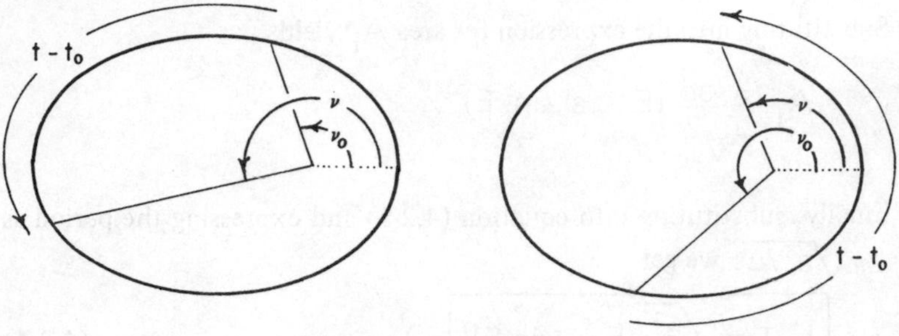

Figure 4.2-4 Time-of-flight between arbitrary points

Suppose we want to find the time-of-flight between a point defined by ν_0 and some general point defined by ν when the initial point is not periapsis. Provided the object does not pass through periapsis enroute from ν_0 to ν (Figure 4.2-4 left), we can say that

$$t - t_0 = (t - T) - (t_0 - T).$$

If the object does pass through periapsis (which is the case whenever ν_0 is greater than ν) then, from Figure 4.2-4 right,

$$t - t_0 = \mathbb{P} + (t - T) - (t_0 - T_0).$$

In general we can say that

$$t - t_0 = \sqrt{\frac{a^3}{\mu}}\,[2k\pi + (E - e \sin E) - (E_0 - e \sin E_o)] \tag{4.2-9}$$

where k is the number of times the object passes through periapsis enroute from ν_0 to ν.

At this point it is instructive to note that this same result can be derived analytically. In this case the eccentric anomaly appears as a convenient variable transformation to permit integration. Only the skeleton of the derivation will be shown here.

From equations in section 1.7.2 we can write

$$\int_T^t h\,dt = \int_0^\nu r^2\,d\nu \tag{4.2-10}$$

or

$$h(t - T) = \int_0^\nu \frac{p^2\,d\nu}{(1 + e\cos\nu)^2} \tag{4.2-11}$$

Now let us introduce the eccentric anomaly as a variable change to make equation (4.2-11) easily integrable. From equation (4.2-8) and geometry the following relationships can be derived:

$$\cos\nu = \frac{e - \cos E}{e\cos E - 1} \tag{4.2-12}$$

$$\sin\nu = \frac{a\sqrt{1 - e^2}}{r}\sin E \tag{4.2-13}$$

$$r = a(1 - e\cos E) \tag{4.2-14}$$

Differentiating equation (4.2-12), we obtain

$$d\nu = \frac{\sin E\,(1 + e\cos\nu)}{\sin\nu\,(1 - e\cos E)}\,dE = \frac{\sin E\,(p/r)}{\sin\nu\,(r/a)}$$

$$= \frac{a\sqrt{1 - e^2}}{r}\,dE\,. \tag{4.2-15}$$

then

$$h(t - T) = \frac{p}{\sqrt{1 - e^2}}\int_0^E r\,dE$$

$$= \frac{pa}{\sqrt{1 - e^2}}\int_0^E (1 - e\cos E)\,dE$$

$$= \frac{pa}{\sqrt{1 - e^2}}(E - e\sin E)\,. \tag{4.2-16}$$

since $h = \sqrt{\mu p}$

$$(t - T) = \sqrt{\frac{a^3}{\mu}}\,(E - e \sin E) \tag{4.2-4}$$

which is identical to the geometrical result.

4.2.2 Time-of-Flight on Parabolic and Hyperbolic Orbits. In a similar manner, the analytical derivation of the parabolic time-of-flight can be shown to be

$$\boxed{t - T = \frac{1}{2\sqrt{\mu}}\left[pD + \frac{1}{3}D^3\right]} \tag{4.2-17}$$

or

$$\boxed{t - t_0 = \frac{1}{2\sqrt{\mu}}\left[\left(pD + \frac{1}{3}D^3\right) - \left(pD_0 + \frac{1}{3}D_0^3\right)\right]} \tag{4.2-18}$$

where $D = \sqrt{p}\,\tan\frac{\nu}{2}$

D is the "parabolic eccentric anomaly."

From either a geometrical or analytical approach the hyperbolic time-of-flight, using the "hyperbolic eccentric anomaly," F, can be derived as

$$\boxed{t - T = \sqrt{\frac{(-a)^3}{\mu}}\,(e \sinh F - F)} \tag{4.2-19}$$

or

$$\boxed{t - t_0 = \sqrt{\frac{(-a)^3}{\mu}}\,[(e \sinh F - F) - (e \sinh F_0 - F_0)]} \tag{4.2-20}$$

where

$$\boxed{\cosh F = \frac{e + \cos\nu}{1 + e\cos\nu}} \tag{4.2-21}$$

or $F = \ln\left[y + \sqrt{y^2 - 1}\right]$

for $y = \cosh F$. Whenever ν is between 0 and π, F should be taken as positive; whenever ν is between π and 2π, F should be taken as negative. Figure 4.2-5 illustrates the hyperbolic variables.

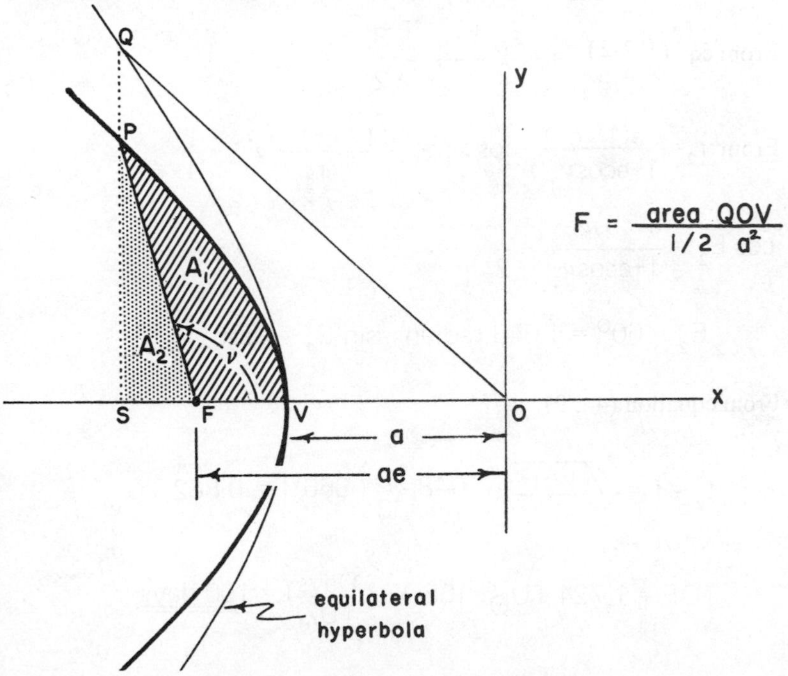

Figure 4.2-5 Hyperbolic eccentric anomaly, F

EXAMPLE PROBLEM. A space probe is in an elliptical orbit around the sun. Perihelion distance is .5 AU and aphelion is 2.5 AU. How many days during each orbit is the probe closer than 1 AU to the sun?

The given information is:

$$r_p = 0.5 \text{ AU}, \quad r_a = 2.5 \text{ AU}$$

$$r_1 = 1 \text{ AU}, \quad r_2 = 1 \text{ AU}$$

Since the portion of the orbit in question is symmetrical, we can compute the time of flight from periapsis to point 2 and then double it.

From eq. (1.7-4)
$$e = \frac{r_a - r_p}{r_a + r_p} = \frac{2}{3}$$

From eq. (1.7-2) $a = \dfrac{r_p + r_a}{2} = \dfrac{3}{2}$

From $r_2 = \dfrac{a(1 - e^2)}{1 + e\cos\nu_2}$, $\cos\nu_2 = \dfrac{a(1 - e^2) - r_2}{er_2} = -\dfrac{1}{4}$

$$\cos E_2 = \frac{e + \cos\nu_2}{1 + e\cos\nu_2} = \frac{1}{2}$$

$$E_2 = 60^\circ = 1.048 \text{ radians}, \sin E_2 = .866$$

From equation (4.2-9)

$$t_2 - T = \sqrt{\frac{(1.5)^3}{1}}\left[1.048 - \frac{2}{3}(.866)\right] = 0.862 \text{ TU}_\odot$$

$$\text{TOF} = 1.724 \text{ TU}_\odot \left(58.133 \frac{\text{Days}}{\text{TU}_\odot}\right) = \underline{\underline{100 \text{ days}}}$$

4.2.3 Loss of Numerical Accuracy for Near-Parabolic Orbits. The Kepler time-of-flight equations suffer from a severe loss in computational accuracy near $e = 1$. The nature of the difficulty can best be illustrated by a numerical example:

Suppose we want to compute the time-of-flight from periapsis to a point where $\nu = 60^\circ$ on an elliptical orbit with $a = 100$ DU and $e = 0.999$. The first step is to compute the eccentric anomaly. From equation (4.2-8),

$$\cos E = \frac{.999 + .5}{1 + .999(.5)} = .99967$$

Therefore $E = 0.02559$, $\sin E = 0.02560$ and $(e \sin E) = 0.02557$.

Substituting these values into equation (4.2-4),

$$t - T = \sqrt{\frac{100^3}{1}}(.02559 - .02557)$$

$$= 1000\,(.00002) = .02 \text{ TU}$$

There is a loss of significant digits in computing E from cos E when E is near zero. There is a further loss in subtracting two nearly equal numbers in the last step. As a result, the answer is totally unreliable.

This loss of computational accuracy near $e = 1$ and the inconvenience of having a different equation for each type of conic orbit will be our principal motivation for developing a universal formulation for time-of-flight in section 4.3.

4.3 A UNIVERSAL FORMULATION FOR TIME-OF-FLIGHT

The classical formulations for time-of-flight involving the eccentric anomalies, E or F, don't work very well for near-parabolic orbits. We have already seen the loss of numerical accuracy that can occur near $e = 1$ in the Kepler equation. Also, solving for E or F when a, e, ν_0 and $t - t_0$ are given is difficult when e is nearly 1 because the trial-and-error solutions converge too slowly or not at all. Both of these defects are overcome in a reformulation of the time-of-flight equations made possible by the introduction of a new auxiliary variable different from the eccentric anomaly. Furthermore, the introduction of this new auxiliary variable allows us to develop a single time-of-flight equation valid for all conic orbits.

The change of variable is known as the "Sundman transformation" and was first proposed in 1912.[6] Only recently, however, has it been used to develop a unified time-of-flight equation. Goodyear[7], Lemmon[8], Herrick[9], Stumpff[10], Sperling[11] and Battin[3] have all presented formulae for computation of time-of-flight via "generalized" or "universal" variables. The original derivation presented below was suggested by Bate[12] and partially makes use of notation introduced by Battin.

4.3.1 Definition of the Universal Variable, x. Angular momentum and energy are related to the geometrical parameters p and a by the familiar equations:

$$h = r^2 \dot{\nu} = \sqrt{\mu p}$$

$$\mathcal{E} = \tfrac{1}{2} v^2 - \frac{\mu}{r} = \frac{-\mu}{2a} \ .$$

If we resolve v into its radial component, $\dot{r}$, and its transverse component, $r\dot{\nu}$, the energy equation can be written as

$$\tfrac{1}{2} \dot{r}^2 + \tfrac{1}{2}(r\dot{\nu})^2 - \frac{\mu}{r} = \frac{-\mu}{2a} \ .$$

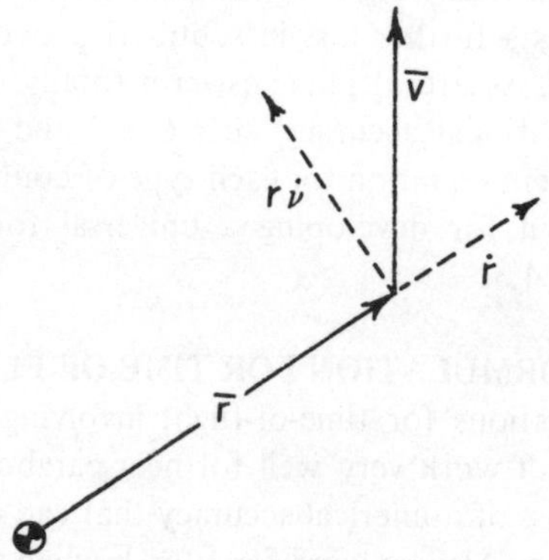

Figure 4.3-1 Radial and transverse components of velocity vector, **v**

Solving for $\dot{r}^2$ and setting $(r\,\dot{\nu})^2 = \frac{\mu p}{r^2}$, we get

$$\dot{r}^2 = \frac{-\mu p}{r^2} + \frac{2\mu}{r} - \frac{\mu}{a}. \tag{4.3-1}$$

Since the solution of this equation is not obvious, we introduce a new independent variable, x, defined as

$$\boxed{\dot{x} = \frac{\sqrt{\mu}}{r}} \tag{4.3-2}$$

First we will develop a general expression for r in terms of x. If we divide equation (4.3-1) by equation (4.3-2) squared, we obtain

$$\left(\frac{dr}{dx}\right)^2 = -p + 2r - \frac{r^2}{a}.$$

Separating the variables yields

$$dx = \frac{dr}{\sqrt{-p+2r-r^2/a}} \tag{4.3-3}$$

For $e \neq 1$ the indefinite integral, calling the constant of integration c_0, is

$$x + c_0 = \sqrt{a}\,\sin^{-1}\frac{(r/a - 1)}{\sqrt{1 - p/a}}$$

But $p = a(1 - e^2)$, so $e = \sqrt{1 - p/a}$ and we may write

$$x + c_o = \sqrt{a} \sin^{-1} \frac{(r/a - 1)}{e}$$

Finally, solving this equation for r gives

$$\boxed{r = a\left(1 + e \sin \frac{x + c_o}{\sqrt{a}}\right)} \quad . \tag{4.3-4}$$

Substituting equation (4.3-4) into the definition of the universal variable, equation (4.3-2), we obtain

$$\sqrt{\mu}\, dt = a\left(1 + e \sin \frac{x + c_o}{\sqrt{a}}\right) dx$$

$$\sqrt{\mu}\, t = ax - ae\sqrt{a}\left(\cos \frac{x + c_o}{\sqrt{a}} - \cos \frac{c_o}{\sqrt{a}}\right) \tag{4.3-5}$$

where we assumed $x = 0$ at $t = 0$.

At this point we have developed equations for both r and t in terms of x. The constant of integration, c_o, has not been evaluated yet. Application of these equations will now be made to a specific problem type.

4.4 THE PREDICTION PROBLEM

With the Kepler time-of-flight equations you can easily solve for the time-of-flight, $t - t_o$, if you are given a, e, ν_o and ν. The inverse problem of finding ν when you are given a, e, ν_o and $t - t_o$ is not so simple, as we shall see. Small[4], in *An Account of the Astronomical Discoveries of Kepler*, relates: "This problem has, ever since the time of Kepler, continued to exercise the ingenuity of the ablest geometers; but no solution of it which is rigorously accurate has been obtained. Nor is there much reason to hope that the difficulty will ever be overcome . . ." This problem classically involves the solution of Kepler's Equation and is often referred to as Kepler's problem.

4.4.1 Development of the Universal Variable Formulation. The prediction problem can be stated as (see Figure 4.4-1):

Given: $\mathbf{r}_o, \mathbf{v}_o, t_o = 0$

Find: **r**, **v** at time t.

We have assumed $x = 0$ at $t = 0$. From equation (4.3-4)

$$e \sin \frac{c_0}{\sqrt{a}} = \frac{r_0}{a} - 1. \tag{4.4-1}$$

Now differentiate equation (4.3-4) with respect to time:

$$\dot{r} = \frac{ae}{\sqrt{a}} \cos\left[\frac{(x + c_0)}{\sqrt{a}}\right]\frac{\sqrt{\mu}}{r}. \tag{4.4-2}$$

Applying the initial conditions to equation (4.4-2) and using the identity $\mathbf{r}\cdot\dot{\mathbf{r}} = r\dot{r}$, we get

$$e \cos \frac{c_0}{\sqrt{a}} = \frac{\mathbf{r}_0 \cdot \mathbf{v}_0}{\sqrt{\mu a}} \tag{4.4-3}$$

Using the trig identity for the cosine of a sum we can write equation (4.3-5) as

$$\sqrt{\mu}\, t = ax - ae\sqrt{a}\left(\cos \frac{x}{\sqrt{a}} \cos \frac{c_0}{\sqrt{a}} - \sin \frac{x}{\sqrt{a}} \sin \frac{c_0}{\sqrt{a}} - \cos \frac{c_0}{\sqrt{a}}\right).$$

Then substituting equations (4.4-1) and (4.4-3) and rearranging:

$$\sqrt{\mu}\, t = a\left(x - \sqrt{a} \sin \frac{x}{\sqrt{a}}\right) + \frac{\mathbf{r}_0 \cdot \mathbf{v}_0}{\sqrt{\mu}}\, a\left(1 - \cos \frac{x}{\sqrt{a}}\right) + r_0\sqrt{a} \sin \frac{x}{\sqrt{a}} \tag{4.4-4}$$

In a similar fashion we can use the trig identity for the sine of a sum to rewrite equation (4.3-4) as

$$r = a + ae\left(\sin \frac{x}{\sqrt{a}} \cos \frac{c_0}{\sqrt{a}} + \cos \frac{x}{\sqrt{a}} \sin \frac{c_0}{\sqrt{a}}\right). \tag{4.4-5}$$

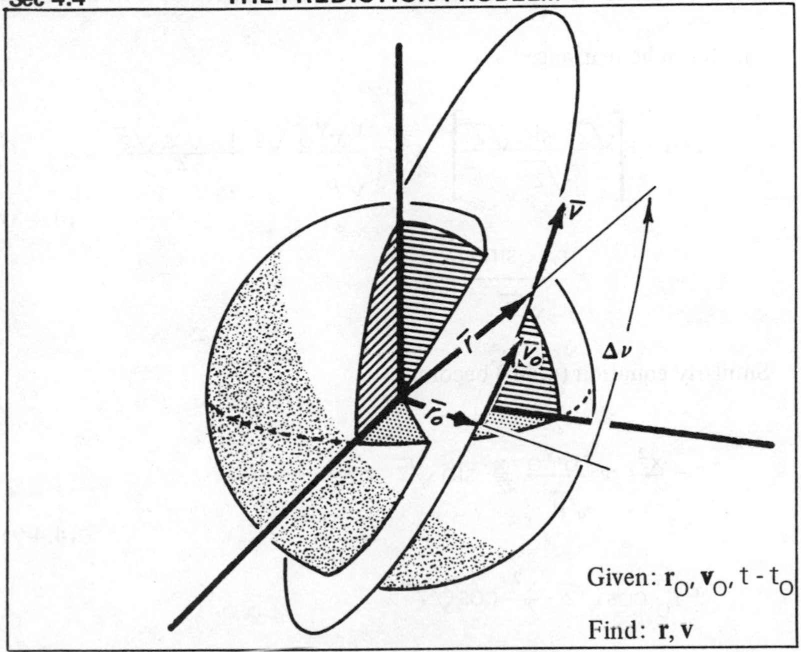

Figure 4.4-1 The Kepler problem

Substituting equations (4.4-1) and (4.4-3) we get:

$$r = a + a\left[\frac{\mathbf{r}_O \cdot \mathbf{v}_O}{\sqrt{\mu a}} \sin \frac{x}{\sqrt{a}} + \left(\frac{r_O}{a} - 1\right) \cos \frac{x}{\sqrt{a}}\right] \tag{4.4-6}$$

At this point let us introduce another new variable

$$z = \frac{x^2}{a}. \tag{4.4-7}$$

Then $a = \frac{x^2}{z}$

Equation (4.4-4) then becomes

$$\sqrt{\mu}\, t = \frac{x^2}{z} \left(x - \frac{x}{\sqrt{z}} \sin\sqrt{z}\right) + \frac{\mathbf{r}_O \cdot \mathbf{v}_O}{\sqrt{\mu}} \frac{x^2}{z} \left(1 - \cos\sqrt{z}\right)$$

$$+ r_O \frac{x}{\sqrt{z}} \sin\sqrt{z}$$

which can be rearranged as

$$\sqrt{\mu}\, t = \left[\frac{\sqrt{z} - \sin\sqrt{z}}{\sqrt{z^3}}\right] x^3 + \frac{\mathbf{r}_0 \cdot \mathbf{v}_0}{\sqrt{\mu}}\, x^2\, \frac{1 - \cos\sqrt{z}}{z} + \frac{r_0 x \sin\sqrt{z}}{\sqrt{z}} \tag{4.4-8}$$

Similarly equation (4.4-6) becomes

$$r = \frac{x^2}{z} + \frac{\mathbf{r}_0 \cdot \mathbf{v}_0}{\sqrt{\mu}}\, \frac{x}{\sqrt{z}} \sin\sqrt{z} + r_0 \cos\sqrt{z} - \frac{x^2}{z}\cos\sqrt{z}. \tag{4.4-9}$$

These equations are indeterminate for $z = 0$. To remedy this we will introduce two very useful functions which can be expressed as a series:

$$C(z) \equiv \frac{1 - \cos\sqrt{z}}{z} = \frac{1 - \cosh\sqrt{-z}}{z} = \frac{1}{2!} - \frac{z}{4!} + \frac{z^2}{6!} - \frac{z^3}{8!} + \cdots$$

$$= \sum_{k=0}^{\infty} \frac{(-z)^k}{(2k+2)!} \tag{4.4-10}$$

$$S(z) = \frac{\sqrt{z} - \sin\sqrt{z}}{\sqrt{z^3}} = \frac{\sinh\sqrt{-z} - \sqrt{-z}}{\sqrt{(-z)^3}}$$

$$= \frac{1}{3!} - \frac{z}{5!} + \frac{z^2}{7!} - \frac{z^3}{9!} + \cdots = \sum_{k=0}^{\infty} \frac{(-z)^k}{(2k+3)!}. \tag{4.4-11}$$

The properties of these functions will be discussed in a later section. Using these functions equations (4.4-8) and (4.4-9) become

$$\sqrt{\mu}\, t = x^3 S + \frac{\mathbf{r}_0 \cdot \mathbf{v}_0}{\sqrt{\mu}}\, x^2 C + r_0 x (1 - zS). \tag{4.4-12}$$

$$r = \sqrt{\mu}\,\frac{dt}{dx} = x^2 C + \frac{\mathbf{r}_0 \cdot \mathbf{v}_0}{\sqrt{\mu}}\, x\,(1 - zS) + r_0\,(1 - zC). \quad (4.4\text{-}13)$$

4.4.2 Solving for x When Time is Known. An intermediate step to finding the radius and velocity vectors at a future time is to find x when time is known. From $\mathbf{r}_0$, $\mathbf{v}_0$ and the energy equation you can obtain the semi-major axis, a. But now we have a problem; since equation (4.4-12) is transcendental in x, we cannot get it by itself on the left of the equal sign. Therefore, a trial-and-error solution is indicated.

Fortunately, the t vs x curve is well-behaved and a Newton iteration technique may be used successfully to solve for x when time-of-flight is given. If we let $t_0 = 0$ and choose a trial value for x–call it x_n, then

$$\sqrt{\mu}\, t_n = \frac{\mathbf{r}_0 \cdot \mathbf{v}_0}{\sqrt{\mu}}\, x_n^2 C + \left(1 - \frac{r_0}{a}\right) x_n^3 S + r_0 x_n \quad (4.4\text{-}14)$$

where t_n is the time-of-flight corresponding to the given $\mathbf{r}_0$, $\mathbf{v}_0$, a and the trial value of x. Equation (4.4-7) has been used to eliminate z. In one sense, C and S should have a subscript of n also because they are functions of the guess of x_n.

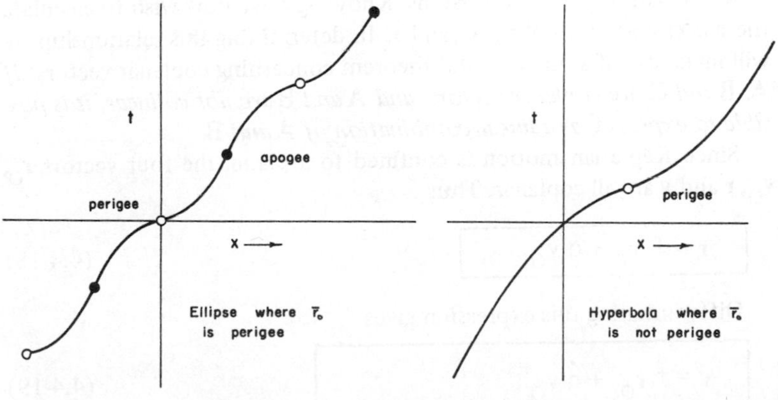

Figure 4.4-2 Typical t vs x plots

A better approximation is then obtained from the Newton iteration algorithm

$$x_{n+1} = x_n + \frac{t - t_n}{dt/dx\big|_{x=x_n}} \tag{4.4-15}$$

where t is the given time-of-flight, and where $dt/dx\big|_{x=x_n}$ is the slope of the t vs x curve at the trial point, x_n.

An analytical expression for the slope may be obtained directly from the definition of x in equation (4.3-2).

$$\frac{dt}{dx} = \frac{1}{\dot{x}} = \frac{r}{\sqrt{\mu}} \tag{4.4-16}$$

Note that the slope of the t vs x curve is directly proportional to r; it will be minimum at periapsis and maximum where r is maximum. Some typical plots of t vs x are illustrated in Figure 4.4-2. Substituting for r in equation (4.4-16) yields

$$\sqrt{\mu}\,\frac{dt}{dx} = x^2 C + \frac{\mathbf{r}_0 \cdot \mathbf{v}_0}{\sqrt{\mu}}\, x\,(1 - zS) + r_0(1 - zC). \tag{4.4-17}$$

When the difference between t and t_n becomes negligible, the iteration may be terminated.

With x known, we must then calculate the corresponding **r** and **v**. To do this we will now develop what is called the f and g expressions in terms of x and z.

4.4.3 The f and g Expressions. Knowing x we now wish to calculate the **r** and **v** in terms of $\mathbf{r}_0$, $\mathbf{v}_0$ and x. In determining this relationship we will make use of a fundamental theorem concerning coplanar vectors: *If* **A**, **B** *and* **C** *are coplanar vectors, and* **A** *and* **B** *are not colinear, it is possible to express* **C** *as a linear combination of* **A** *and* **B**.

Since Keplerian motion is confined to a plane, the four vectors $\mathbf{r}_0$, $\mathbf{v}_0$, **r** and **v** are all coplanar. Thus

$$\mathbf{r} = f\,\mathbf{r}_0 + g\,\mathbf{v}_0\,. \tag{4.4-18}$$

Differentiating this expression gives

$$\mathbf{v} = \dot{f}\,\mathbf{r}_0 + \dot{g}\,\mathbf{v}_0\,. \tag{4.4-19}$$

where f, g, $\dot f$ and $\dot g$ are time dependent scalar quantities. The main purpose of this section will be to determine expressions for these scalars in terms of the universal variable, x.

First, however, we will derive an interesting relationship between f, g, $\dot f$ and $\dot g$. Crossing equation (4.4-18) into equation (4.4-19) gives

$$\mathbf{r}\times\mathbf{v} = (f\,\mathbf{r}_0 + g\,\mathbf{v}_0)\times(\dot f\,\mathbf{r}_0 + \dot g\,\mathbf{v}_0)$$

$$\overbrace{\mathbf{r}\times\mathbf{v}}^{\mathbf{h}} = f\dot f\,\overbrace{\mathbf{r}_0\times\mathbf{r}_0}^{0} + f\dot g\,\overbrace{\mathbf{r}_0\times\mathbf{v}_0}^{\mathbf{h}} - \dot f g\,\overbrace{\mathbf{r}_0\times\mathbf{v}_0}^{\mathbf{h}}$$

$$+\, g\dot g\,\overbrace{\mathbf{v}_0\times\mathbf{v}_0}^{0}.$$

Equating the scalar components of $\mathbf{h}$ on both sides of the equation yields

$$\boxed{1 = f\dot g - \dot f g} \tag{4.4-20}$$

This equation shows that f, g, $\dot f$ and $\dot g$ are not independent; if you know any three you can determine the fourth from this useful identity.

We will now develop the f and g expressions in terms of perifocal coordinates. We can isolate the scalar, f, in equation (4.4-18) by crossing the equation into $\mathbf{v}_0$:

$$\mathbf{r}\times\mathbf{v}_0 = f(\overbrace{\mathbf{r}_0\times\mathbf{v}_0}^{h\mathbf{W}}) + g(\overbrace{\mathbf{v}_0\times\mathbf{v}_0}^{0})$$

Since $\mathbf{r} = x_\omega\,\mathbf{P} + y_\omega\,\mathbf{Q}$ and $\mathbf{v}_0 = \dot x_{\omega_0}\,\mathbf{P} + \dot y_{\omega_0}\,\mathbf{Q}$, the left side of the equation becomes

$$\mathbf{r}\times\mathbf{v}_0 = \begin{vmatrix} \mathbf{P} & \mathbf{Q} & \mathbf{W} \\ x_\omega & y_\omega & 0 \\ \dot x_{\omega_0} & \dot y_{\omega_0} & 0 \end{vmatrix} = (x_\omega \dot y_{\omega_0} - \dot x_{\omega_0} y_\omega)\mathbf{W}$$

Equating the scalar components of **W** and solving for f,

$$f = \frac{x_\omega \dot{y}_{\omega_0} - \dot{x}_{\omega_0} y_\omega}{h} . \tag{4.4-21}$$

We can isolate g in a similar manner by crossing $\mathbf{r}_0$ into equation (4.4-18):

$$\mathbf{r}_0 \times \mathbf{r} = f(\mathbf{r}_0 \times \mathbf{r}_0) + g(\mathbf{r}_0 \times \mathbf{v}_0) = gh\mathbf{W}$$

$$\mathbf{r}_0 \times \mathbf{r} = \begin{vmatrix} \mathbf{P} & \mathbf{Q} & \mathbf{W} \\ x_{\omega_0} & y_{\omega_0} & 0 \\ x_\omega & y_\omega & 0 \end{vmatrix} = (x_{\omega_0} y_\omega - x_\omega y_{\omega_0})\mathbf{W}$$

$$g = \frac{x_{\omega_0} y_\omega - x_\omega y_{\omega_0}}{h} . \tag{4.4-22}$$

To obtain $\dot{f}$ and $\dot{g}$ we only need to differentiate the expressions for f and g (or we could cross $\mathbf{r}_0$ into equation (4.4-19) to get $\dot{g}$ and then cross equation (4.4-19) into $\mathbf{v}_0$ to get $\dot{f}$):

$$\dot{f} = \frac{\dot{x}_\omega \dot{y}_{\omega_0} - \dot{x}_{\omega_0} \dot{y}_\omega}{h} \tag{4.4-23}$$

$$\dot{g} = \frac{x_{\omega_0} \dot{y}_\omega - \dot{x}_\omega y_{\omega_0}}{h} . \tag{4.4-24}$$

To get the f and g expressions in terms of x, we need to relate the perifocal coordinates to x. From the standard conic equation we obtain

$$re \cos \nu = a(1 - e^2) - r . \tag{4.4-25}$$

Combining equations (4.3-4) and (4.4-25),

$$x_\omega = r \cos \nu = -a\left(e + \sin \frac{x + c_0}{\sqrt{a}}\right) . \tag{4.4-26}$$

Since $\quad y_\omega{}^2 = r^2 - x_\omega{}^2$

we obtain

$$y_\omega = a\sqrt{1-e^2}\cos\frac{x+c_0}{\sqrt{a}} \quad . \tag{4.4-27}$$

Now by differentiating equations (4.4-26) and (4.4-27) and using the definition of the universal variable, equation (4.3-2),

$$\dot{x}_\omega = -\frac{\sqrt{\mu a}}{r}\cos\frac{x+c_0}{\sqrt{a}} \tag{4.4-28}$$

$$\dot{y}_\omega = -\frac{h}{r}\sin\frac{x+c_0}{\sqrt{a}} \quad . \tag{4.4-29}$$

Substituting equations (4.4-26) through (4.4-29) into equation (4.4-21)

$$f = \frac{1}{h}[a(e+\sin\frac{x+c_0}{\sqrt{a}})\frac{h}{r_0}\sin\frac{c_0}{\sqrt{a}} +$$

$$(\frac{\sqrt{\mu a}}{r_0}\cos\frac{c_0}{\sqrt{a}})\,a\sqrt{1-e^2}\cos\frac{x+c_0}{\sqrt{a}}] \; .$$

Recall that $x=0$ at $t=0$. Using the trig identities for sine and cosine of a sum,

$$f = \frac{a}{r_0}(e\sin\frac{c_0}{\sqrt{a}} + \cos\frac{x}{\sqrt{a}}). \tag{4.4-30}$$

Using the definitions of z, $C(z)$ and equation (4.4-1), equation (4.4-30) becomes

$$f = 1 - \frac{a}{r_0}(1-\cos\frac{x}{\sqrt{a}}) = 1 - \frac{x^2}{r_0}C. \tag{4.4-31}$$

We can derive the expression for g in a similar manner:

$$g = \frac{1}{h}\left[-a\left(e + \sin\frac{c_0}{\sqrt{a}}\right) a\sqrt{1-e^2}\cos\frac{x + c_0}{\sqrt{a}}\right.$$

$$\left.+ a\sqrt{1-e^2}\left(\cos\frac{c_0}{\sqrt{a}}\right) a\left(e + \sin\frac{x + c_0}{\sqrt{a}}\right)\right]$$

$$= \frac{a^2}{\sqrt{\mu a}}\left[e\left(\cos\frac{c_0}{\sqrt{a}} - \cos\frac{x}{\sqrt{a}}\cos\frac{c}{\sqrt{a}}\right.\right.$$

$$\left.\left. + \sin\frac{x}{\sqrt{a}}\sin\frac{c_0}{\sqrt{a}}\right) + \sin\frac{x}{\sqrt{a}}\right]$$

Using equations (4.4-1) and (4.4-3)

$$g = \frac{a^2}{\sqrt{\mu a}}\left[\frac{\mathbf{r}_0 \cdot \mathbf{v}_0}{\sqrt{\mu a}}\left(1 - \cos\frac{x}{\sqrt{a}}\right) + \frac{r_0}{a}\sin\frac{x}{\sqrt{a}}\right]$$

Then

$$\sqrt{\mu}\, g = x^2\frac{\mathbf{r}_0 \cdot \mathbf{v}_0}{\sqrt{\mu}} C + r_0 x\,(1 - zS). \qquad (4.4\text{-}32)$$

Comparing this to equation (4.4-14) we see that

$$\sqrt{\mu}\, g = \sqrt{\mu}\, t - x^3 S \qquad (4.4\text{-}33)$$

and

$$\boxed{g = t - \frac{x^3}{\sqrt{\mu}} S\,.} \qquad (4.4\text{-}34)$$

In a similar fashion we can show that

$$\boxed{\dot{g} = 1 - \frac{a}{r} + \frac{a}{r}\cos\frac{x}{\sqrt{a}} = 1 - \frac{x^2}{r}C} \qquad (4.4\text{-}35)$$

and

$$\boxed{\dot{f} = -\frac{\sqrt{\mu a}}{r_0 r}\sin\frac{x}{\sqrt{a}} = \frac{\sqrt{\mu}}{r_0 r}\, x\,(zS - 1).} \qquad (4.4\text{-}36)$$

In computation, note that equation (4.4-20) can be used as a check on the accuracy of the f and g expressions. Also, in any of the equations where z appears, its definition, x^2/a, can be used. Note also that if t_o were not zero, the expression $(t - t_o)$ would replace t.

4.4.4 Algorithm for Solution of the Kepler Problem.

1. From $\mathbf{r}_o$ and $\mathbf{v}_o$ determine r_o and a.
2. Given $t - t_o$ (usually t_o is assumed to be zero), solve the universal time-of-flight equation for x using a Newton iteration scheme.
3. Evaluate f and g from equations (4.4-31) and (4.4-34); then compute $\mathbf{r}$ and r from equation (4.4-18).
4. Evaluate $\dot{f}$ and $\dot{g}$ from equations (4.4-35) and (4.4-36); then compute $\mathbf{v}$ from equation (4.4-19).

The advantages of this method over other methods are that only one set of equations works for all conic orbits and accuracy and convergence for nearly parabolic orbits is better.

4.5 IMPLEMENTING THE UNIVERSAL VARIABLE FORMULATION

In this section several aspects of the universal variable formulation will be presented which will increase your understanding and facilitate its computer implementation.

4.5.1 The Physical Significance of x and z. Up to now we have developed universally valid expressions for r and $t - t_o$ in terms of the auxiliary variables x and z; but we have not said what x and z represent. Obviously, if we are going to use equation (4.4-12) above we must know how x and z are related to the physical parameters of the orbit.

Let's compare the expression for r in terms of x with the corresponding expression for r in terms of eccentric anomaly:

$$r = a\left(1 + e \sin \frac{x + c_o}{\sqrt{a}}\right) \tag{4.3-4}$$

$$r = a\,(1 - e \cos E). \tag{4.5-1}$$

We can conclude that

$$\sin \frac{x + c_o}{\sqrt{a}} = - \cos E\,.$$

But

$$\sin \frac{x + c_o}{\sqrt{a}} = \cos\left[\frac{\pi}{2} - \frac{x + c_o}{\sqrt{a}}\right] = - \cos\left[\frac{\pi}{2} + \frac{x + c_o}{\sqrt{a}}\right],$$

so $$E = \frac{\pi}{2} + \frac{x + c_o}{\sqrt{a}}. \tag{4.5-2}$$

If we compare equations (4.3-4) and (4.5-1) at time t_o when $x = 0$, $r = r_o$, and $E = E_o$, we get

$$E_o = \frac{\pi}{2} + \frac{c_o}{\sqrt{a}} \tag{4.5-3}$$

Subtracting equation (4.5-3) from (4.5-2) yields

$$\boxed{x = \sqrt{a}\,(E - E_o).} \tag{4.5-4}$$

Using the identity $F = iE$, we can conclude that

$$\boxed{x = \sqrt{-a}\,(F - F_o)} \tag{4.5-5}$$

whenever a is negative.

To determine what x represents on the parabolic orbit let's look at the general expression for r in terms of x:

$$r = x^2 C + \frac{\mathbf{r}_o \cdot \mathbf{v}_o}{\sqrt{\mu}}\, x\,(1 - zS) + r_o(1 - zC). \tag{4.4-13}$$

For the parabolic orbit $z = 0$ and $C = 1/2$, since $a = \infty$, so

$$r = \frac{1}{2}x^2 + \frac{\mathbf{r}_o \cdot \mathbf{v}_o}{\sqrt{\mu}}\, x + r_o .$$

We will show later in equation (4.6-6) that $\mathbf{r}_o \cdot \mathbf{v}_o = \sqrt{\mu}\, D_o$ and in equation (4.6-5) that $r = 1/2\,(p + D^2)$ and $r_o = 1/2\,(p + D_o^2)$, therefore

$$\frac{1}{2}(p + D^2) = \frac{1}{2}x^2 + D_o x + \frac{1}{2}(p + D_o^2) .$$

If we solve this quadratic for x, we get

$$\boxed{x = D - D_o.} \tag{4.5-6}$$

Which is valid for parabolic orbits.

Obviously x is related to the change in eccentric anomaly that occurs between $\mathbf{r}_o$ and $\mathbf{r}$. Since $z = x^2/a$,

$$\boxed{z = (E - E_o)^2} \tag{4.5-7}$$

when z is positive. If z is negative it can only mean that E and E_o are imaginary, so

$$-z = (F - F_o)^2. \tag{4.5-8}$$

When z is zero, either a is infinite or the change in eccentric anomaly is zero.

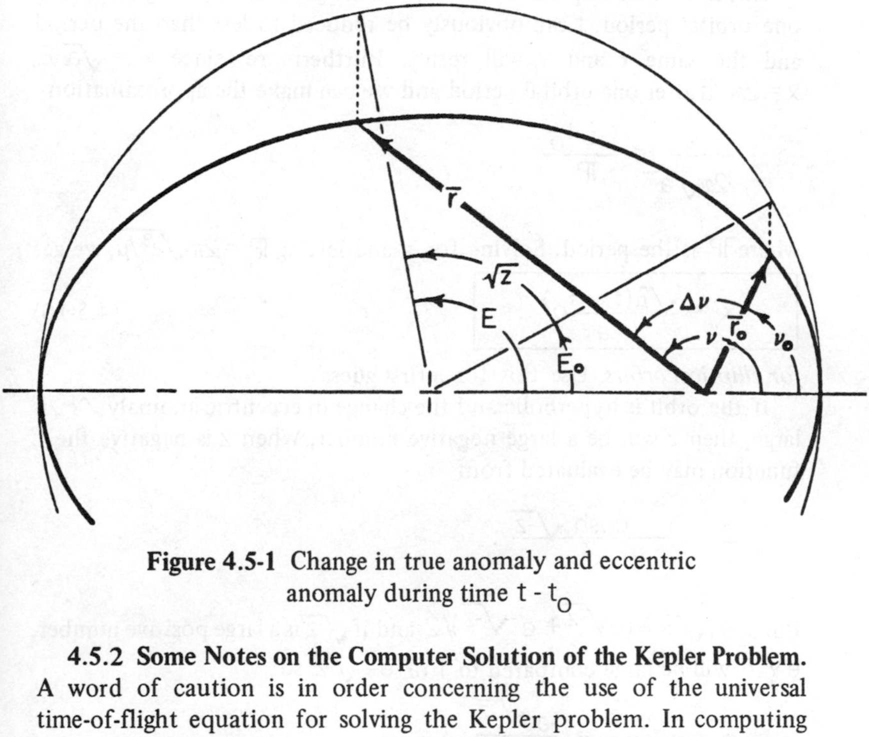

Figure 4.5-1 Change in true anomaly and eccentric anomaly during time $t - t_o$

4.5.2 Some Notes on the Computer Solution of the Kepler Problem. A word of caution is in order concerning the use of the universal time-of-flight equation for solving the Kepler problem. In computing the semi-major axis, a, from $\mathbf{r}_o$ and $\mathbf{v}_o$ and the energy equation we get

$$a = \frac{-\mu}{v_o^2 - 2\mu/r_o}.$$

If the orbit is parabolic the denominator of this expression is zero and an error finish would result if the computation is performed on a digital computer. Therefore, the reciprocal of a should be computed and stored instead:

$$\alpha = \frac{1}{a} = \frac{2\mu/r_o - v_o^2}{\mu}. \tag{4.5-9}$$

All equations should then be modified by replacing $1/a$, wherever it occurs, with α.

The number of iterations required to compute x to any desired degree of accuracy depends mainly on the initial trial value of x; if the initial estimate of x is close to the correct value, convergence will be extremely rapid.

In the case of elliptical orbits, where the given time-of-flight exceeds one orbital period, t can obviously be reduced to less than the period and the same **r** and **v** will result. Furthermore, since $x = \sqrt{a}\Delta E$, $x = 2\pi\sqrt{a}$ after one orbital period and we can make the approximation

$$\frac{x}{2\pi\sqrt{a}} \approx \frac{t - t_0}{\mathbb{P}}$$

where $\mathbb{P}$ is the period. Solving for x and letting $\mathbb{P} = 2\pi\sqrt{a^3/\mu}$, we get

$$\boxed{x \approx \frac{\sqrt{\mu}(t - t_0)}{a}\ .} \tag{4.5-10}$$

for elliptical orbits. Use this for a first guess.

If the orbit is hyperbolic and the change in eccentric anomaly, ΔF, is large, then z will be a large negative number. When z is negative the C function may be evaluated from

$$C = \frac{1 - \cosh\sqrt{-z}}{z}\ .$$

But $\cosh\sqrt{-z} = (e^{\sqrt{-z}} + e^{-\sqrt{-z}})/2$ and if $\sqrt{-z}$ is a large positive number, $e^{\sqrt{-z}}$ will be large compared to 1 or $e^{-\sqrt{-z}}$, so

$$C \approx \frac{-e^{\sqrt{-z}}}{2z} = \frac{-ae^{\sqrt{-z}}}{2x^2}\ .$$

Similarly we can say that

$$S = \frac{\sinh\sqrt{-z} - \sqrt{-z}}{\sqrt{(-z)^3}}$$

But $\sinh\sqrt{-z} = (e^{\sqrt{-z}} - e^{-\sqrt{-z}})/2$, and

$$\sqrt{(-z)^3} = \pm\, x^3/(-a\sqrt{-a}),\ \text{so}\ S \approx \frac{e^{\sqrt{-z}}}{2\sqrt{(-z)^3}} = \frac{-a\sqrt{-a}\,e^{\sqrt{-z}}}{\pm\, 2x^3}\ .$$

The ± sign can be resolved easily since we know (section 4.5.3) that the S function is positive for all values of z. Therefore, if x is positive we should take the + sign and if x is negative we should take the - sign in the above expression. Anytime $t - t_0$ is positive, x will be positive and vice versa, so

$$S \approx \frac{-a\sqrt{-a}\, e^{\sqrt{-z}}}{\text{sign}(t - t_0) 2x^3}$$

Substituting these approximate values for C and S into the universal time-of-flight equation and neglecting the last term, we get

$$t - t_0 \approx -a\frac{(\mathbf{r}_0 \cdot \mathbf{v}_0)}{2\mu}\, e^{\sqrt{-z}} - \text{sign}(t - t_0)\frac{a\sqrt{-a}}{2\sqrt{\mu}}\left(1 - \frac{r_0}{a}\right) e^{\sqrt{-z}}$$

Solving for $e^{\sqrt{-z}}$ yields

$$e^{\sqrt{-z}} \approx \frac{-2\mu(t - t_0)}{a[(\mathbf{r}_0 \cdot \mathbf{v}_0) + \text{sign}(t - t_0)\sqrt{-\mu a}\,(1 - \frac{r_0}{a})]}$$

$$\sqrt{-z} = \pm\frac{x}{\sqrt{-a}}$$

$$\approx \ln\left[\frac{-2\mu(t - t_0)}{a[(\mathbf{r}_0 \cdot \mathbf{v}_0) + \text{sign}(t - t_0)\sqrt{-\mu a}\,(1 - \frac{r_0}{a})]}\right]$$

We can resolve the ± sign in the same way as before, recognizing that x will be positive when $(t - t_0)$ is positive. Thus, *for hyperbolic orbits:*

$$x \approx \text{sign}(t - t_0)\sqrt{-a}\,\ln\left[\frac{-2\mu\,(t - t_0)}{a[(\mathbf{r}_0 \cdot \mathbf{v}_0) + \text{sign}(t - t_0)\sqrt{-\mu a}\,(1 - \frac{r_0}{a})]}\right] \qquad (4.5\text{-}11)$$

The use of these approximations, where appropriate, for selecting the first trial value of x will greatly speed convergence.

4.5.3 Properties of C(z) and S(z). The functions C(z) and S(z) are defined as

$$C = \frac{1 - \cos\sqrt{z}}{z} \tag{4.4-10}$$

$$S = \frac{\sqrt{z} - \sin\sqrt{z}}{\sqrt{z^3}} \tag{4.4-11}$$

We can write equation (4.4-10) as

$$C = \frac{1 - \cos i\sqrt{-z}}{z}$$

where $i = \sqrt{-1}$. But $\cos i\theta = \cosh\theta$, so an *equivalent expression for* C is

$$C = \frac{1 - \cosh\sqrt{-z}}{z}. \tag{4.5-12}$$

Similarly, we can write equation (4.4-11) as

$$S = \frac{i\sqrt{-z} - \sin i\sqrt{-z}}{-i\sqrt{(-z)^3}} = \frac{-i\sin i\sqrt{-z} - \sqrt{-z}}{\sqrt{(-z)^3}}.$$

But $-i\sin i\theta = \sinh\theta$, so an *equivalent expression for* S is

$$S = \frac{\sinh\sqrt{-z} - \sqrt{-z}}{\sqrt{(-z)^3}} \tag{4.5-13}$$

To evaluate C and S when z is positive, use equations (4.4-10) and (4.4-11); if z is negative, use equations (4.5-12) and (4.5-13). If z is near zero, the power series expansions of the functions may be used to evaluate C and S. The series expansions are easily derived from the power series for $\sin\theta$ and $\cos\theta$:

$$\cos\theta = 1 - \frac{\theta^2}{2!} + \frac{\theta^4}{4!} - \frac{\theta^6}{6!} + \ldots$$

$$\sin\theta = \theta - \frac{\theta^3}{3!} + \frac{\theta^5}{5!} - \frac{\theta^7}{7!} + \ldots$$

Substituting these series into the definitions of C and S, we get

$$C = \frac{1}{z}\left[1 - (1 - \frac{z}{2!} + \frac{z^2}{4!} - \frac{z^3}{6!} + \ldots)\right]$$

$$\boxed{C = \frac{1}{2!} - \frac{z}{4!} + \frac{z^2}{6!} - \ldots} \tag{4.5-14}$$

$$S = \frac{1}{\sqrt{z^3}}\left[\sqrt{z} - \left(\sqrt{z} - \frac{\sqrt{z^3}}{3!} + \frac{\sqrt{z^5}}{5!} - \frac{\sqrt{z^7}}{7!} + \ldots\right)\right]$$

$$\boxed{S = \frac{1}{3!} - \frac{z}{5!} + \frac{z^2}{7!} - \ldots} \tag{4.5-15}$$

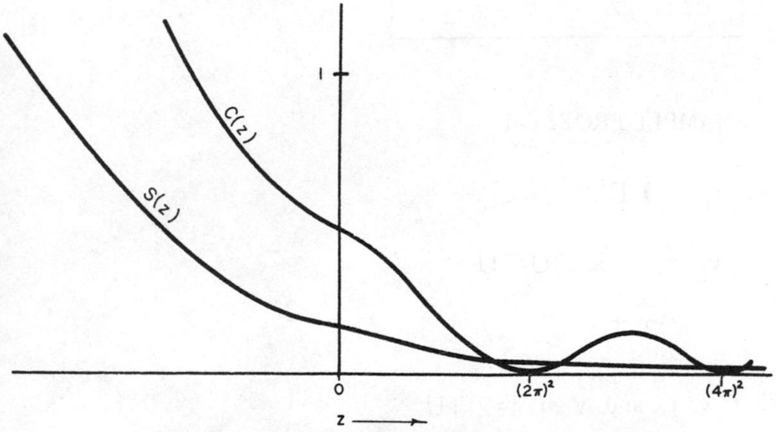

Figure 4.5-2 Plot of S(z) and C(z) versus z

Both the C and S functions approach infinity as z approaches minus infinity. The S function is 1/6 when $z = 0$ and decreases asymptotically to zero as z approaches plus infinity. The C function is 1/2 when $z = 0$ nd decreases to zero at $z = (2\pi)^2, (4\pi)^2, (6\pi)^2$, etc.

Since the C and S functions are defined by means of the cos and sin functions, it is not surprising that the derivatives, dC/dz and dS/dz, can be expressed in terms of the functions themselves. Differentiating the definition of C, we get

$$\frac{dC}{dz} = \frac{1}{2z}\left[\frac{\sin\sqrt{z}}{\sqrt{z}} - \frac{2(1 - \cos\sqrt{z})}{z}\right]$$

$$\boxed{\frac{dC}{dz} = \frac{1}{2z}(1 - zS - 2C).} \tag{4.5-16}$$

Differentiating the definition of S, we get

$$\frac{dS}{dz} = \frac{1}{2z}\left[\frac{1 - \cos\sqrt{z}}{z} - \frac{3(\sqrt{z} - \sin\sqrt{z})}{\sqrt{z^3}}\right]$$

$$\boxed{\frac{dS}{dz} = \frac{1}{2z}(C - 3S).} \tag{4.5-17}$$

EXAMPLE PROBLEM.

Given $\mathbf{r}_o = \mathbf{I}$ DU

$\mathbf{v}_o = 1.1\mathbf{K}$ DU/TU

$t = 2$ TU

Find x, $\mathbf{r}$, and $\mathbf{v}$ at $t=2$ TU.

$$\mathbf{r}_o = \mathbf{I},\ \mathbf{v}_o = 1.1\mathbf{K},\ t=2\ \text{TU}_\oplus$$

$$\mathbf{r}_o \cdot \mathbf{v}_o = 0,\ v_o > \sqrt{\frac{\mu}{r_o}}$$

Therefore $\mathbf{r}_o = \mathbf{r}_p$

$$\mathcal{E} = \frac{1.1^2}{2} - \frac{1}{1} = -.395,\quad a = \frac{-1}{2\mathcal{E}} = 1.266$$

Since $x = \sqrt{a}\Delta E$ and $\Delta E = 2\pi$ for one complete orbit, we can make a first approximation for x:

$$\frac{x}{2\pi\sqrt{a}} \cong \frac{t-t_0}{TP} \quad \text{or} \quad x \cong \frac{\sqrt{\mu}(t-t_0)}{a}$$

$$\therefore x_1 = \frac{2}{1.266} = 1.58$$

$$z_1 = \frac{(1.58)^2}{a} = 1.973$$

Using equation (4.4-14) and (4.4-17) to find t_1 and dt/dx_1

$$\sqrt{\mu}t_1 = \frac{\mathbf{r}_0 \cdot \mathbf{v}_0}{\sqrt{\mu}} x_1^2 C_1 + \left(1 - \frac{r_0}{a}\right) x_1^3 S_1 + r_0 x_1 = 1.705$$

$$\sqrt{\mu}\frac{dt}{dx_1} = x_1^2 C_1 + \frac{\mathbf{r}_0 \cdot \mathbf{v}_0}{\sqrt{\mu}} x_1 (1 - z_1 S_1) + r_0 (1 - z_1 C_1)$$
$$= 1.222$$

Inserting these figures in the Newton iteration:

$$x_2 = x_1 + \frac{2-1.705}{1.222} = 1.58 + \frac{.295}{1.222} = 1.58 + .241 = 1.821$$

Repeating the process, solving for t_2, $\frac{dt}{dx_2}$, x_3, etc., we can construct a table and see how the iteration process drives our successive values for t_n toward $t = 2TU$.

x_n	t_n	$\frac{dt}{dx_n}$	x_{n+1}
1.58	1.705	1.222	1.821
1.821	2.007	1.279	1.816
1.816	2.000	1.277	1.816

After three iterations, we have found the value of x for a time-of-flight of 2 TU accurate to three decimal places, using a slide rule. Using a ligital computer this accuracy can be improved to 11 decimal places if ecessary.

Then from the definitions of $f, \dot{f}, g, \dot{g}$, $f = -0.321$, $g = 1.124$, $\dot{f} = -0.8801$ and $\dot{g} = -0.035$ Thus

$$\mathbf{r} = f\mathbf{r}_0 + g\mathbf{v}_0 = \underline{\underline{-0.321\mathbf{I} + 1.236\mathbf{K}}}$$

$$\mathbf{v} = \dot{f}\mathbf{r}_0 + \dot{g}\mathbf{v}_0 = \underline{\underline{-0.8801\mathbf{I} - 0.039\mathbf{K}}}$$

4.6 CLASSICAL FORMULATIONS OF THE KEPLER PROBLEM

In the interest of relating to the historical development of the solution of the Kepler problem we will briefly summarize the solution using the various eccentric anomalies. But first some useful identities will be presented.

4.6.1 Some Useful Identities Involving D, E and F. We have developed time-of-flight equations for the parabola, ellipse and hyperbola which involve the auxiliary variables D, E and F. You have already seen how these eccentric anomalies relate to the true anomaly, ν. Now let's look at the relationship between these auxiliary variables and some other physical parameters of the orbit.

Taking each of the eccentric anomalies in turn, we will derive expressions for x_ω, y_ω and r in terms of D, E or F. Then we will relate the eccentric anomalies to the dot product $\mathbf{r} \cdot \mathbf{v}$. Finally, we will examine the very interesting and fundamental relationship between E and F.

In order to simplify Barker's equation (4.2-17), we introduced the parabolic eccentric anomaly, D, as

$$D = \sqrt{p}\, \tan \frac{\nu}{2}\ . \tag{4.6-1}$$

From Figure 4.6-1 we can see that

$$x_\omega = r \cos \nu\ .$$

But, for the parabola, since $e = 1$,

$$r = \frac{p}{1 + \cos \nu}\ ,$$

$$\text{so } x_\omega = \frac{p \cos \nu}{1 + \cos \nu}\ . \tag{4.6.2}$$

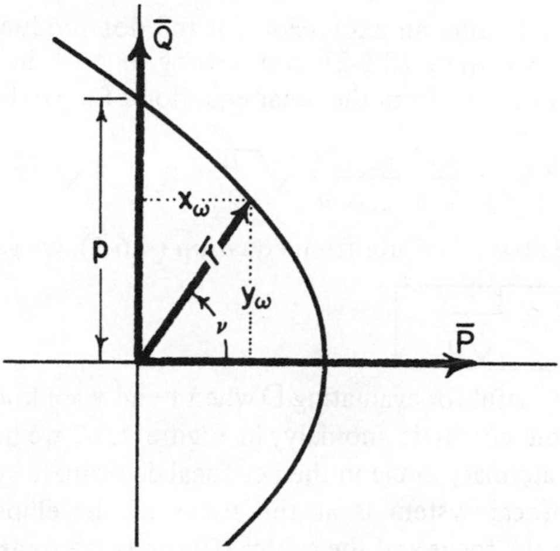

Figure 4.6-1 Perifocal components of **r**

Now substitute $\cos\nu = \cos^2\frac{\nu}{2} - \sin^2\frac{\nu}{2}$ in the numerator of equation (4.6-2) and substitute $\cos\nu = 2\cos^2\frac{\nu}{2} - 1$ in the denominator:

$$x_\omega = \frac{p}{2}\left(1 - \tan^2\frac{\nu}{2}\right).$$

If we substitute from equation (4.6-1), the expression for x_ω reduces to

$$\boxed{x_\omega = \frac{1}{2}\left(p - D^2\right).} \tag{4.6-3}$$

The expression for y_ω is much simpler:

$$y_\omega = r\sin\nu = \frac{p\sin\nu}{1+\cos\nu} = p\tan\frac{\nu}{2}$$

$$\boxed{y_\omega = \sqrt{p}\,D.} \tag{4.6-4}$$

Since x_ω and y_ω are the rectangular components of the vector **r** in the perifocal coordinate system,

$$r^2 = x_\omega^2 + y_\omega^2$$

$$\boxed{r = \frac{1}{2}\left(p + D^2\right).} \tag{4.6-5}$$

Now let's find an expression for the dot product, $\mathbf{r} \cdot \mathbf{v}$. Using $\mathbf{r} \cdot \dot{\mathbf{r}} = r\dot{r}$, and equation (2.5-2) and setting $e = 1$ in this equation and substituting for r from the polar equation of a parabola yields

$$\mathbf{r} \cdot \mathbf{v} = \frac{p}{1 + \cos \nu} \sqrt{\frac{\mu}{p}} \sin \nu = \sqrt{\mu p} \tan \frac{\nu}{2}.$$

If we now substitute from equation (4.6-1), we get

$$\boxed{D = \frac{\mathbf{r} \cdot \mathbf{v}}{\sqrt{\mu}}} \tag{4.6-6}$$

which is useful for evaluating D when $\mathbf{r}$ and $\mathbf{v}$ are known.

For the eccentric anomaly, in Figure 4.6-2 we have drawn an ellipse with its auxiliary circle in the perifocal coordinate system. The origin of the perifocal system is at the focus of the ellipse and the distance between the focus and the center (0) is just $c = ae$ as shown.

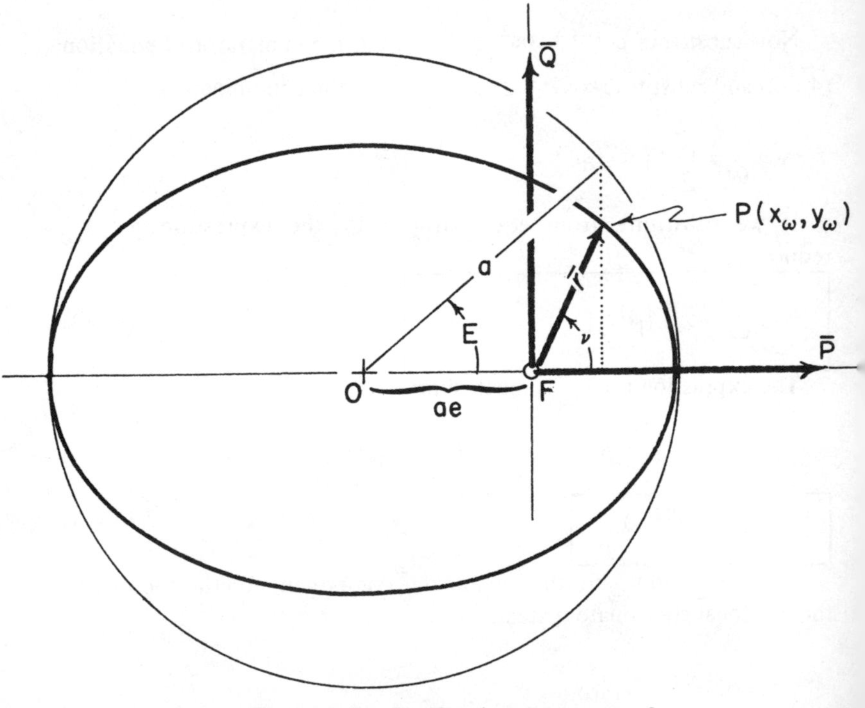

Figure 4.6-2 Perifocal components of r

From Figure 4.6-2, we see that

$$x_\omega = a \cos E - ae$$

or

$$\boxed{x_\omega = a\,(\cos E - e).} \tag{4.6-7}$$

From geometry and the relationship between the y_ω-ordinates of the ellipse and circle, we get

$$y_\omega = \frac{b}{a}\,(a \sin E)$$

But, since $b^2 = a^2 - c^2$ for the ellipse and $c = ae$,

$$\boxed{y_\omega = a\sqrt{1 - e^2}\,\sin E.} \tag{4.6-8}$$

Just as we did for the parabola, we can say that

$$r^2 = x_\omega^2 + y_\omega^2\,.$$

Substituting from equations (4.6-7) and (4.6-8) and simplifying, we obtain

$$\boxed{r = a(1 - e \cos E).} \tag{4.6-9}$$

If we solve this equation for $e \cos E$ we get another useful expression:

$$\boxed{e \cos E = 1 - \frac{r}{a}.} \tag{4.6-10}$$

We will find it convenient to have an expression for $e \sin E$ also. To get it we need to differentiate equation (4.6-9).

$$\dot{r} = (ae \sin E)\dot{E}\,.$$

To find $\dot{E}$ we could differentiate the Kepler time-of-flight equation:

$$t - T = \sqrt{\frac{a^3}{\mu}}\,(E - e \sin E)\,. \tag{4.2-4}$$

Thus

$$1 = \sqrt{\frac{a^3}{\mu}}\,(1 - e\cos E)\dot{E}\,.$$

Solving this equation for $\dot{E}$ and substituting for $e \cos E$ from equation (4.6-10), we get

$$\dot{E} = \frac{1}{r}\sqrt{\frac{\mu}{a}}\,. \tag{4.6-11}$$

And so

$$\dot{r} = \sqrt{\mu a}\,\frac{e \sin E}{r}\,.$$

Finally, solving this equation for $e \sin E$ and again noting that $r\dot{r} = \mathbf{r} \cdot \mathbf{v}$, we obtain

$$e \sin E = \frac{\mathbf{r} \cdot \mathbf{v}}{\sqrt{\mu a}} \tag{4.6-12}$$

which is particularly useful since the term $e \sin E$ appears in the Kepler time-of-flight equation.

For the hyperbolic eccentric anomaly we will exploit the relationship between E and F to arrive at a set of identities involving F which are analogous to the ones involving E.

The relationship between the eccentric anomalies and the true anomaly, ν, is given by

$$\cos E = \frac{e + \cos\nu}{1 + e\cos\nu} \tag{4.2-8}$$

$$\cosh F = \frac{e + \cos\nu}{1 + e\cos\nu}\,. \tag{4.2-21}$$

From which we may conclude that

$$\cosh F = \cos E$$

Using the identity $\cosh\theta = \cos i\theta$ we see that apparently

$$\boxed{E = \pm iF} \tag{4.6-13}$$

In other words, when E is a real number, F is imaginary; when F is a real number, E is imaginary. The $\pm$ sign is a result of defining E in the range from 0 to 2π while F is defined from minus infinity to plus infinity. The proper sign can always be determined from physical reasoning.

From equation (4.6-7) we can write

$$\boxed{x_\omega = a\,(\cos E - e)}$$

$$= a\,(\cos iF - e).$$

But $\cos iF = \cosh F$, so

$$\boxed{x_\omega = a(\cosh F - e).} \tag{4.6-14}$$

Similarly,

$$y_\omega = a\sqrt{1-e^2}\,\sin E$$

$$= ai\sqrt{e^2-1}\,\sin iF,$$

But $i \sin iF = -\sinh F$, so

$$\boxed{y_\omega = -a\sqrt{e^2-1}\,\sinh F.} \tag{4.6-15}$$

The following identities are obtained in analogous fashion:

$$\boxed{r = a(1 - e\cosh F)} \tag{4.6-16}$$

$$\boxed{e\cosh F = 1 - \frac{r}{a}} \tag{4.6-17}$$

$$\boxed{e\sinh F = \frac{\mathbf{r}\cdot\mathbf{v}}{\sqrt{-\mu a}}} \tag{4.6-18}$$

This last identity is particularly useful since the expression $e \sinh F$ appears in the hyperbolic time-of-flight equation.

4.6.2 The f and g Expressions in Terms of $\Delta\nu$. In Figure 4.6-3 we have drawn an orbit in the perifocal system. Although an ellipse is shown we need to make no assumption concerning the type of conic.

The rectangular components of a general position vector, **r**, may be written as

$$x_\omega = r\cos\nu \tag{4.6-19}$$

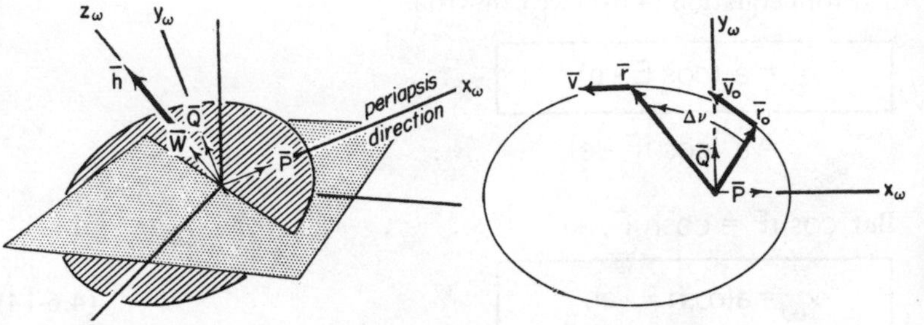

Figure 4.6-3 Perifocal components of position and velocity

$$y_\omega = r \sin \nu \tag{4.6-20}$$

From equation (2.5-4), the rectangular components of the velocity vector, **v**, are

$$\dot{x}_\omega = -\sqrt{\frac{\mu}{p}} \sin \nu \tag{4.6-21}$$

$$\dot{y}_\omega = \sqrt{\frac{\mu}{p}} (e + \cos \nu) \ . \tag{4.6-22}$$

If we substitute these expressions into equation (4.4-21), and note that $h = \sqrt{\mu p}$, we get

$$f = \frac{(r \cos \nu)\sqrt{\frac{\mu}{p}}(e + \cos \nu_0) + \sqrt{\frac{\mu}{p}} \sin \nu_0 (r \sin \nu)}{\sqrt{\mu p}}$$

But $\cos \nu \cos \nu_0 + \sin \nu \sin \nu_0 = \cos \Delta\nu$, so

$$\boxed{f = 1 - \frac{r}{p}(1 - \cos \Delta\nu)} \tag{4.6-23}$$

where $\Delta\nu = \nu - \nu_0$.

Similarly we obtain

$$\boxed{g = \frac{r r_0 \sin \Delta\nu}{\sqrt{\mu p}}} \tag{4.6-24}$$

$$\boxed{\dot{g} = 1 - \frac{r_0}{p}(1 - \cos \Delta\nu)} \tag{4.6-25}$$

and

$$\boxed{\dot{f} = \sqrt{\frac{\mu}{p}} \tan \frac{\Delta\nu}{2} \left(\frac{1 - \cos \Delta\nu}{p} - \frac{1}{r} - \frac{1}{r_0}\right).} \tag{4.6-26}$$

4.6.3 The f and g Expressions in Terms of Eccentric Anomaly. From equations (4.6-7) and (4.6-8), the rectangular components of velocity may be obtained directly by differentiation. Thus

$$\dot{x}_\omega = -a\dot{E}\sin E .$$

But, using equation (4.6-11)

$$\dot{x}_\omega = -\frac{1}{r}\sqrt{\mu a}\sin E. \tag{4.6-27}$$

Differentiating the expression for y_ω above yields

$$\dot{y}_\omega = a\sqrt{1-e^2}\,\dot{E}\cos E .$$

Therefore

$$\dot{y}_\omega = \frac{1}{r}\sqrt{\mu a(1-e^2)}\cos E . \tag{4.6-28}$$

If we now substitute into equation (4.4-21) and recognize that $h = \sqrt{\mu a(1-e^2)}$, we get

$$f = \frac{a(\cos E - e)\sqrt{\mu a(1-e^2)}\cos E_0}{r_0\sqrt{\mu a(1-e^2)}} + \frac{\sqrt{\mu a}\sin E_0\, a\sqrt{1-e^2}\sin E}{r_0\sqrt{\mu a(1-e^2)}}$$

$$= \frac{a}{r_0}(\cos E\cos E_0 + \sin E\sin E_0 - e\cos E_0).$$

But $\cos E\cos E_0 + \sin E\sin E_0 = \cos\Delta E$ and using equation (4.6-10),

$$f = 1 - \frac{a}{r_0}(1-\cos\Delta E). \tag{4.6-29}$$

Similarly,

$$g = (t-t_0) - \sqrt{\frac{a^3}{\mu}}(\Delta E - \sin\Delta E) \tag{4.6-30}$$

$$\dot{f} = -\frac{\sqrt{\mu a}\sin\Delta E}{r\,r_0} \tag{4.6-31}$$

and

$$\dot{g} = 1 - \frac{a}{r}(1-\cos\Delta E) . \tag{4.6-32}$$

As before we will use the relationship that $\Delta E = i\Delta F$ and the identities relating the circular and hyperbolic functions to derive the hyperbolic expressions directly from the ones involving ΔE.

From equation (4.6-29) we get

$$f = 1 - \frac{a}{r_0}(1 - \cos i\Delta F)$$

But $\cos i\Delta F = \cosh \Delta F$ so

$$f = 1 - \frac{a}{r_0}(1 - \cosh \Delta F) \,. \tag{4.6-33}$$

In the same manner

$$g = (t - t_0) - \sqrt{\frac{(-a)^3}{\mu}}(\sinh \Delta F - \Delta F). \tag{4.6-34}$$

$$\dot{f} = -\frac{\sqrt{-\mu a}\,\sinh \Delta F}{r\, r_0} \,. \tag{4.6-35}$$

and

$$\dot{g} = 1 - \frac{a}{r}(1 - \cosh \Delta F). \tag{4.6-36}$$

4.6.4 Kepler Problem Algorithm. In classical terms, the Kepler problem is basically the solution of the equation

$$M = E - e \sin E \tag{4.2-6}$$

where M is known as $n(t - T)$. M can be obtained from

$$M = \sqrt{\frac{\mu}{a^3}}(t - t_0) - 2k\pi + M_0 \,.$$

Even though equation (4.2-6) is one equation in one unknown, it is transcendental in E; there is no way of getting E by itself on the left of the equal sign. Kepler himself realized this, of course, and Small[4] tells us: "But, with respect to the direct solution of the problem–from the mean anomaly given to find the true–[Kepler] tells us that he found it impracticable, and that he did not believe there was any geometrical or rigorous method of attaining to it."

The first approximate solution for E was quite naturally made by Kepler himself. The next was by Newton in the *Principia*; from a graphical construction involving the cycloid he was able to find an approximate solution for the eccentric anomaly. A very large number of analytical and graphical solutions have since been discovered because nearly every prominent mathematician since Newton has given some attention to the problem. We will resort again to the Newton iteration method.

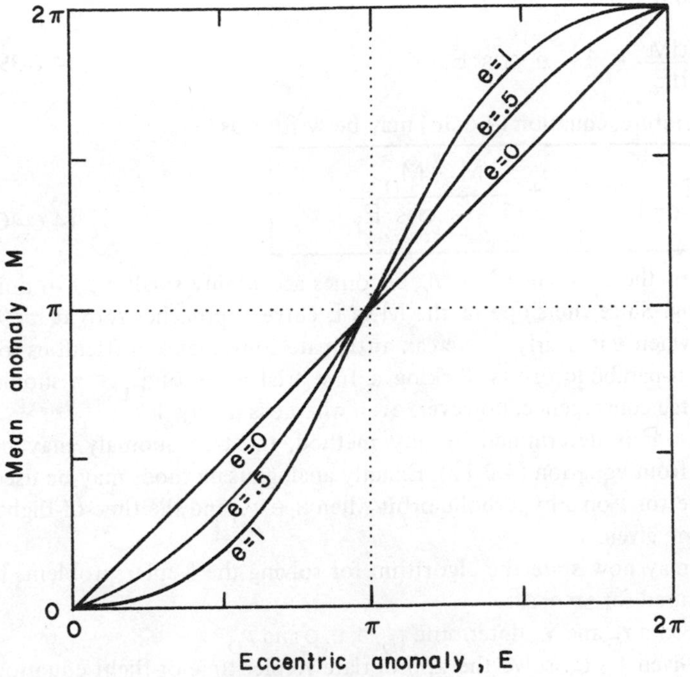

Figure 4.6-4 M vs E plot

It would, of course, be possible to graph Kepler's equation as we have done in Figure 4.6-4 and then determine what value of E corresponds to a known value of M; but this is not very accurate. Since we can derive an analytical expression for the slope of the M vs E curve, we can formulate a Newton iteration scheme as follows: first select a trial value for E–call it E_n. Next, compute the mean anomaly, M_n, that results from this trial value.

$$M_n = E_n - e \sin E_n \,. \tag{4.6-37}$$

Now, select a new trial value, E_{n+1}, from

$$E_{n+1} = E_n + \frac{M - M_n}{dM/dE\big|_{E=E_n}} \tag{4.6-38}$$

where $dM/dE\big|_{E=E_n}$ is the slope of the M vs E curve at the trial value, E_n.

The slope expression is obtained by differentiating Kepler's equation.

$$\frac{dM}{dE} = 1 - e \cos E. \tag{4.6-39}$$

Therefore, equation (4.6-36) may be written as

$$\boxed{E_{n+1} = E_n + \frac{M - M_n}{1 - e \cos E_n}.} \tag{4.6-40}$$

When the difference $M - M_n$ becomes acceptably small we can quit iterating. Since the slope of the M vs E curve approaches zero at $E=0$ or 2π when e is nearly 1, we can anticipate convergence difficulties for the near-parabolic orbits. Picking a first trial value of $E_1 = \pi$ should guarantee convergence, however, even when e is nearly 1.

Once E is determined by any method, the true anomaly may be found from equation (4.2-12). Exactly analogous methods may be used to solve for ν on a hyperbolic orbit when a, e, ν_0 and the time-of-flight, $t - t_0$, are given.

We may now state the algorithm for solving the Kepler problem. It can be used for $\Delta\nu$ or ΔE.

1. From $\mathbf{r}_0$ and $\mathbf{v}_0$ determine r_0, a, e, p and ν_0.
2. Given $t - t_0$, solve the appropriate Kepler time-of-flight equation for E or F using a trial-and-error method such as the Newton iteration. Solve for ν if needed.
3. Solve for r from the polar equation of a conic or equation (4.2-14) or the similar expression for the hyperbola.
4. Evaluate the f and g expressions above using r, r_0, p and $\Delta\nu$ (or ΔE or ΔF).
5. Determine $\mathbf{r}$ and $\mathbf{v}$ from equations (4.4-18) and (4.4-19).

The algorithm using ΔE (or ΔF) is shorter than using $\Delta\nu$ since neither p or ν need to be calculated.

EXERCISES

4.1 The equation of a body in earth orbit is

$$r = \frac{1.5}{1 + .5 \cos \nu}\ \text{DU}$$

Calculate the time of flight from one end of the minor axis out to apogee.
(Ans. TOF = 5.85 TU)

4.2 In deriving TOF on an ellipse, it was stated that the area beneath the ellipse was to the area beneath the auxiliary circle as b/a, i.e.,

$$\frac{\text{area PSV}}{\text{area QSV}} = \frac{b}{a}$$

as a result of the fact that

$$\frac{Y_{\text{ellipse}}}{Y_{\text{circle}}} = \frac{b}{a}$$

Explain why this must be so.

4.3 Given that $\mathbf{r}_o = \mathbf{I} + \mathbf{J}$ DU and $\mathbf{v}_o = 2\mathbf{K}$ DU/TU, find **r** and **v** for $\Delta\nu = 60^o$.
(Partial Ans. $\mathbf{v} = -.348\mathbf{I} - .348\mathbf{J} + 1.5\mathbf{K}$ DU/TU)

4.4 If, in a computer solution for position and velocity on an ellipse, given $\mathbf{r}_o$, $\mathbf{v}_o$, and t, one modifies t by subtracting an integer number of periods to make the $t < TP$, how should the area of search for x be limited, to reduce iterations to a minimum?

4.5 A radar ship at 150° W on the equator picks up an object directly overhead. Returns indicate a position and velocity of:

$$\mathbf{r}_o = 1.2\mathbf{I} \text{ DU}_\oplus$$

$$\mathbf{v}_o = .1\mathbf{I} + \mathbf{J} \text{ DU}_\oplus/\text{TU}_\oplus$$

Four hours later another ship at 120° W on the equator spots the same object directly overhead. Find the values of f, g, $\dot{f}$ and $\dot{g}$ that could be used to calculate position and velocity at the second sighting.
(Partial Ans. $\dot{f} = -.625$)

4.6 For the data given in problem 4.3, find the universal variable, x, corresponding to $\Delta\nu$ of 60°.

4.7 The text, in equations (4.5-10) and (4.5-11), gives analytic expressions to use as a first guess for x in an iterative solution for either the elliptical or hyperbolic trajectory. Develop, and give your analytic reasoning behind, an expression for a first guess for x for the parabolic trajectory.

4.8 Why is the slope of the t vs x curve a minimum at a point corresponding to periapsis? If the slope equals zero at that point, what type of conic section does the curve represent? Draw the family of t vs x curves for ellipses with the same period but different eccentricities (show $e = 0$, $e = .5$, $e = .99$).

4.9 At burnout, a space probe has the following position and velocity:

$$\mathbf{r}_{bo} = 1.1\mathbf{J}\ \text{DU}$$

$$\mathbf{v}_{bo} = \sqrt{2}\mathbf{I}\ \text{DU/TU}$$

How long will it take for the probe to cross the x-axis?
(Ans. TOF = 2.22 TU)

4.10 A satellite is in a polar orbit, with a perigee above the north pole. $r_p = 1.5$ DU, $r_a = 2.5$ DU. Find the time required to go from a point above 30° N latitude to a point above 30° S latitude.
(Ans. 2.73 TU)

4.11 For problem 4.9 compare the calculations using classical and universal variable methods. Do the same for problem 4.10. Which method would be most convenient to program on a computer?

4.12 Construct a flow chart for an algorithm that will read in values for $\mathbf{r}_o$, $\mathbf{v}_o$ and $\Delta\nu$ and will solve for f, g, $\dot{f}$ and $\dot{g}$.

4.13 Construct a flow chart for an algorighm that will read in values for $\mathbf{r}_o$, $\mathbf{v}_o$ and $t - t_o$ and solve for $\mathbf{r}$ and $\mathbf{v}$.

4.14 Any continuous time-varying function can be expressed as a Taylor Series Expansion about a starting value, i.e., if $x = x(t)$, then

$$x = x_0 + \frac{(t - t_0)}{1!}\dot{x}_0 + \frac{(t - t_0)^2}{2!}\ddot{x}_0 + \frac{(t - t_0)^3}{3!}\dddot{x}_0 + \ldots$$

Where $\dot{x}_0 = \left.\frac{dx}{dt}\right|_{x = x_0}$

By defining $U \triangleq -\frac{\mu}{r^3}$ and using it in our equation of motion

$$\ddot{\mathbf{r}} = -\frac{\mu}{r^3}\mathbf{r} = U\mathbf{r}$$

Expand **r** and **v** in Taylor Series and substitute the equation of motion to derive series expressions for f, g, $\dot{f}$ and $\dot{g}$ in terms of $(t - t_0)$ and derivatives of U_0. Find the first three terms of each expression.

4.15 Derive analytically the expression for time-of-flight on a parabola, equation (4.2-18).

4.16 Verify the results expressed in equations (4.4-35) and (4.4-36).

4.17 Derive the expression for g, $\dot{f}$ and $\dot{g}$ in terms of the eccentric anomaly, ΔE. See equations (4.6-30) through (4.6-32).

***4.18** A lunar probe is given just escape speed at burnout altitude of .2 $DU_\oplus$ and flight-path angle of 45°. How long will it take to get to the vicinity of the moon (r_2 = 60 DU) disregarding the Moon's gravity? (Ans. TOF = 219.6 TU)

LIST OF REFERENCES

1. Einstein, Albert. Introduction to *Johannes Kepler: Life and Letters.* New York, NY, Philosophical Library, 1951.

2. Koestler, Arthur. *The Sleepwalkers.* New York, NY, The Macmillan Company, 1959.

3. Battin, Richard H. *Astronautical Guidance.* New York, NY, McGraw-Hill Book Company, 1964.

4. Small, Robert. *An Account of the Astronomical Discoveries of Kepler.* A reprinting of the 1804 text with a forward by William D. Stahlman. Madison, Wisc, The University of Wisconsin Press, 1963.

5. Moulton, Forest Ray. *An Introduction to Celestial Mechanics.* New York, The MacMillan Company, 1914.

6. Sundman, K. F. "Memoire sur le probleme des trois corps," *Acta Mathmatica.* Vol 36, pp 105-179, 1912.

7. Goodyear, W. H. "Completely General Closed Form Solution for Coordinates and Partial Derivatives of the Two-body Problem," *Astron. J.* Vol 70, pp 189-192, 1965.

8. Lemmon, W. W. and J. E. Brooks. "A Universal Formulation for Conic Trajectories—Basic Variables and Relationships." Report 3400-6019-TU000 TRW/Systems, Redondo Beach, California, 1965.

9. Herrick, S. H. "Universal Variables," *Astron. J.* Vol 70, pp 309-315, 1965.

10. Stumpff, K. "Neue Formeln and Hilfstafeln zur Ephemeridenrechung," *Astr. Nach.* Vol 275, pp 108-128, 1947.

11. Sperling, H. "Computation of Keplerian Conic Sections," *ARS J.* Vol 31, pp 660-661, 1961.

12. Bate, Roger R. Dept of Astronautics and Computer Science, United States Air Force Academy. Unpublished notes.

CHAPTER 5

ORBIT DETERMINATION FROM TWO POSITIONS AND TIME

Probably all mathematicians today regret that Gauss was deflected from his march through darkness by "a couple of clods of dirt which we call planets"–his own words–which shone out unexpectedly in the night sky and led him astray. Lesser mathematicians than Gauss–Laplace for instance–might have done all that Gauss did in computing the orbits of Ceres and Pallas, even if the problem of orbit determination was of a sort which Newton said belonged to the most difficult in mathematical astronomy. But the brilliant success of Gauss in these matters brought him instant recognition as the first mathematician in Europe and thereby won him a comfortable position where he could work in comparative peace; so perhaps those wretched lumps of dirt were after all his lucky stars.

—Eric Temple Bell[1]

5.1 HISTORICAL BACKGROUND

The most brilliant chapter in the history of orbit determination was written by Carl Fredrich Gauss, a 24-year-old German mathematician, in the first year of the 19th century. Ever since Sir William Herschel had discovered the seventh planet, Uranus, in 1781, astronomers had been looking for further members of the solar system–especially since Bode's law predicted the existence of a planet between the orbits of Mars and Jupiter. A plan was formed dividing the sky into several areas which were to be searched for evidence of a new planet. But, before the search operation could begin one of the prospective participants,

Giuseppe Piazzi of Palermo, on New Year's day of 1801, observed what he first mistook for a small comet approaching the sun. The object turned out to be Ceres, the first of the swarm of asteroids or minor planets circling the sun between Mars and Jupiter.

It is ironic that the discovery of Ceres coincided with the publication of the famous philosopher Hegel of a vitriolic attack on astronomers for wasting their time in search for an eighth planet. If they paid some attention to philosophy, Hegel asserted, they would see immediately that there can be precisely seven planets, no more, no less. This slight lapse on Hegel's part no doubt has been explained by his disciples even if they cannot explain away the hundreds of minor planets which mock his philosophic ban.

To understand why computing the orbit of Ceres was such a triumph for Gauss, you must appreciate the meager data which was available in the case of sighting a new object in the sky. Without radar or any other means of measuring the distance or velocity of the object, the only information astronomers had to work with was the line-of-sight direction at each sighting. To compound the difficulty in the case of Ceres, Piazzi was only able to observe the asteroid for about 1 month before it was lost in the glare of the sun. The challenge of rediscovering the insignificant clod of dirt when it reappeared from behind the sun seduced the intellect of Gauss and he calculated as he had never calculated before. Ceres was rediscovered on New Year's day in 1802, exactly 1 year later, precisely where the ingenious and detailed calculations of the young Gauss had predicted she must be found.[1]

The method which Gauss used is just as pertinent today as it was in 1802, but for a different reason. The data that Gauss used to determine the orbit of Ceres consisted of the right ascension and declination at three observation times. His method is much simplified if the original data consists of two position vectors and the time-of-flight between them. The technique of determining an orbit from two positions and time is of considerable interest to modern astrodynamics since it has direct application in the solution of intercept and rendezvous or ballistic missile targeting problems. Because of its importance, and for convenience in referring to it later, we will formally define the problem of orbit determination from two positions and time and give it a name—"the Gauss problem."

5.2 THE GAUSS PROBLEM—GENERAL METHODS OF SOLUTION

We may define the Gauss problem as follows: Given $\mathbf{r}_1$, $\mathbf{r}_2$, the time-of-flight from $\mathbf{r}_1$ to $\mathbf{r}_2$ which we will call t, and the direction of motion, find $\mathbf{v}_1$ and $\mathbf{v}_2$.

By "direction of motion" we mean whether the satellite is to go from $\mathbf{r}_1$ to $\mathbf{r}_2$ the "short way," through an angular change $(\Delta\nu)$ of less than π radians, or the "long way," through an angular change greater than π.

Obviously, there are an infinite number of orbits passing through $\mathbf{r}_1$ and $\mathbf{r}_2$, but only two which have the specified time-of-flight–one for each possible direction of motion.

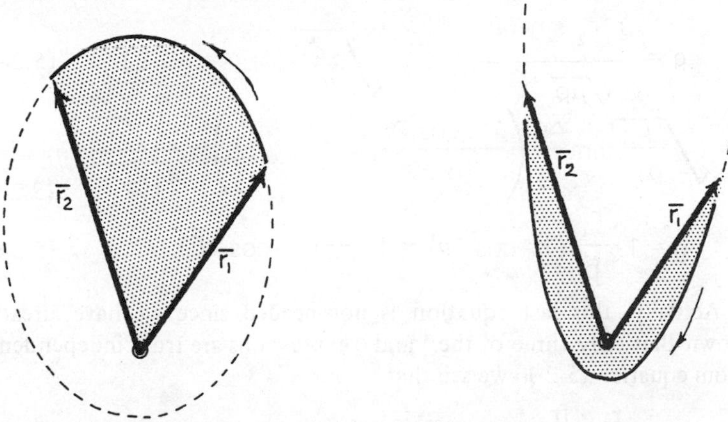

Figure 5.2-1 Short-way and long-way trajectories with same time-of-flight

One thing is immediately obvious from Figure 5.2-1; the two vectors $\mathbf{r}_1$ and $\mathbf{r}_2$ uniquely define the plane of the transfer orbit. If the vectors $\mathbf{r}_1$ and $\mathbf{r}_2$ are collinear and in opposite directions $(\Delta\nu = \pi)$, the plane of the transfer orbit is not determined and a unique solution for $\mathbf{v}_1$ and $\mathbf{v}_2$ is not possible. If the two position vectors are collinear and in the same direction $(\Delta\nu = 0$ or $2\pi)$, the orbit is a degenerate conic, but a unique solution is possible for $\mathbf{v}_1$ and $\mathbf{v}_2$. In the latter case, the method of solution may have to be modified as there may be a mathematical singularity in the equations used, particularly if the parameter, p, appears in the denominator of any expression.

The relationship between the four vectors $\mathbf{r}_1$, $\mathbf{r}_2$, $\mathbf{v}_1$ and $\mathbf{v}_2$ is contained in the f and g expressions which were developed in Chapter 4. It is not surprising, therefore, that nearly every known method for solving the Gauss problem may be derived from the f and g relations. We will rewrite them below making an obvious change in notation to be consistent with the definition of the Gauss problem:

$$\mathbf{r}_2 = f\mathbf{r}_1 + g\mathbf{v}_1 \tag{5.2-1}$$

$$\mathbf{v}_2 = \dot{f}\mathbf{r}_1 + \dot{g}\mathbf{v}_1 \tag{5.2-2}$$

where

$$f = 1 - \frac{r_2}{p}(1 - \cos\Delta\nu) = 1 - \frac{a}{r_1}(1 - \cos\Delta E) \tag{5.2-3}$$

$$g = \frac{r_1 r_2 \sin\Delta\nu}{\sqrt{\mu p}} = t - \sqrt{\frac{a^3}{\mu}}(\Delta E - \sin\Delta E) \tag{5.2-4}$$

$$\dot{f} = \sqrt{\frac{\mu}{p}} \tan\frac{\Delta\nu}{2}\left(\frac{1 - \cos\Delta\nu}{p} - \frac{1}{r_1} - \frac{1}{r_2}\right) = \frac{-\sqrt{\mu a}}{r_1 r_2} \sin\Delta E \tag{5.2-5}$$

$$\dot{g} = 1 - \frac{r_1}{p}(1 - \cos\Delta\nu) = 1 - \frac{a}{r_2}(1 - \cos\Delta E). \tag{5.2-6}$$

Actually, this last equation is not needed since we have already shown that only three of the f and g expressions are truly independent. From equation (5.2-1) we see that

$$\mathbf{v}_1 = \frac{\mathbf{r}_2 - f\mathbf{r}_1}{g} \tag{5.2-7}$$

Since equations (5.2-7) and (5.2-2) express the vectors $\mathbf{v}_1$ and $\mathbf{v}_2$ in terms of $f, g, \dot{f}, \dot{g}$ and the two known vectors $\mathbf{r}_1$ and $\mathbf{r}_2$, the solution to the Gauss problem is reduced to evaluating the scalars $f, g, \dot{f}$ and $\dot{g}$.

Consider equations (5.2-3), (5.2-4) and (5.2-5). There are seven variables–$\mathbf{r}_1$, $\mathbf{r}_2$, $\Delta\nu$, t, p, a and ΔE; but the first four are known, so what we have is three equations in three unknowns. The only trouble is that the equations are transcendental in nature, so a trial-and-error solution is necessary. We may outline the general method of solution as follows:

1. Guess a trial value for one of the three unknowns, p, a or ΔE directly or indirectly by guessing some other parameter of the transfer orbit which in turn establishes p, a or ΔE.

2. Use equations (5.2-3) and (5.2-5) to compute the remaining two unknowns.

3. Test the result by solving equation (5.2-4) for t and check it against the given value of time-of-flight.

4. If the computed value of t does not agree with the given value, adjust the trial value of the iteration variable and repeat the procedure until it does agree.

This last step is perhaps the most important of all, since the method used to adjust the trial value of the iteration variable is what determines how quickly the procedure converges to a solution. This point is frequently overlooked by authors who suggest a method of accomplishing the first three steps but give no guidance on how to adjust the iterative variable.

Several methods for solving the Gauss problem will be discussed in this chapter, including the original method suggested by Gauss. In each case, a scheme for adjusting the trial value of the iterative variable will be suggested and the relative advantages of one method over another will be discussed. What we are referring to as the Gauss problem can also be stated in terms of the Lambert theorem. Solution using the Lambert theorem will not be treated here because the universal variable method avoids much of the awkwardness of special cases which must be treated when applying Lambert's theorem.

5.3 SOLUTION OF THE GAUSS PROBLEM VIA UNIVERSAL VARIABLES

In discussing general methods for solving the Gauss problem earlier in this chapter, we indicated that the f and g expressions provide us with three independent equations in three unknowns, p, a and ΔE or ΔF. Later we will show that a trial-and-error solution based on guessing a value of p can be formulated. A direct iteration on the variable a is more difficult since picking a trial value for a does not determine a unique value for p or ΔE. A solution based on guessing a trial value of ΔE or ΔF would work, however, and would enable us to use the universal variables, x and z, introduced in Chapter 4, since $z = \Delta E^2$ and $-z = \Delta F^2$.

To see how such a scheme might work, let's write the expressions for f, g, $\dot{f}$ and $\dot{g}$ in terms of the universal variables:

$$f = 1 - \frac{r_2}{p}(1 - \cos \Delta\nu) = 1 - \frac{x^2}{r_1}C \tag{5.3-1}$$

$$g = \frac{r_1 r_2 \sin \Delta\nu}{\sqrt{\mu p}} = t - \frac{x^3}{\sqrt{\mu}}S. \tag{5.3-2}$$

$$\dot{f} = \sqrt{\frac{\mu}{p}}\,\frac{(1-\cos\Delta\nu)}{\sin\Delta\nu}\left(\frac{1-\cos\Delta\nu}{p}-\frac{1}{r_1}-\frac{1}{r_2}\right) = \frac{-\sqrt{\mu}}{r_1 r_2}\,x(1-zS) \tag{5.3-3}$$

$$\dot{g} = 1 - \frac{r_1}{p}(1-\cos\Delta\nu) = 1 - \frac{x^2}{r_2}C. \tag{5.3-4}$$

Solving for x from equation (5.3-1), we get

$$x = \sqrt{\frac{r_1 r_2 (1-\cos\Delta\nu)}{pC}}\,. \tag{5.3-5}$$

Substituting for x in equation (5.3-3) and cancelling $\sqrt{\mu/p}$ from both sides, yields

$$\frac{1-\cos\Delta\nu}{\sin\Delta\nu}\left(\frac{1-\cos\Delta\nu}{p}-\frac{1}{r_1}-\frac{1}{r_2}\right) = -\sqrt{\frac{1-\cos\Delta\nu}{r_1 r_2}}\,\frac{(1-zS)}{\sqrt{C}}\,.$$

If we multiply both sides by $r_1 r_2$ and rearrange, we obtain

$$\frac{r_1 r_2(1-\cos\Delta\nu)}{p} = r_1 + r_2 - \frac{\sqrt{r_1 r_2}\,\sin\Delta\nu}{\sqrt{1-\cos\Delta\nu}}\,\frac{(1-zS)}{\sqrt{C}}\,. \tag{5.3-6}$$

We can write this equation more compactly if we define a constant, A, as

$$\boxed{A = \frac{\sqrt{r_1 r_2}\,\sin\Delta\nu}{\sqrt{1-\cos\Delta\nu}}\,.} \tag{5.3-7}$$

We will also find it convenient to define another auxiliary variable, y, such that

$$y = \frac{r_1 r_2(1-\cos\Delta\nu)}{p}\,. \tag{5.3-8}$$

Using these definitions of A and y, equation (5.3-6) may be written more compactly as

$$\boxed{y = r_1 + r_2 - A\,\frac{(1-zS)}{\sqrt{C}}\,.} \tag{5.3-9}$$

We can also express x in equation (5.3-5) more concisely:

$$\boxed{x = \sqrt{\frac{y}{C}}} \tag{5.3-10}$$

If we now solve for t from equation (5.3-2), we get

$$\sqrt{\mu}\, t = x^3 S + \frac{r_1 r_2 \sin \Delta\nu}{\sqrt{p}} \tag{5.3-11}$$

But, by using equations (5.3-7) and (5.3-8), the last term of this expression may be simplified, so that

$$\boxed{\sqrt{\mu}\, t = x^3 S + A\sqrt{y}\,.} \tag{5.3-12}$$

The simplification of the equations resulting from the introduction of the constant, A, and the auxiliary variable, y, can be extended to the f and g expressions themselves. From equations (5.3-1), (5.3-2) and (5.3-4), we can obtain the following simplified expressions:

$$\boxed{f = 1 - \frac{y}{r_1}} \tag{5.3-13}$$

$$\boxed{g = A\sqrt{\frac{y}{\mu}}} \tag{5.3-14}$$

$$\boxed{\dot{g} = 1 - \frac{y}{r_2}\,.} \tag{5.3-15}$$

Since $\mathbf{r}_2 = f\,\mathbf{r}_1 + g\,\mathbf{v}_1$, we can compute $\mathbf{v}_1$ from

$$\boxed{\mathbf{v}_1 = \frac{\mathbf{r}_2 - f\mathbf{r}_1}{g}\,.} \tag{5.3-16}$$

The velocity, $\mathbf{v}_2$, may be expressed as

$$\mathbf{v}_2 = \dot{f}\,\mathbf{r}_1 + \dot{g}\,\mathbf{v}_1$$

Substituting for $\mathbf{v}_1$ from equation (5.3-16) and using the identity $f\dot{g} - \dot{f}g = 1$ from equation (4.4-20), this last expression simplifies to

$$\boxed{\mathbf{v}_2 = \frac{\dot{g}\,\mathbf{r}_2 - \mathbf{r}_1}{g}\,.} \tag{5.3-17}$$

A simple algorithm for solving the Gauss problem via universal variables may now be stated as follows:

1. From $\mathbf{r}_1$, $\mathbf{r}_2$ and the "direction of motion," evaluate the constant, A, using equation (5.3-7).

2. Pick a trial value for z. Since $z = \Delta E^2$ and $-z = \Delta F^2$, this amounts to guessing the change in eccentric anomaly. The usual range for z is from minus values to $(2\pi)^2$. Values of z greater than $(2\pi)^2$ correspond to changes in eccentric anomaly of more than 2π and can occur only if the satellite passes back through $\mathbf{r}_1$ enroute to $\mathbf{r}_2$.

3. Evaluate the functions S and C for the selected trial value of z using equations (4.4-10) and (4.4-11).

4. Determine the auxiliary variable, y, from equation (5.3-9).

5. Determine x from equation (5.3-10).

6. Check the trial value of z by computing t from equation (5.3-12) and compare it with the desired time-of-flight. If it is not nearly the same, adjust the trial value of z and repeat the procedure until the desired value of t is obtained. A Newton iteration scheme for adjusting z will be discussed in the next section.

7. When the method has converged to a solution, evaluate f, g and $\dot{g}$ from equations (5.3-13), (5.3-14) and (5.3-15), then compute $\mathbf{v}_1$ and $\mathbf{v}_2$ from equations (5.3-16) and (5.3-17).

5.3.1 Selecting A New Trial Value of z. Although any iterative scheme, such as a Bolzano bisection technique or linear interpolation, may be used successfully to pick a better trial value for z, a Newton iteration, which converges more rapidly, may be used if we can determine the slope of the t vs z curve at the last trial point.

The derivative, dt/dz, necessary for a Newton iteration can be determined by differentiating equation (5.3-12) for t:

$$\sqrt{\mu}\, t = x^3 S + A\sqrt{y} \tag{5.3-12}$$

$$\sqrt{\mu}\,\frac{dt}{dz} = 3x^2\frac{dx}{dz}S + x^3\frac{dS}{dz} + \frac{A}{2\sqrt{y}}\frac{dy}{dz}. \tag{5.3-18}$$

Differentiating equation (5.3-10) for x yields

$$\frac{dx}{dz} = \frac{1}{2xC}\left(\frac{dy}{dz} - x^2\frac{dC}{dz}\right). \tag{5.3-19}$$

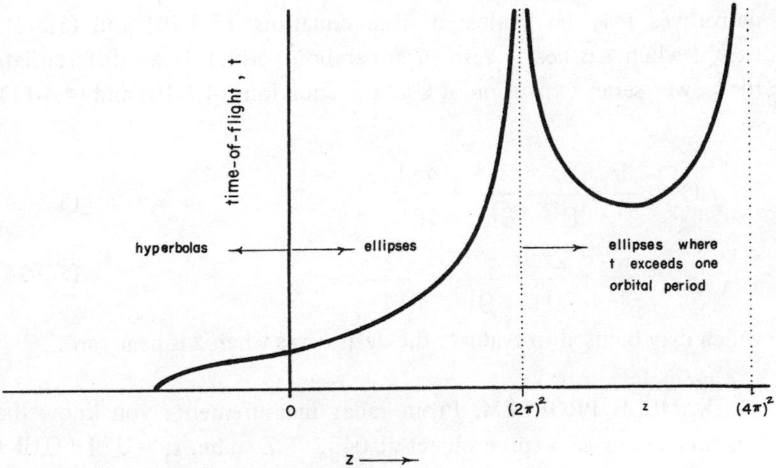

Figure 5.3-1 Typical t vs Z plot for a fixed $\mathbf{r}_1$ and $\mathbf{r}_2$

Differentiating equation (5.3-9) for y, we get

$$\frac{dy}{dz} = -\frac{A}{C}\left[\sqrt{C}\left(-S - z\frac{dS}{dz}\right) - \frac{(1 - zS)}{2\sqrt{C}}\ \frac{dC}{dz}\right]$$

But, in section 4.5.3, we showed that

$$\frac{dS}{dz} = \frac{1}{2z}\ (C - 3S) \tag{5.3-20}$$

$$\frac{dC}{dz} = \frac{1}{2z}\ (1 - zS - 2C) \tag{5.3-21}$$

so, the expression for dy / dz reduces to

$$\frac{dy}{dz} = \frac{A}{4}\sqrt{C} \tag{5.3-22}$$

If we now substitute equations (5.3-19) and (5.3-22) into equation (5.3-18), we get

$$\boxed{\sqrt{\mu}\frac{dt}{dz} = x^3\left(S' - \frac{3SC'}{2C}\right) + \frac{A}{8}\left(\frac{3S\sqrt{y}}{C} + \frac{A}{x}\right)} \tag{5.3-23}$$

where S' and C' are the derivatives of S and C with respect to z. These derivatives may be evaluated from equations (5.3-20) and (5.3-21) except when z is nearly zero (near-parabolic orbit). If we differentiate the power series expansion of C and S, equations (4.4-10) and (4.4-11), we get

$$C' = \frac{1}{4!} + \frac{2z}{6!} - \frac{3z^2}{8!} + \frac{4z^3}{10!} - \dots \tag{5.3-24}$$

$$S' = \frac{1}{5!} + \frac{2z}{7!} - \frac{3z^2}{9!} + \frac{4z^3}{11!} - \dots \tag{5.3-25}$$

which may be used to evaluate the derivatives when z is near zero.

EXAMPLE PROBLEM. From radar measurements you know the position vector of a space object at 0432:00Z to be: $\mathbf{r}_1 = 0.5\mathbf{I} + 0.6\mathbf{J} + 0.7\mathbf{K}$ DU. At 0445:00Z the position of the object was: $\mathbf{r}_2 = 0.0\mathbf{I} + 1.0\mathbf{J} + 0.0\mathbf{K}$ DU. Using universal variables, determine on what two paths the object could have moved from position one to position two. Assume the object occupies each position only once during this time period.

There is only one path for a given direction of motion. Thus the two paths sought are the "short way" and "long way." We first find $\mathbf{v}_1$ and $\mathbf{v}_2$ for each direction by solving the Gauss problem and then use $\mathbf{r}_1$, $\mathbf{v}_1$ and the method of Chapter 2 to solve for the orbital elements.

To facilitate our solution let us define an integer quantity DM, "direction of motion":

$$DM \triangleq \text{sign}\,(\pi - \Delta\nu). \tag{5.3-26}$$

For DM = -1 we have "long way"($\Delta\nu > \pi$); DM = +1 we have "short way" ($\Delta\nu > \pi$).

For ease of numerical solutions, a convenient form of equation (5.3-7) is

$$A = DM\sqrt{r_1 r_2 (1 + \cos \Delta\nu)}. \tag{5.3-27}$$

which does not present the problem that (5.3-7) does when $\Delta\nu$ is small.

Now following the algorithm of section 5.3:

Step 1: $\mathbf{r}_1 = 1.0488088482$ DU

$\mathbf{r}_2 = 1.0000000000$ DU

$\Delta\nu$ short way = the smallest angle between $\mathbf{r}_1$ and $\mathbf{r}_2$ = .962 radians

$\Delta\nu$ long way = 2π - $\Delta\nu$ short way = 5.321 radians

Using (5.3-27),

$$A_{\text{long way}} = -1.28406$$

$$A_{\text{short way}} = 1.28406$$

Step 2 and Step 3:

Let the first estimate be $Z = 0$; from equations (4.5-14) and (4.5-15)

$$S(Z)\Big|_{Z=0} = \frac{1}{6}, \quad C(Z)\Big|_{Z=0} = \frac{1}{2}$$

Step 4:

Using (5.3-9),

$$y_{\text{long way}} = 1.0488088482 + 1.0000 + 1.28406\left(\frac{1}{\sqrt{1/2}}\right)$$
$$= 3.86474323$$

$$y_{\text{short way}} = 1.0488088482 + 1.0000$$
$$-1.28406\left(\frac{1}{\sqrt{1/2}}\right) = 0.23287446$$

Here we must insure that we heed the caution concerning the sign of y for the short way.

Step 5:

Using equation (5.3-10),

$$x_{\text{long way}} = \sqrt{\frac{y_{\text{long}}}{C(O)}} = \sqrt{\frac{3.86474323}{1/2}} = 2.78019540$$

$$x_{\text{short way}} = \sqrt{\frac{y_{\text{short}}}{C(O)}} = \sqrt{\frac{0.23287446}{1/2}} = 0.68245800$$

Step 6:

Now using equation (5.3-12) solve for t:

$$t_{\text{long}} = \sqrt{\mu}\, X^3_{\text{long}}\, S + A_{\text{long}} \sqrt{y_{\text{long}}}$$

$$t_{\text{long}} = 21.489482706 \left(\frac{1}{6}\right) - 1.28406 \sqrt{3.86474323}$$

$$= 1.05725424 \text{ TU}$$

$$t_{\text{short}} = 0.317854077 \left(\frac{1}{6}\right) + 1.28406 \sqrt{0.23287446}$$

$$= 0.67262516 \text{ TU}$$

Now, from the given data, the desired time of flight is:

$$\Delta t = 0445z - 0432z = \frac{13 \text{ min}}{13.444689 \text{ min/TU}}$$

$$= 0.9667663 \text{ TU}$$

We see that our computed values of t using $z = 0$ are too far off, so we adjust the value of z using a Newton iteration and repeat steps 2 through 6. Care must be taken to insure that the next trial value of z does not cause y to be negative. (The iterations *must* be performed separately for long way and short way.)

Convergence criteria should be chosen with the size of Δt taken into consideration. Since Δt for this problem is less than 1.0 we do not normalize.

When $\Delta t - t \leqslant 10^{-4}$ (within 0.1 seconds) we consider the problem solved.

We have:

Long way (after 2 iterations)	Short way (after 2 iterations)
$z = -3.61856634$	$z = 0.83236253$
$x = 2.66224213$	$x = 0.94746230$
$A = -1.28406$	$A = 1.28406$
$y = 4.74994739$	$y = 0.41856019$
$t = 0.96681012$	$t = 0.96670788$

Step 7:

Using equations (5.3-13), (5.3-14), (5.3-15), (5.3-16) and (5.3-17) we get:

Long way:

$f = -3.52889714 \qquad v_1 = 1.554824$ DU/TU

$g = -2.79852734 \qquad v_2 = 1.584516$ DU/TU

$\dot{g} = -3.74994739$

$\mathbf{v}_1 = -0.6304918096\mathbf{I} - 1.1139209665\mathbf{J} - 0.8826885334\mathbf{K}$

$\mathbf{v}_2 = 0.1786653974\mathbf{I} + 1.5544139777\mathbf{J} - 0.2501315563\mathbf{K}$

Short way:

$f = 0.60091852 \qquad v_1 = 0.989794$ DU/TU

$g = 0.83073807 \qquad v_2 = 1.035746$ DU/TU

$\dot{g} = 0.58143981$

$\mathbf{v}_1 = -0.36167749\mathbf{I} + 0.76973587\mathbf{J} - 0.50634848\mathbf{K}$

$\mathbf{v}_2 = -0.60187442\mathbf{I} - 0.02234181\mathbf{J} - 0.84262419\mathbf{K}$

Then using the given **r** vectors and these **v** vectors we have specified the two paths.

Notice that the long way trajectory is a hyperbola:

Energy	$= 0.2553\ DU^2/TU^2$
Eccentricity	$= 3.96$
and perigee radius	$= 0.0191\ DU$

while the short way is an ellipse:

Energy	$= -4.636\ DU^2/TU^2$
Eccentricity	$= 0.076832$
and perigee radius	$= 0.9958\ DU$

Since the hyperbola passes through the earth between the two positions, and the ellipse intersects the earth (but only after it passes the second position), the ellipse is the only trajectory a real object could travel on. This result illustrates the beauty of the universal variables approach to this problem: with one set of equations we solved problems involving two different types of conics.

There is one pitfall in the solution of the Gauss problem via universal variables which you should be aware of. For "short way" trajectories, where $\Delta\nu$ is less than π, the t vs z curve crosses the $t = 0$ axis at some negative value of z as shown in Figure 5.3-1. In other words, there is a negative lower limit for permissable values of z when $\Delta\nu < \pi$. The reason for this may be seen by examining equations (5.3-8) and (5.3-9).

From equation (5.3-8) it is obvious that y cannot be negative, yet equation (5.3-9) will result in a negative value for y if $\Delta\nu < \pi$ and z is too large a negative number. This is apparent from the fact that A is positive whenever $\Delta\nu < \pi$ and negative whenever $\pi < \Delta\nu < 2\pi$. Because both C and S become large positive numbers when z is large and negative (see Figure 4.5-2), the expression

$$\frac{A(1 - zS)}{\sqrt{C}}$$

can become a large positive number if A is positive. Whenever

$$\frac{A(1 - zS)}{\sqrt{C}} > r_1 + r_2$$

the value of y will be negative and x will be imaginary in equation (5.3-10).

Any computational algorithm for solving the Gauss problem via universal variables should include a check to see if y is negative prior to evaluating x. This check is only necessary when A is positive. For "long way" trajectories y is positive for all values of z and the t vs z curve approaches zero asymptotically as z approaches minus infinity.

5.4 THE p-ITERATION METHOD

The next method of solving the Gauss problem that we will look at could be called a direct p-iteration technique. It differs from the p-iteration method first proposed by Herrick and Liu in 1959[6] since it does not directly involve eccentricity. The method consists of guessing a trial value of p from which we can compute the other two unknowns, a and ΔE. The trial values are checked by solving for t and comparing it with the given time-of-flight.

The p-iteration method presented below is unusual in that it will permit us to develop an analytical expression for the slope of the t vs p curve; hence, a Newton iteration scheme is possible for adjusting the trial value of p.

5.4.1 Expressing p as a Function of ΔE. From equation (5.2-5), if we cancel $\sqrt{\mu}$ from both sides and write $\tan(\Delta\nu/2)$ as $(1-\cos\Delta\nu)/\sin\Delta\nu$, we obtain

$$\frac{1-\cos\Delta\nu}{\sqrt{p}\,\sin\Delta\nu}\left(\frac{1-\cos\Delta\nu}{p}-\frac{1}{r_1}-\frac{1}{r_2}\right)=\frac{-\sqrt{a}\,\sin\Delta E}{r_1 r_2} \tag{5.4-1}$$

From equation (5.2-3), we can solve for a and get

$$a=\frac{r_1 r_2(1-\cos\Delta\nu)}{p(1-\cos\Delta E)} \tag{5.4-2}$$

Substituting this expression for a into equation (5.4-1) and rearranging yields

$$\frac{1-\cos\Delta\nu}{p}-\frac{1}{r_1}-\frac{1}{r_2}=\frac{-1}{\sqrt{r_1 r_2}}\;\frac{\sin\Delta\nu}{\sqrt{1-\cos\Delta\nu}}\;\frac{\sin\Delta E}{\sqrt{1-\cos\Delta E}}$$

Using the trigonometric identity, $(\sin x)/\sqrt{1-\cos x}=\sqrt{2}\cos\frac{x}{2}$, and solving for p, we get

$$p = \frac{r_1 r_2 (1 - \cos \Delta\nu)}{r_1 + r_2 - 2\sqrt{r_1 r_2} \cos \frac{\Delta\nu}{2} \cos \frac{\Delta E}{2}} . \tag{5.4-3}$$

5.4.2 Expressing a as a Function of p. The first step in the solution is to find an expression for a as a function of p and the given information. We will find it convenient to define three constants which may be determined from the given information:

$$\begin{aligned} k &= r_1 r_2 (1 - \cos \Delta\nu) \\ \ell &= r_1 + r_2 \\ m &= r_1 r_2 (1 + \cos \Delta\nu) . \end{aligned} \tag{5.4-4}$$

Using these definitions, a may be written as

$$a = \frac{k}{p(1 - \cos \Delta E)} = \frac{k}{2p \sin^2 \frac{\Delta E}{2}} = \frac{k}{2p(1 - \cos^2 \frac{\Delta E}{2})} \tag{5.4-5}$$

Using these same definitions, and noting that $\sqrt{2 r_1 r_2} \cos \frac{\Delta\nu}{2} = \pm \sqrt{r_1 r_2 (1 + \cos \Delta\nu)}$, we can rewrite equation (5.4-3) as

$$p = \frac{k}{\ell \pm \sqrt{2m} \cos \frac{\Delta E}{2}} . \tag{5.4-6}$$

Solving for $\cos \frac{\Delta E}{2}$, we get

$$\cos \frac{\Delta E}{2} = \frac{k - \ell p}{\pm \sqrt{2m}\, p}$$

$$\cos^2 \frac{\Delta E}{2} = \frac{(k - \ell p)^2}{2mp^2} \tag{5.4-7}$$

If we substitute this last expression into equation (5.4-5) and simplify, we obtain

$$a = \frac{mkp}{(2m - \ell^2)p^2 + 2k\ell p - k^2} \tag{5.4-8}$$

where k, ℓ and m are all constants that can be determined from $\mathbf{r}_1$ and $\mathbf{r}_2$. It is clear from this equation that once p is specified a unique value of a is determined.

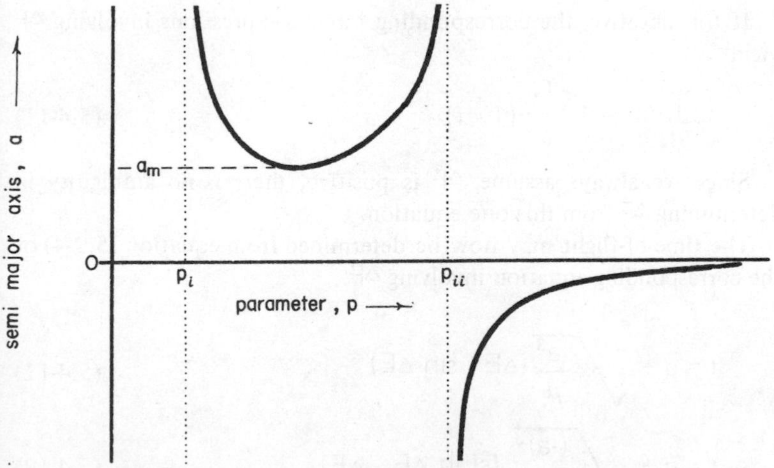

Figure 5.4-1 Typical plot of a vs p for a fixed $\mathbf{r}_1$ and $\mathbf{r}_2$

In Figure 5.4-1 we have drawn a typical plot of a vs p for a fixed $\mathbf{r}_1$, $\mathbf{r}_2$ and $\Delta\nu$ less than π. Notice that for those orbits where a is positive (ellipses), a may not be smaller than some minimum value, a_m. The point where a is a minimum corresponds to the "minimum energy ellipse" joining $\mathbf{r}_1$ and $\mathbf{r}_2$.

The two points where a approaches infinity correspond to parabolic orbits joining $\mathbf{r}_1$ and $\mathbf{r}_2$. The values of p that specify the parabolic orbits are labeled p_i and p_{ii} and will be important to us later.

Those orbits where a is negative are hyperbolic. The limiting case where p approaches infinity and a approaches zero is the straight-line orbit connecting $\mathbf{r}_1$ and $\mathbf{r}_2$. It would require infinity energy and have a time-of-flight of zero.

5.4.3 Checking the Trial Value of p. Once we have selected a trial value of p and computed a from equation (5.4-8), we are ready to solve for t and check it against the given time-of-flight. First, however, we need to determine ΔE (or ΔF in case a is negative).

From the trial value of p and the known information, we can compute f, g and $\dot{f}$ from equations (5.2-3), (5.2-4) and (5.2-5). If a is positive, we can determine ΔE from equations (5.2-3) and (5.2-5):

$$\cos \Delta E = 1 - \frac{r_1}{a}(1 - f) \tag{5.4-9}$$

$$\sin \Delta E = \frac{-r_1 r_2 \dot{f}}{\sqrt{\mu a}} \tag{5.4-10}$$

If a is negative, the corresponding f and g expressions involving ΔF yield

$$\cosh \Delta F = 1 - \frac{r_1}{a}(1 - f) \tag{5.4-11}$$

Since we always assume ΔF is positive, there is no ambiguity in determining ΔF from this one equation.

The time-of-flight may now be determined from equation (5.2-4) or the corresponding equation involving ΔF:

$$t = g + \sqrt{\frac{a^3}{\mu}}(\Delta E - \sin \Delta E) \tag{5.4-12}$$

$$t = g + \sqrt{\frac{(-a)^3}{\mu}}(\sinh \Delta F - \Delta F). \tag{5.4-13}$$

5.4.4 The t vs p Curve. Before discussing the method of selecting a new trial value of p, we need to understand what the t vs p curve looks like. To get a feeling for the problem, let's look at the family of orbits that can be drawn between a given $\mathbf{r}_1$ and $\mathbf{r}_2$. In Figure 5.4-2 we have drawn $\mathbf{r}_1$ and $\mathbf{r}_2$ to be of equal length. The conclusions we will reach from examining this illustration also apply to the more general case where $\mathbf{r}_1 \neq \mathbf{r}_2$.

First, let's consider the orbits that permit traveling from $\mathbf{r}_1$ to $\mathbf{r}_2$ the "short way." The quickest way to get from $\mathbf{r}_1$ to $\mathbf{r}_2$ is obviously along the straight line from $\mathbf{r}_1$ to $\mathbf{r}_2$. This is the limiting case where p is infinite and t is zero. As we take longer to make the journey from $\mathbf{r}_1$ to $\mathbf{r}_2$, the trajectories become more "lofted" until we approach the limiting case where we try to go through the open end of the parabola joining $\mathbf{r}_1$ and $\mathbf{r}_2$. Notice that, although t approaches infinity for this orbit, p approaches a finite minimum value which we will call p_i.

If we look at "long way" trajectories, we see that the limiting case of zero time-of-flight is achieved on the degenerate hyperbola that goes straight down $\mathbf{r}_1$ to the focus and then straight up $\mathbf{r}_2$ and has p equal to zero. As we choose trajectories with longer times-of-flight, p increases until we reach the limiting case where we try to travel through the open end of the other parabola that joins $\mathbf{r}_1$ and $\mathbf{r}_2$. For this case, t approaches infinity as p approaches a finite limiting value which we will call p_{ii}.

From the discussion above, we can construct a typical plot of t vs p for a fixed $\mathbf{r}_1$ and $\mathbf{r}_2$. In Figure 5.4-3 the solid line represents "short way" trajectories and the dashed line represents "long way" trajectories.

For $\Delta\nu$ less than π, p must lie between p_i and infinity; for $\Delta\nu$ greater than π, p must lie between 0 and p_{ii}. Since it is important that the first trial value, as well as all subsequent guesses for p, lie within the prescribed limits, we should first compute p_i or p_{ii}.

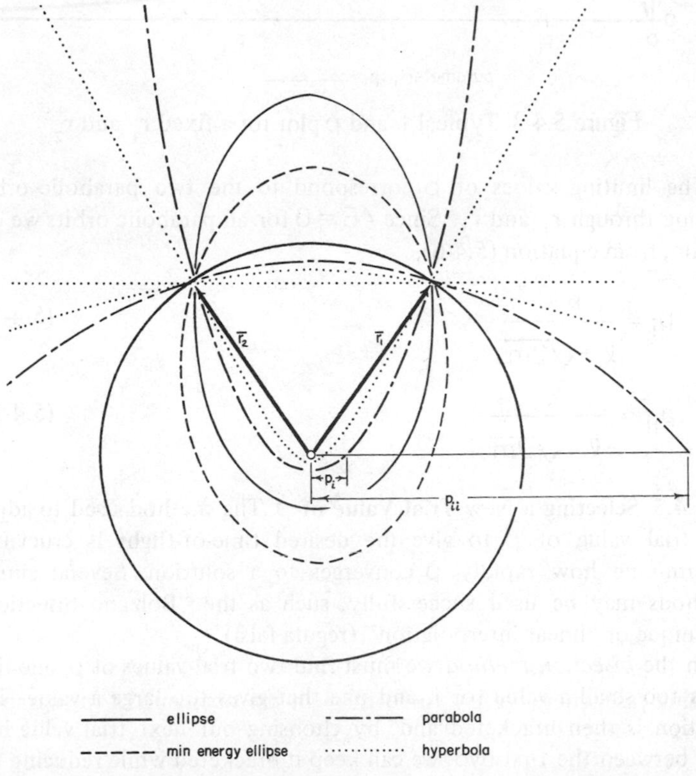

Figure 5.4-2 Family of possible transfer orbits connecting $\mathbf{r}_1$ and $\mathbf{r}_2$

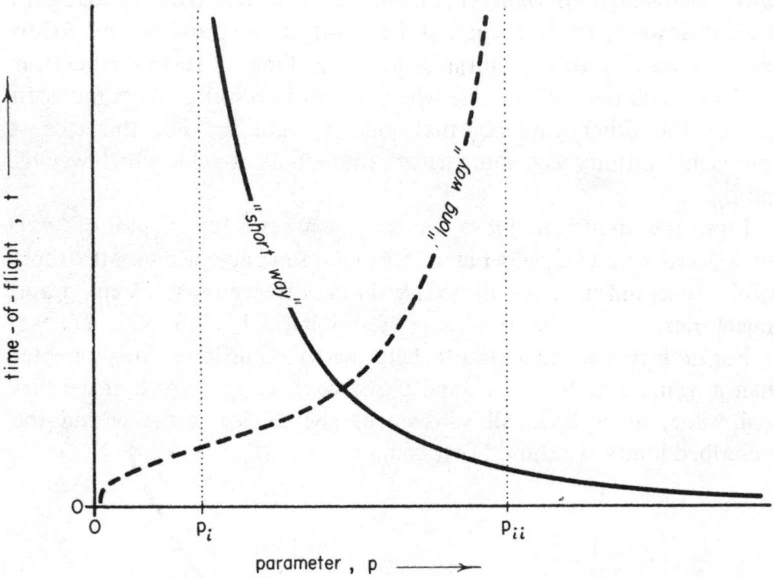

Figure 5.4-3 Typical t and p plot for a fixed $\mathbf{r}_1$ and $\mathbf{r}_2$

The limiting values of p correspond to the two parabolic orbits passing through $\mathbf{r}_1$ and $\mathbf{r}_2$. Since $\Delta E = 0$ for all parabolic orbits we can obtain, from equation (5.4-6),

$$p_i = \frac{k}{\ell + \sqrt{2m}} \tag{5.4-14}$$

$$p_{ii} = \frac{k}{\ell - \sqrt{2m}} \tag{5.4-15}$$

5.4.5 Selecting a New Trial Value of p. The method used to adjust the trial value of p to give the desired time-of-flight is crucial in determining how rapidly p converges to a solution. Several simple methods may be used successfully, such as the "Bolzano bisection" technique or "linear interpolation" (regula falsi).

In the *bisection method* we must find two trial values of p; one that gives too small a value for t, and one that gives too large a value. The solution is then bracketed and, by choosing our next trial value half way between the first two, we can keep it bracketed while reducing the interval of uncertainty to some arbitrarily small value.

In the *linear interpolation* method we choose two trial values of p—call them p_{n-1} and p_n. If t_{n-1} and t_n are the times-of-flight corresponding to these trial values of p, then we select a new value from

$$p_{n+1} = p_n + \frac{(t - t_n)(p_n - p_{n-1})}{(t_n - t_{n-1})} \tag{5.4-16}$$

This scheme can be repeated, always retaining the latest two trial values of p and their corresponding times-of-flight for use in computing a still better trial value from equation (5.4-16). It is not necessary that the initial two trial values bracket the answer.

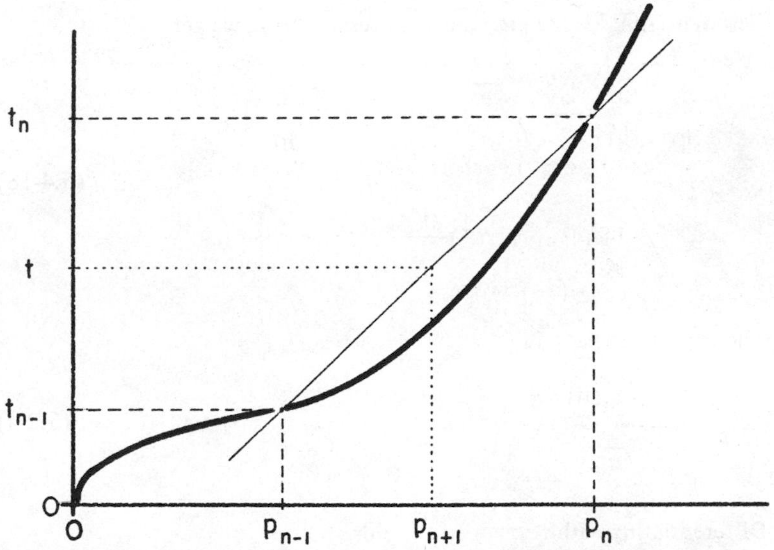

Figure 5.4-4 Selecting a new trial value of p by linear interpolation

An even faster method is a *Newton iteration*, but it requires that we compute the slope of the t vs p curve at the last trial value of p. If this last trial value is called p_n and t_n is the time-of-flight that results from it, then a better estimate of p may be obtained from

$$p_{n+1} = p_n + \frac{t - t_n}{dt/dp\big|_{p=p_n}} \tag{5.4-17}$$

where t is the desired time-of-flight and $dt/dp\big|_{p=p_n}$ is the slope at the last trial point. We must now obtain an expression for the derivative in equation (5.4-17).

From the f and g expressions we can write

$$t = g + \sqrt{\frac{a^3}{\mu}}\,(\Delta E - \sin \Delta E). \tag{5.4-12}$$

Differentiating this expression with respect to p, we get

$$\frac{dt}{dp} = \frac{dg}{dp} + \frac{3}{2}\sqrt{\frac{a}{\mu}}(\Delta E - \sin \Delta E)\frac{da}{dp} + \sqrt{\frac{a^3}{\mu}}\,(1 - \cos \Delta E)\frac{d\Delta E}{dp} \tag{5.4-18}$$

The expression for g is

$$g = \frac{r_1 r_2 \sin \Delta\nu}{\sqrt{\mu p}} \tag{5.2-4}$$

Differentiating with respect to p yields

$$\frac{dg}{dp} = \frac{-r_1 r_2 \sin \Delta\nu}{2p\sqrt{\mu p}}$$

$$\frac{dg}{dp} = \frac{-g}{2p} \tag{5.4-19}$$

The derivative da / dp comes directly from differentiating the expression for a:

$$a = \frac{mkp}{(2m - \ell^2)p^2 + 2k\ell p - k^2} \qquad (5.4\text{-}8)$$

$$\frac{da}{dp} = \frac{mk(2m - \ell^2)p^2 + 2mk^2\ell p - mk^3 - 2mk(2m - \ell^2)p^2 - 2mk^2\ell p}{[(2m - \ell^2)p^2 + 2k\ell p - k^2]^2}$$

which simplifies to

$$\frac{da}{dp} = \frac{-a^2 [k^2 + (2m - \ell^2)p^2]}{mkp^2} . \qquad (5.4\text{-}20)$$

From equation (5.4-7) we can write

$$\cos^2 \frac{\Delta E}{2} = \frac{(k - \ell p)^2}{2mp^2} . \qquad (5.4\text{-}7)$$

Differentiating both sides of this equation yields

$$(-\tfrac{1}{2})\ 2 \cos \frac{\Delta E}{2} \sin \frac{\Delta E}{2} \frac{d\Delta E}{dp} = \frac{-4mp^2 (k - \ell p)\ell - (k - \ell p)^2 4mp}{4m^2 p^4}$$

which simplifies to

$$\tfrac{1}{2} \sin \frac{\Delta E}{2} \frac{d\Delta E}{dp} = \frac{k(k - \ell p)}{mp^3}$$

Solving for $d\Delta E / dp$ and noting, from equation (5.4-7), that $mp^2 = (k - \ell p)^2 / (1 + \cos \Delta E)$, we obtain

$$\frac{d\Delta E}{dp} = \frac{2k(1 + \cos \Delta E)}{p(k - \ell p) \sin \Delta E} \tag{5.4-21}$$

We are now ready to substitute into equation (5.4-18) from (5.4-19), (5.4-20) and (5.4-21):

$$\frac{dt}{dp} = \frac{-g}{2p} - \frac{3a}{2}\sqrt{\frac{a^3}{\mu}}\,(\Delta E - \sin \Delta E)\left[\frac{k^2 + (2m - \ell^2)p^2}{mkp^2}\right]$$

$$+ \sqrt{\frac{a^3}{\mu}}\,\frac{(1 - \cos \Delta E)(1 + \cos \Delta E)2k}{\sin \Delta E\, p(k - \ell p)}$$

which simplifies to

$$\boxed{\frac{dt}{dp} = \frac{-g}{2p} - \frac{3}{2}a(t - g)\left(\frac{k^2 + (2m - \ell^2)p^2}{mkp^2}\right) + \sqrt{\frac{a^3}{\mu}}\,\frac{2k \sin \Delta E}{p(k - \ell p)}} \tag{5.4-22}$$

which is valid for the elliptical portion of the t vs p curve.

By an analogous derivation starting with the equation

$$t = g + \sqrt{\frac{(-a)^3}{\mu}}\,(\sin \Delta F - \Delta F)$$

we can arrive at the following slope expression which is valid for the hyperbolic portion of the t vs p curve:

$$\boxed{\begin{aligned} \frac{dt}{dp} &= \frac{-g}{2p} - \frac{3}{2}a(t - g)\left(\frac{k^2 + (2m - \ell^2)p^2}{mkp^2}\right) \\ &- \sqrt{\frac{(-a)^3}{\mu}}\,\frac{2k \sinh \Delta F}{p(k - \ell p)}. \end{aligned}} \tag{5.4-23}$$

To evaluate the slope at p_n, the values of g, a, t and ΔE or ΔF obtained from the trial value, p_n, should be used in equation (5.4-22) or (5.4-23).

We can summarize the steps involved in solving the Gauss problem via the p-iteration technique as follows:

1. Evaluate the constants k, ℓ and m from r_1, r_2 and $\Delta\nu$ using equations (5.4-4).
2. Determine the limits on the possible values of p by evaluating p_i and p_{ii} from equations (5.4-14) and (5.4-15).
3. Pick a trial value of p within the appropriate limits.
4. Using the trial value of p, solve for a from equation (5.4-8). The type conic orbit will be known from the value of a.
5. Solve for f, g and $\dot{f}$ from equations (5.2-3), (5.2-4) and (5.2-5).
6. Solve for ΔE or ΔF, as appropriate, using equations (5.4-9) and (5.4-10) or equation (5.4-11).
7. Solve for t from equation (5.4-12) or (5.4-13) and compare it with the desired time-of-flight.
8. Adjust the trial value of p using one of the iteration methods discussed above until the desired time-of-flight is obtained.
9. Evaluate $\dot{g}$ from equation (5.2-6) and then solve for $\mathbf{v}_1$ and $\mathbf{v}_2$ using equations (5.2-7) and (5.2-2).

The p-iteration method converges in all cases except when $\mathbf{r}_1$ and $\mathbf{r}_2$ are collinear. Its main disadvantage is that separate equations are used for the ellipse and hyperbola. This defect may be overcome by using the universal variables x and z introduced in Chapter 4 and discussed earlier in this chapter.

5.5 THE GAUSS PROBLEM USING THE f AND g SERIES

In this section we will develop another method for solving the Gauss problem. Instead of using the f and g expressions, we will develop and use the f and g *series*. As stated earlier, the motion of a body in a Keplerian orbit is in a plane which contains the radius vector from the center of force to the body and the velocity vector. If we know the position $\mathbf{r}_o$ and the velocity $\mathbf{v}_o$ at some time t_o then we know that the position vector at any time t can be expressed as a linear combination of $\mathbf{r}_o$ and $\mathbf{v}_o$ because it always lies in the same plane as $\mathbf{r}_o$ and $\mathbf{v}_o$. The coefficients of $\mathbf{r}_o$ and $\mathbf{v}_o$ in this linear combination will be functions of the time and will depend upon the vectors $\mathbf{r}_o$ and $\mathbf{v}_o$.

$$\boxed{\mathbf{r} = f(\mathbf{r}_0, \mathbf{v}_0, t)\ \mathbf{r}_0 + g(\mathbf{r}_0, \mathbf{v}_0, t)\ \mathbf{v}_0\ .} \tag{5.5-1}$$

We can determine the functions f and g by expanding **r** in a Taylor series expansion around $t = t_0$

$$\mathbf{r} = \sum_{n=0}^{\infty} \frac{(t - t_0)^n}{n!}\ \mathbf{r}_0^{(n)} \tag{5.5-2}$$

where

$$\mathbf{r}_0^{(n)} \equiv \left.\frac{d^n\mathbf{r}}{dt^n}\right|_{t=t_0} \tag{5.5-3}$$

Since the motion is in a plane all the time derivatives of **r** must lie in the plane of **r** and **v**. Therefore, we can write in general

$$\mathbf{r}^{(n)} = F_n\,\mathbf{r} + G_n\,\mathbf{v}. \tag{5.5-4}$$

Differentiating with respect to time

$$\mathbf{r}^{(n+1)} = \dot{F}_n\,\mathbf{r} + F_n\,\mathbf{v} + \dot{G}_n\,\mathbf{v} + G_n\,\ddot{\mathbf{r}} \tag{5.5-5}$$

but $\ddot{\mathbf{r}} = -\frac{\mu}{r^3}\mathbf{r}$ and if we define $u \equiv \frac{\mu}{r^3}$ we can write

$$\mathbf{r}^{(n+1)} = (\dot{F}_n - u\,G_n)\,\mathbf{r} + (F_n + \dot{G}_n)\,\mathbf{v} \tag{5.5-6}$$

Comparing with equation (5.5-4) we have the following recursion formulas

$$\boxed{F_{n+1} = \dot{F}_n - u\,G_n} \tag{5.5-7}$$

$$\boxed{G_{n+1} = F_n + \dot{G}_n\,.} \tag{5.5-8}$$

In order to determine F_0 and G_0 so that we can start the recursion, we write equation (5.5-4) for $n = 0$.

$$\mathbf{r}^{(0)} = \mathbf{r} = F_0\,\mathbf{r} + G_0\,\dot{\mathbf{r}} \tag{5.5-9}$$

From which it is obvious that $F_0 = 1$ and $G_0 = 0$.

5.5.1 Development of the Series Coefficients. Before continuing with the development of the functions F_n and G_n it is convenient to digress at this point to introduce and discuss three quantities u, p and q which will be useful later. We define them as follows:

$$u \equiv \frac{\mu}{r^3} \tag{5.5-10}$$

$$p \equiv \frac{1}{r^2}(\mathbf{r} \cdot \mathbf{v}) \quad (\textit{not} \text{ the semi-latus rectum}) \tag{5.5-11}$$

$$q \equiv \frac{1}{r^2}(v)^2 - u\,. \tag{5.5-12}$$

These quantities can all be determined if the position **r** and the velocity **v** are known. These quantities are useful because their time derivatives can be expressed in terms of the quantities u, p, q, as we shall demonstrate.

$$\dot{u} = \frac{d}{dt}\left(\frac{\mu}{r^3}\right) = -\frac{3\mu}{r^4}\dot{r}\,. \tag{5.5-13}$$

Using the relationship $\mathbf{r} \cdot \mathbf{v} = r\dot{r}$ and the definition of p, this can be written:

$$\dot{u} = -\frac{3\mu}{r^5}(\mathbf{r} \cdot \dot{\mathbf{r}}) = -3up \tag{5.5-14}$$

$$\dot{p} = \frac{d}{dt}\left[\frac{1}{r^2}(\mathbf{r} \cdot \dot{\mathbf{r}})\right] = \frac{1}{r^2}(v)^2 + \frac{1}{r^2}(\mathbf{r} \cdot \ddot{\mathbf{r}}) - \frac{2}{r^3}\dot{r}(\mathbf{r} \cdot \mathbf{v})\,. \tag{5.5-15}$$

Using the relationship $\mathbf{r} \cdot \mathbf{v} = r\dot{r}$ and the equation of motion $\ddot{\mathbf{r}} = -u\mathbf{r}$, this can be written:

$$\dot{p} = \frac{1}{r^2}(v)^2 - u - \frac{2}{r^4}(\mathbf{r} \cdot \mathbf{v})^2 \tag{5.5-16}$$

Using the definitions of q and p we have

$$\dot{p} = q - 2p^2 \tag{5.5-17}$$

$$\dot{q} = \frac{d}{dt}\left[\frac{1}{r^2}(v)^2 - u\right] = \frac{2}{r^2}(\mathbf{v} \cdot \ddot{\mathbf{r}}) - \frac{2}{r^3}\dot{r}(v)^2 - \dot{u}. \qquad (5.5\text{-}18)$$

Similarly,

$$\dot{q} = -\frac{2}{r^2}u(\mathbf{r} \cdot \dot{\mathbf{r}}) - \frac{2}{r^4}(\mathbf{r} \cdot \dot{\mathbf{r}})(v)^2 - \dot{u}. \qquad (5.5\text{-}19)$$

Using the definition of p, q and the expression for $\dot{u}$ we have

$$\dot{q} = -2p\,(u + q + u) + 3up \qquad (5.5\text{-}20)$$

$$\dot{q} = -p(u + 2q). \qquad (5.5\text{-}21)$$

Summarizing:

$$\begin{array}{ll} u \equiv \dfrac{\mu}{r^3} & \dot{u} = -3up \\ p \equiv \dfrac{1}{r^2}(\mathbf{r} \cdot \mathbf{v}) & \dot{p} = q - 2p^2 \\ q \equiv \dfrac{1}{r^2}(v)^2 - u & \dot{q} = -p(u + 2q) \end{array} \qquad (5.5\text{-}22)$$

After this digression we are now in a position to carry out the recursion of the F_n and G_n. We have already seen that

$$F_0 = 1 \text{ and } G_0 = 0.$$

Applying the recursion formulas (5.5-7) and (5.5-8) we obtain

$$F_1 = \dot{F}_0 - u\,G_0 = 0$$

$$G_1 = F_0 + \dot{G}_0 = 1$$

$$F_2 = \dot{F}_1 - u\,G_1 = -u$$

$$G_2 = F_1 + \dot{G}_1 = 0$$

It may be appropriate to stop at this point and check that these results make sense. Writing equation (5.5-4) for $n = 0, 1, 2$ we have:

$$\mathbf{r}^{(0)} = \mathbf{r} = F_0 \mathbf{r} + G_0 \dot{\mathbf{r}} = \mathbf{r}$$

$$\mathbf{r}^{(1)} = \dot{\mathbf{r}} = F_1 \mathbf{r} + G_1 \dot{\mathbf{r}} = \dot{\mathbf{r}}$$

$$\mathbf{r}^{(2)} = \ddot{\mathbf{r}} = F_2 \mathbf{r} + G_2 \mathbf{r} = - u\mathbf{r} = - \frac{\mu}{r^3} \mathbf{r}$$

Since everything seems to be working we shall continue.

$$F_3 = \dot{F}_2 - u\, G_2 = - \dot{u} = 3up$$

$$G_3 = F_2 + \dot{G}_2 = - u$$

In the next step we shall see the value of u, p and q.

$$F_4 = \dot{F}_3 - u\, G_3 = 3\dot{u}p + 3u\dot{p} + u^2$$

$$= 3p\,(-3up) + 3u(q - 2p^2) + u^2$$

$$= u(u - 15p^2 + 3q)$$

$$G_4 = F_3 + \dot{G}_3 = 3up + 3up = 6up$$

Continuing,

$$F_5 = \dot{F}_4 - uG_4 = \dot{u}\,(u - 15p^2 + 3q)$$

$$+ u(\dot{u} - 30p\dot{p} + 3\dot{q}) - 6u^2 p$$

$$= - 3up\,(u - 15p^2 + 3q) + u\left[-3up - 30p(q - 2p^2)\right.$$

$$\left. - 3p(u + 2q)\right] - 6u^2 p$$

$$= - 15up(u - 7p^2 + 3q)$$

$$G_5 = F_4 + \dot{G}_4 = u(u - 15p^2 + 3q) + 6u\dot{p} + 6\dot{u}p$$

$$= u(u - 15p^2 + 3q) + 6p(q - 2p^2) + 6p(-3up)$$

$$= u(u - 45p^2 + 9q)$$

To continue much farther is obviously going to become tedious and laborious; therefore, the algebra was programmed on a computer to get further terms which are indicated in Table 5.5-1, but we shall use our results so far to determine the functions f and g to terms in t^5. Referring to equations (5.5-2) and (5.5-4)

$$\mathbf{r} = \sum_{n=0}^{\infty} \frac{(t - t_0)^n}{n!} \mathbf{r}_0^{(n)}$$

$$= \sum_{n=0}^{\infty} \frac{(t - t_0)^n}{n!} \left[(F_n \mathbf{r} + G_n \mathbf{v}) \right]_{t=t_0}$$

$$= \left(\sum_{n=0}^{\infty} \frac{\tau^n}{n!} F_n \right) \mathbf{r}_0 + \left(\sum_{n=0}^{\infty} \frac{\tau^n}{n!} G_n \right) \mathbf{v}_0 \qquad (5.5\text{-}23)$$

where $\tau \equiv t - t_0$. Comparing with equation (5.5-1)

$$f(\mathbf{r}_0, \mathbf{v}_0, t) = \sum_{n=0}^{\infty} \frac{\tau^n}{n!} [F_n]_{t=t_0} \qquad (5.5\text{-}24)$$

$$g(\mathbf{r}_0, \mathbf{v}_0, t) = \sum_{n=0}^{\infty} \frac{\tau^n}{n!} [G_n]_{t=t_0} \qquad (5.5\text{-}25)$$

Using the results for F_n and G_n we have previously derived

$$f = 1 - \frac{1}{2} u_0 \tau^2 + \frac{1}{2} u_0 p_0 \tau^3 + \frac{1}{24} u_0 (u_0 - 15p_0^2 + 3q_0) \tau^4$$
$$+ \frac{1}{8} u_0 p_0 (7 p_0^2 - u_0 - 3q_0) \tau^5 + \ldots \qquad (5.5\text{-}26)$$

$F_0 = 1,\quad F_1 = 0,\quad F_2 = -u,\quad F_3 = 3up$

$F_4 = u(-15p^2 + 3q + u),\quad F_5 = 15up(7p^2 - 3q - u)$

$F_6 = 105up^2(-9p^2 + 6q + 2u) - u(45q^2 + 24up + u^2)$

$F_7 = 315up^3(33p^2 - 30q - 10u) + 63up(25q^2 + 14up + u^2)$

$F_8 = 10395up^4(-13p^5 + 15q + 5u) - 315up^2(15q + 7u)(9q + u)$

$\quad + u(1575q^3 + 1107uq^2 + 117u^2q + u^3)$

$F_9 = 135135up^5(15p^2 - 21q - 7u) + 3465up^3(315q^2 + 186uq + 19u^2)$

$\quad - 15up(6615q^3 + 4959uq^2 + 729u^2q + 17u^3)$

$F_{10} = 675675up^6(-15p^2 + 84q + 28u) - 1891890up^4(15q^2 + 9uq + u^2)$

$\quad + 660up^2(6615q^3 + 5184uq^2 + 909u^2q + 32u^3)$

$\quad - u(99225q^4 + 85410uq^3 + 15066u^2q^2 + 498u^3q + u^4)$

$G_0 = 0,\quad G_1 = 1,\quad G_2 = 0,\quad G_3 = -u,\quad G_4 = 6up$

$G_5 = u(-45p^2 + 9q + u),\quad G_6 = 30up(14p^2 - 6q - u)$

$G_7 = 315up^2(-15p^2 + 10q + 2u) - u(225q^2 + 54uq + u^2)$

$G_8 = 630up^3(99p^2 - 90q - 20u) + 126up(75q^2 + 24\,uq + u^2)$

$G_9 = 10395up^4(-91p^2 + 105q + 25u) - 945up^2(315q^2 + 118up + 7u^2)$

$\quad + u(11025q^3 + 4131uq^2 + 243u^2q + u^3)$

$G_{10} = 810810up^5(20p^2 - 28q - 7u) + 13860up^3(630q^2 + 261uq + 19u^2)$

$\quad - 30up(26460q^3 + 12393uq^2 + 1170u^2q + 17u^3)$

Table 5.5-1

$$g = \tau - \frac{1}{6} u_0 \tau^3 + \frac{1}{4} u_0 p_0 \tau^4 + \frac{1}{120} u_0 (u_0 - 45 p_0^2 + 9 q_0) \tau^5 + \ldots\ldots \qquad (5.5\text{-}27)$$

where u_0, p_0 and q_0 are the values of u, p, q at $t = t_0$.

The f and g series will be used in a later section to determine an orbit from sighting directions only.

5.5.2 Solution of the Gauss Problem. The f and g series may be used to solve Gauss' problem if the time interval between the two measurements is not too large. This method has the advantage of not having a quadrant ambiguity as some other methods. We assume that we are given the positions $\mathbf{r}_1$ and $\mathbf{r}_2$ at times t_1 and t_2 and we wish to find $\mathbf{v}$. From equation (5.5-1)

$$\mathbf{r}_2 = f(\mathbf{r}_1, \mathbf{v}_1, t_2 - t_1)\, \mathbf{r}_1 + g(\mathbf{r}_1, \mathbf{v}_1, t_2 - t_1)\, \mathbf{v}_1 \qquad (5.5\text{-}28)$$

from which we find

$$\mathbf{v}_1 = \frac{\mathbf{r}_2 - f(\mathbf{r}_1, \dot{\mathbf{r}}_1, \tau)\, \mathbf{r}_1}{g(\mathbf{r}_1, \dot{\mathbf{r}}_1, \tau)} \qquad (5.5\text{-}29)$$

If we guess a value of $\mathbf{v}_1$ we can compute f and g and then can use equation (5.5-29) to compute a new value of $\mathbf{v}_1$. This method of successive approximations can be continued until $\mathbf{v}_1$ is determined with sufficient accuracy. This method converges very rapidly if τ is not too large.

5.6 THE ORIGINAL GAUSS METHOD

In the interest of its historic and illustrative value we will examine the method which was originally proposed by Gauss in 1809[4]. Although we will assume that the transfer orbit connecting $\mathbf{r}_1$ and $\mathbf{r}_2$ is an ellipse, the extension of the method to cover hyperbolic orbits will be obvious. The derivation of the necessary equations "from scratch" is long and tedious and may be found in Escobal[3] or Moulton[5]. Since all of the relationships we need are contained in the f and g expressions, we will present a very compact and concise development of the Gauss method using only equations (5.2-3), (5.2-4) and (5.2-5).

5.6.1 Ratio of Sector to Triangle. In going from $\mathbf{r}_1$ to $\mathbf{r}_2$ the radius vector sweeps out the shaded area shown in Figure 5.6-1. In Chapter 1 we showed that area is swept out at a constant rate:

$$dt = \frac{2}{h}\, dA. \tag{1.7-7}$$

Since $h = \sqrt{\mu p}$, the area of the shaded sector, A_s, becomes

$$A_s = \tfrac{1}{2}\sqrt{\mu p}\; t$$

where t is the time-of-flight from $\mathbf{r}_1$ to $\mathbf{r}_2$.

The area of the triangle formed by the two radii and the subtended chord is just one-half the base times the altitude; so

$$A_t = \tfrac{1}{2}\, r_1 r_2 \sin \Delta\nu .$$

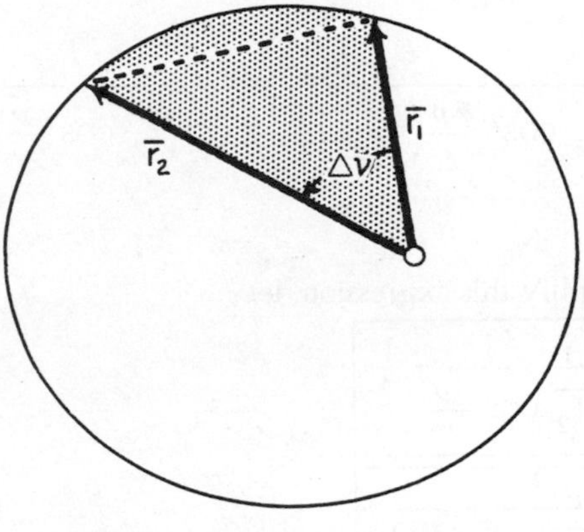

Figure 5.6-1 Sector and triangle area

Gauss called the *ratio of sector to triangle* area y, thus

$$\boxed{y = \frac{\sqrt{\mu p}\; t}{r_1 r_2 \sin \Delta\nu}} \,. \tag{5.6-1}$$

The Gauss method is based on obtaining two independent equations relating y and the change in eccentric anomaly, ΔE. A trial value of y (usually $y \approx 1$) is selected and the first equation is solved for ΔE. This value of ΔE is then used in the second equation to compute a better trial value of y. This technique of successive approximations will converge rapidly if y is nearly one, but fails completely if the radius vector spread is large.

The first equation of Gauss will be obtained by substituting for p in equation (5.6-1) an expression which contains ΔE as the only unknown.

5.6.2 The First Equation of Gauss. If we square the expression for the sector-to-triangle ratio, we obtain

$$y^2 = \frac{\mu p t^2}{(r_1 r_2 \sin \Delta\nu)^2}$$

Substituting for p from equation (5.4-3) and using the identity, $(\sin^2 x)/(1 - \cos x) = 2\cos^2 \frac{x}{2}$, this expression becomes

$$y^2 = \frac{\mu t^2}{2r_1 r_2 \cos^2 \frac{\Delta\nu}{2}\left(r_1 + r_2 - 2\sqrt{r_1 r_2}\cos\frac{\Delta\nu}{2}\cos\frac{\Delta E}{2}\right)}.$$

In order to simplify this expression, let

$$s = \frac{r_1 + r_2}{4\sqrt{r_1 r_2}\cos\frac{\Delta\nu}{2}} - \frac{1}{2} \tag{5.6-2}$$

$$w = \frac{\mu t^2}{\left(2\sqrt{r_1 r_2}\cos\frac{\Delta\nu}{2}\right)^3}. \tag{5.6-3}$$

Note that s and w are constants that may be evaluated from the given information.

A little trigonometric manipulation will prove that y^2 may be expressed compactly as

$$y^2 = \frac{w}{s + \frac{1}{2}\left(1 - \cos\frac{\Delta E}{2}\right)} \tag{5.6-4}$$

which is known as "the first equation of Gauss."

5.6.3 The Second Equation of Gauss. Another completely independent expression for y involving ΔE as the only unknown may be derived from equations (5.2-4) and (5.6-1). From the first of these equations, we see that

$$\frac{r_1 r_2 \sin \Delta\nu}{\sqrt{\mu p}} = t - \sqrt{\frac{a^3}{\mu}}\,(\Delta E - \sin \Delta E).$$

But $r_1 r_2 \sin \Delta\nu/\sqrt{\mu p} = t/y$, so

$$1 - \frac{1}{y} = \frac{1}{t}\sqrt{\frac{a^3}{\mu}}\,(\Delta E - \sin \Delta E)\,. \tag{5.6-5}$$

We still need to eliminate a from this expression. Using the identity, $\sin \Delta\nu = 2 \sin \frac{\Delta\nu}{2} \cos \frac{\Delta\nu}{2}$, equation (5.6-1) becomes

$$y = \frac{\sqrt{\mu p}\, t}{2 r_1 r_2 \sin \frac{\Delta\nu}{2} \cos \frac{\Delta\nu}{2}}\,. \tag{5.6-7}$$

From equation (5.2-3) we can write

$$1 - \cos \Delta\nu = \frac{ap}{r_1 r_2}\,(1 - \cos \Delta E)$$

$$2 \sin^2 \frac{\Delta\nu}{2} = 2\,\frac{ap}{r_1 r_2} \sin^2 \frac{\Delta E}{2}$$

$$\sin \frac{\Delta\nu}{2} = \sqrt{\frac{ap}{r_1 r_2}} \sin \frac{\Delta E}{2}\,.$$

Substituting this last expression into equation (5.6-7) eliminates $\sqrt{p}$ in favor of $\sqrt{a}$:

$$y = \frac{\sqrt{\mu}\, t}{2\sqrt{a r_1 r_2} \sin \frac{\Delta E}{2} \cos \frac{\Delta\nu}{2}} \tag{5.6-8}$$

If we now cube this equation and multiply it by equation (5.6-5), a will be eliminated and we end up with

$$y^3\left(1 - \frac{1}{y}\right) = \frac{\mu t^2}{\left(2\sqrt{r_1 r_2} \cos \frac{\Delta\nu}{2}\right)^3}\,\frac{(\Delta E - \sin \Delta E)}{\sin^3 \frac{\Delta E}{2}}\,.$$

Recognizing the first factor as w, we may write, more compactly

$$y^2\,(y-1) = w\,\frac{(\Delta E - \sin \Delta E)}{\sin^3 \frac{\Delta E}{2}}$$

Substituting for y^2 from (5.6-4) and solving for y, we get

$$y = 1 + \left(\frac{\Delta E - \sin \Delta E}{\sin^3 \frac{\Delta E}{2}}\right)\left(s + \frac{1 - \cos \frac{\Delta E}{2}}{2}\right) \qquad (5.6\text{-}9)$$

which is known as "the second equation of Gauss."

5.6.4 Solution of the Equations. To review what we have done so far, recall that we started with three equations, (5.2-3), (5.2-4) and (5.2-5), in three unknowns, p, a and ΔE. We then added another independent equation, (5.6-1), and one more unknown, y. By a process of eliminating p and a between these four equations, we now have reduced the set to two equations in two unknowns, y and ΔE. Unfortunately, equations (5.6-4) and (5.6-9) are transcendental, so a trial-and-error solution is necessary.

The first step is to evaluate the constants, s and w, from $\mathbf{r}_1$, $\mathbf{r}_2$, $\Delta\nu$ and t. Next, pick a trial value for y; since this method only works well if $\Delta\nu$ is less than about 90°, a good first guess is $y \approx 1$.

We can now solve Gauss' first equation for ΔE, using the trial value of y:

$$\cos \frac{\Delta E}{2} = 1 - 2\left(\frac{w}{y^2} - s\right). \qquad (5.6\text{-}10)$$

If we assume that ΔE is less than 2π (which will always be the case unless the satellite passes back through $\mathbf{r}_1$ enroute to $\mathbf{r}_2$), there is no problem determining the correct quadrant for ΔE.

We are now ready to use this approximate value for ΔE to compute a better approximation for y from Gauss' second equation. This better value of y is then used in equation (5.6-10) to compute a still better value of ΔE, and so on, until two successive approximations for y are nearly identical.

When convergence has occurred, the parameter p may be computed from equation (5.4-3) and the f and g expressions evaluated. The determination of $\mathbf{v}_1$ and $\mathbf{v}_2$ from equations (5.2-7) and (5.2-2) completes the solution.

Since the equations above involve ΔE, they are valid only if the transfer orbit from $\mathbf{r}_1$ to $\mathbf{r}_2$ is elliptical. The extension of Gauss' method to include hyperbolic and parabolic orbits is the subject of the next section.

5.6.5 Extension of Gauss' Method to Any Type of Conic Orbit. If the given time-of-flight is short, the right-hand side of equation (5.6-10) may become greater than one, indicating that ΔE is imaginary. Since we already know that when ΔE is imaginary, ΔF is real, we can conclude that the transfer orbit is hyperbolic when this occurs. Noting that $\Delta E = i\Delta F$ and $\cos i\Delta F = \cosh \Delta F$, equation (5.6-10) may also be written as

$$\cosh \frac{\Delta F}{2} = 1 - 2\left(\frac{w}{y^2} - s\right) \qquad (5.6\text{-}11)$$

whenever the right side is greater than one.

Using the identity, $-i \sin i\Delta F = \sinh \Delta F$, equation (5.6-9) becomes

$$y = 1 + \left(\frac{\sinh \Delta F - \Delta F}{\sinh^3 \frac{\Delta F}{2}}\right)\left(s + \frac{1 - \cosh \frac{\Delta F}{2}}{2}\right). \qquad (5.6\text{-}12)$$

These equations may be used exactly as equations (5.6-9) and (5.6-10) to determine y.

If the transfer orbit being sought happens to be parabolic, then ΔE and ΔF will be zero and both equations (5.6-9) and (5.6-12) become indeterminate. For this reason, difficulties may be anticipated any time ΔE or ΔF are close to zero. Gauss solved this problem by defining two auxiliary variables, x (not to be confused with the universal variable of Chapter 4) and X as follows:

$$x = \frac{1}{2}\left(1 - \cos \frac{\Delta E}{2}\right)$$

$$X = \frac{\Delta E - \sin \Delta E}{\sin^3 \frac{\Delta E}{2}}.$$

The first equation of Gauss may then be written, as

$$y^2 = \frac{w}{s + x}$$

$$\boxed{x = \frac{w}{y^2} - s.} \tag{5.6-13}$$

The second equation of Gauss may be written as

$$\boxed{y = 1 + X\ (s + x).} \tag{5.6-14}$$

Now, it is possible to expand the function X as a power series in x. This may be accomplished by first writing the power series expansion for X in terms of ΔE, and then expressing ΔE as a power series in x. The result, which is developed by Moulton, is

$$\boxed{X = \frac{4}{3}\left(1 + \frac{6}{5}x + \frac{6\cdot 8}{5\cdot 7}x^2 + \frac{6\cdot 8\cdot 10}{5\cdot 7\cdot 9}x^3 + \ldots\right).} \tag{5.6-15}$$

We may now reformulate the algorithm for *solving the Gauss problem via the Gauss method* as follows:

1. Compute the constants, s and w, from $\mathbf{r}_1$, $\mathbf{r}_2$, $\Delta\nu$ and t using equations (5.6-2) and (5.6-3).
2. Assume $y \approx 1$ and compute x from equation (5.6-13).
3. Determine X from equation (5.6-15) and use it to compute a better approximation to y from equation (5.6-14). Repeat this cycle until y converges to a solution.
4. The type of conic orbit is determined at this point, the orbit being an ellipse, parabola, or hyperbola according to whether x is positive, zero, or negative. Depending on the type of conic, determine ΔE or ΔF from equation (5.6-10) or (5.6-11).
5. Determine p from equation (5.4-3), replacing $\cos\frac{\Delta E}{2}$ with $\cosh\frac{\Delta F}{2}$ in the case of the hyperbolic orbit.
6. Evaluate f, g, $\dot{f}$ and $\dot{g}$ from equations (5.2-3), (5.2-4), (5.2-5) and (5.2-6).
7. Solve for $\mathbf{v}_1$ and $\mathbf{v}_2$ from equations (5.2-7) and (5.2-2).

The method outlined above is perhaps the most accurate and rapid technique known for solving the Gauss problem when $\Delta\nu$ is less than 90^0; the iteration to determine y fails to converge shortly beyond this point.

5.7 PRACTICAL APPLICATIONS OF THE GAUSS PROBLEM–INTERCEPT AND RENDEZVOUS

A fundamental problem of astrodynamics is that of getting from one point in space to another in a predetermined time. Usually, we would like to know what velocity is required at the first point in order to coast along a conic orbit and arrive at the destination at the prescribed time. If the object of the mission is to rendezvous with some other satellite, then we may also be interested in the velocity we will have upon arrival at the destination.

Applications of the Gauss problem are almost limitless and include interplanetary transfers, satellite intercept and rendezvous, ballistic missile targeting, and ballistic missile interception. The subject of ballistic missile targeting is covered in Chapter 6 and interplanetary trajectories are covered in Chapter 8. The Gauss problem is also applicable to lunar trajectories which is the subject of Chapter 7.

Oribt determination from two positions and time is usually part of an even larger problem which we may call "mission planning." Mission planning includes determining the optimum timing and sequence of maneuvers for a particular mission and the cost of the mission in terms of Δv or "characteristic velocity."

In all but the simplest cases, the problem of determining the optimum sequencing of velocity changes to give the minimum total Δv defies analytical solution, and we must rely on a computer analysis to establish suitable "launch windows" for a particular mission. To avoid generalities, let's define a hypothetical mission and show how such an analysis is performed.

Let's assume that we have the position and velocity of a target satellite at some time t_0 and we wish to intercept this target satellite from a ground launch site. We will assume that a single impulse is added to the launch vehicle to give it its launch velocity. The problem is to establish the optimum launch time and time-of-flight for the interceptor.

To make the problem even more specific, we will assume that the target satellite is in a nearly circular orbit inclined 65° to the equator and at time, t_0, it is over the Aleutian Islands heading southeastward. Our launch site will be at Johnston Island in the Pacific. The situation at time t_0 is illustrated in Figure 5.7-2. We will assume that time t_0 is 1200 GMT.

The Δv required to intercept the target depends on two parameters

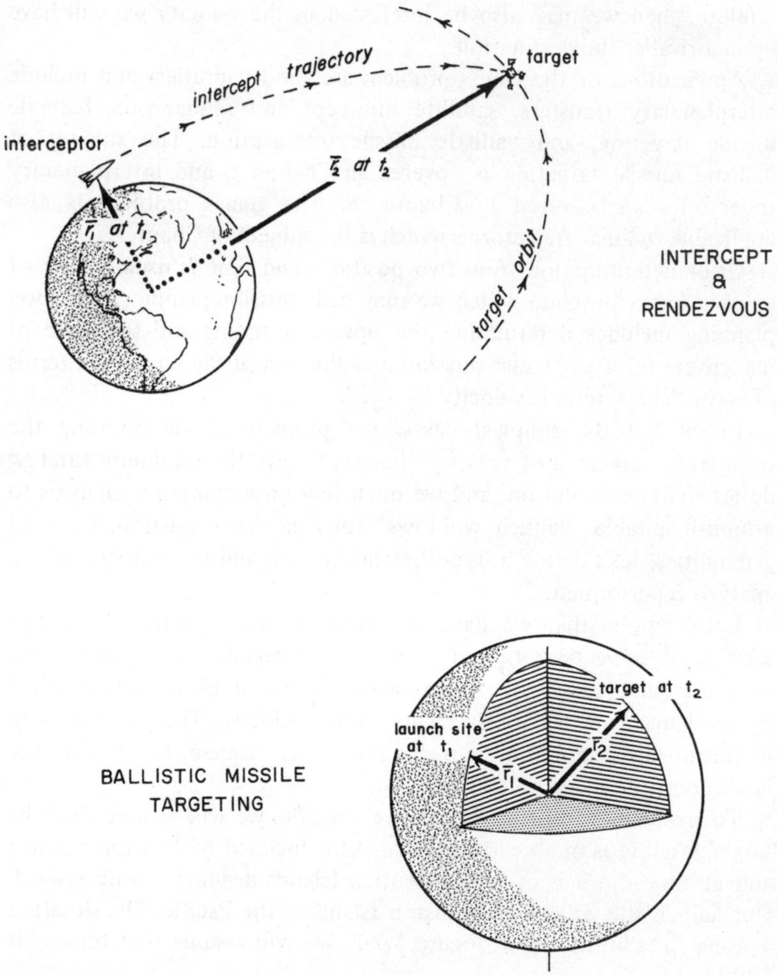

Figure 5.7-1 Important practical applications of the Gauss problem

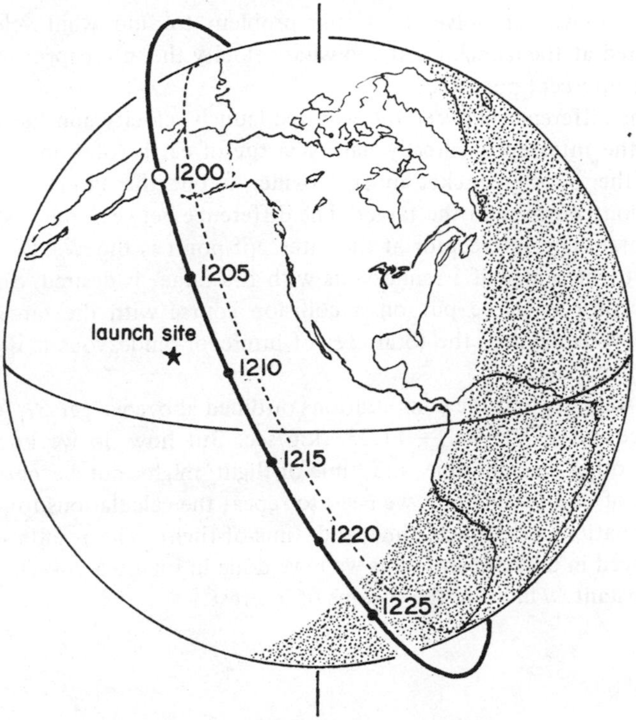

Figure 5.7-2 Example intercept problem

which we are free to choose arbitrarily, the launch time and the time-of-flight of the interceptor. Suppose we pick a launch time of 1205 and a time-of-flight of 5 minutes. The first step in determining Δv is to find the position and velocity of the interceptor at 1205. Since it is stationary on the launch pad at Johnston Island, we need to find the position vector from the center of the Earth to the launch site at 1205. We have already discussed this problem in Chapter 2. The velocity of the interceptor is due solely to earth rotation and is in the eastward direction at the site.

The next step is to determine where the target will be at 1210, which is when the intercept will occur. Since we know the position and velocity of the target at 1200, we can update **r** and **v** to 1210 by solving the Kepler problem which we discussed in Chapter 4.

We now have two position vectors and the time-of-flight between them, so we can solve the Gauss problem to find what velocity is required at the launch point and what velocity the interceptor will have at the intercept point.

The difference between the required launch velocity and the velocity that the interceptor already has by virtue of earth rotation is the Δv_1 that the booster rocket must provide to put the interceptor on a collision course with the target. The difference between the velocity of the target and interceptor at the intercept point is the Δv_2 that would have to be added if a rendezvous with the target is desired. Since the interceptor must be put on a collision course with the target for a rendezvous mission, the total cost of intercept-rendezvous is the scalar sum of Δv_1 and Δv_2.

If we carry out the computations outlined above we get $\Delta v_1 = 4.121$ km/sec and $\Delta v_1 + \Delta v_2 = 11.238$ km/sec. But how do we know that some other launch time and time-of-flight might not be cheaper in terms of Δv? To find out, we need to repeat the calculations for various combinations of launch time and time-of-flight. The results may be displayed in tabular form or as we have done in Figure 5.7-4 where lines of constant Δv indicate the regions of interest.

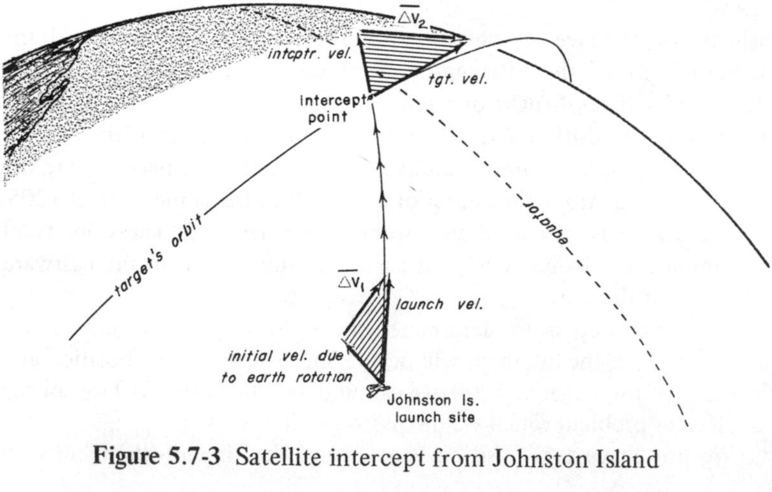

Figure 5.7-3 Satellite intercept from Johnston Island

5.7.1 Interpretation of the Δv Plot. A great deal may be learned about a particular mission just by studying a Δv plot. In Figure 5.7-4 the shaded area represents those conditions that result in the interceptor striking the earth enroute to the target. This is most likely to occur when the intercept point is located far from the launch site and time-of-flight is short. The lowest Δv's for intercept are likely to occur when the intercept point is close to the launch site. From Figure 5.7-2 we can see that this will first occur when the intercept time is about 1210 (launch at 1205 plus 5 min time-of-flight). Keep in mind that the Earth is rotating and, although the target satellite returns to the same position in space after one orbital period, the launch site will have moved eastward.

The target satellite in our example will again pass close to Johnston Island at about 1340 GMT and another good launch opportunity will present itself. Because of earth rotation, the next pass at 1510 will not bring the target satellite nearly so close to the launch site. After 12 hours the launch site will again pass through the plane of the target's orbit and another good series of launch windows should occur.

In summary, we can say that the most important single factor in determining the Δv required for intercept is the choice of where the intercept is made relative to the launch site; the closer the intercept point is to the launch site, the better.

A similar Δv plot for intercept-plus-rendezvous would show that the lowest Δv's occur when the transfer orbit is coplanar with the target's orbit. This can only occur if the launch takes place at the exact time the launch site is passing through the target's orbital plane and only occurs twice every 24 hours at most.

5.7.2 Definition of "Optimum Launch Conditions." It would be a mistake to assume that those launch conditions which minimize Δv are the optimum for a particular mission. We have to know more about the mission before we can properly interpret a Δv plot.

Suppose the mission is to intercept the target as quickly as possible with an interceptor that has a fixed Δv capability. In this case "optimum" launch conditions are those that result in the earliest intercept time without exceeding the Δv limitations of the interceptor.

There are two common instances where we would be trying to minimize Δv. One is where we have a fixed payload and are trying to minimize the size of the launch vehicle required to accomplish the mission. The other is where we have fixed the launch vehicle and are

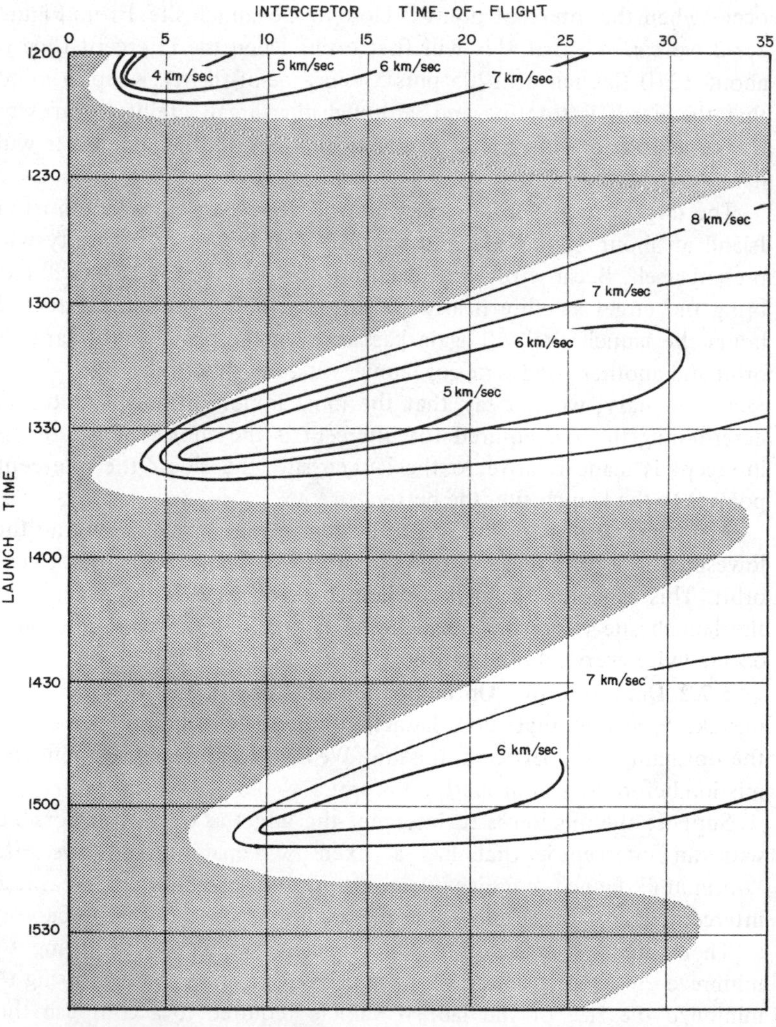

Figure 5.7-4 Δv plot for example intercept problem

trying to maximize the payload it can carry.

Although we have examined a very specific example, the technique of producing and interpreting a Δv plot as an aid in mission planning is generally applicable to all types of mission analysis.

5.8 DETERMINATION OF ORBIT FROM SIGHTING DIRECTIONS AT STATION

Though one method of orbit determination from optical sightings was presented in Chapter 2, now that we have developed the f and g series, another method will be developed.

Let us assume that we can measure the *right ascension* and *declination* of an Earth satellite from some station on the Earth at three times t_1, t_2, t_3. The unit vector $\mathbf{L}_i$ pointing in the direction of the satellite from the station is

$$\mathbf{L}_i = \mathbf{I} \cos\delta_i \cos\alpha_i + \mathbf{J} \cos\delta_i \sin\alpha_i + \mathbf{K} \sin\delta_i \,. \tag{5.8-1}$$

The vector $\mathbf{r}$ from the center of the Earth is

$$\mathbf{r} = \mathbf{R} + \rho\,\mathbf{L} \,. \tag{5.8-2}$$

We expand the vector $\mathbf{r}$ in terms of the f and g series evaluated at t_2 so that

$$\mathbf{r} = f\,(\mathbf{r}_2, \mathbf{v}_2, t - t_2)\,\mathbf{r}_2 + g\,(\mathbf{r}_2, \mathbf{v}_2, t - t_2)\,\mathbf{v}_2 \,. \tag{5.8-3}$$

then letting $f_i = f\,(\mathbf{r}_2, \mathbf{v}_2, t_i - t_2)$ and $g_i = g\,(\mathbf{r}_2, \mathbf{v}_2, t_i - t_2)$

$$f_1\,\mathbf{r}_2 + g_1\,\mathbf{v}_2 = \mathbf{R}_1 + \rho_1\,\mathbf{L}_1 \tag{5.8-4}$$

$$\mathbf{r}_2 = \mathbf{R}_2 + \rho_2\,\mathbf{L}_2 \tag{5.8-5}$$

$$f_3\,\mathbf{r}_2 + g_3\,\mathbf{v}_2 = \mathbf{R}_3 + \rho_3\,\mathbf{L}_3 \,. \tag{5.8-6}$$

This is a set of nine equations in the nine unknowns $\mathbf{r}_2, \mathbf{v}_2, \rho_1, \rho_2, \rho_3$, i.e. the three components of $\mathbf{r}_2$, the three components of $\mathbf{v}_2$ and the three quantities ρ_1, ρ_2, ρ_3.

We can eliminate the ρ_i by cross multiply the ith equation by $\mathbf{L}_i$ to obtain

$$f_1 \mathbf{L}_1 \times \mathbf{r}_2 + g_1 \mathbf{L}_1 \times \mathbf{v}_2 = \mathbf{L}_1 \times \mathbf{R}_1 \tag{5.8-7}$$

$$\mathbf{L}_2 \times \mathbf{r}_2 = \mathbf{L}_2 \times \mathbf{R}_2 \tag{5.8-8}$$

$$f_3 \mathbf{L}_3 \times \mathbf{r}_2 + g_3 \mathbf{L}_3 \times \mathbf{v}_2 = \mathbf{L}_3 \times \mathbf{R}_3 \,. \tag{5.8-9}$$

Although it appears that there are now nine equations in six unknowns, i.e. the components of $\mathbf{r}_2$ and $\mathbf{v}_2$, in fact only six of the equations are independent. Letting the Cartesian components of $\mathbf{r}_2$ be x, y and z, the components of $\mathbf{v}_2$ be $\dot{x}$, $\dot{y}$ and $\dot{z}$, and eliminating the **K** components, these equations are:

$$\begin{aligned}
&f_1 L_{1z}x - f_1 L_{1x}z + g_1 L_{1z}\dot{x} - g_1 L_{1x}\dot{z} = R_{1x}L_{1z} - R_{1z}L_{1x}\\
&f_1 L_{1z}y - f_1 L_{1y}z + g_1 L_{1z}\dot{y} - g_1 L_{1y}\dot{z} = R_{1y}L_{1z} - R_{1z}L_{1y}\\
&L_{2z}x - L_{2x}z = R_{2x}L_{2z} - R_{2z}L_{2x}\\
&L_{2z}y - L_{2y}z = R_{2y}L_{2z} - R_{2z}L_{2y}\\
&f_3 L_{3z}x - f_3 L_{3x}z + g_3 L_{3z}\dot{x} - g_3 L_{3x}\dot{z} = R_{3x}L_{3z} - R_{3z}L_{3x}\\
&f_3 L_{3z}y - f_3 L_{3y}z + g_3 L_{3z}\dot{y} - g_3 L_{3y}\dot{z} = R_{3y}L_{3z} - R_{3z}L_{3y}
\end{aligned} \tag{5.8-10}$$

A procedure which may be used to solve this set of equations is as follows:

a. Estimate the magnitude of $\mathbf{r}_2$.

b. Using this estimate compute $u_2 = \frac{\mu}{r_2^{\,3}}$.

c. Compute the values of f_1, g_1, f_3 and g_3 using the terms of equations (5.5-26) and (5.5-27) which are independent of p_2 and q_2.

d. Substitute these values of f_1, g_1, f_3, g_3 and the known values of the components of $\mathbf{L}_1$, $\mathbf{L}_2$, $\mathbf{L}_3$, $\mathbf{R}_1$, $\mathbf{R}_2$, $\mathbf{R}_3$ into equations (5.8-10) and solve the resulting linear algebraic equations for the six unknowns x, y,

$z, \dot{x}, \dot{y}, \dot{z}$, which are the components of $\mathbf{r}_2$ and $\mathbf{v}_2$.

e. Compute new values of u_2, p_2, q_2 from this $\mathbf{r}_2$ and $\mathbf{v}_2$ using their definitions equations (5.5-22). Then compute new values of f_1, g_1, f_3 and g_3 from equations (5.5-26) and (5.5-27), using as many terms as are necessary to obtain required accuracy.

f. Repeat steps d and e until process converges to correct values of $\mathbf{r}_2$ and $\mathbf{v}_2$.

This process converges very rapidly if the time intervals $t_3 - t_2$ and $t_2 - t_1$ are not too large, but it is probably too tedious for hand computation because of the time required to solve the system of six linear algebraic equations several times. The process is well suited to solution on a digital computer.

EXERCISES

5.1 Several methods for solving the Gauss problem have been developed in this chapter. Discuss and rank each method for each of the following criteria:

a. Limitations.
b. Ease of computation.
c. Accuracy.

5.2 As a mission planner you could calculate Δv's for various combinations of reaction time and time of flight to the target for a given interceptor location. What relative orientation between launch site and target would minimize the Δv required for intercept?

5.3 Verify equations (5.3-13) through (5.3-15) by developing them from equations (5.3-1), (5.3-2), and (5.3-9).

5.4 Verify the development of equation (5.4-8).

5.5 Make a plot similar to Figure 5.4-1 for specific values of $\mathbf{r}_1$ and $\mathbf{r}_2$ (such as $\mathbf{r}_1 = 2\mathbf{I}$, $\mathbf{r}_2 = \mathbf{I} + \mathbf{J}$ DU). Using specific numbers, verify and amplify the descriptive statements used in discussing Figure 5.4-1.

5.6 Derive equation (5.4-23) for hyperbolic orbits.

5.7 Given two position vectors $\mathbf{r}_1$, $\mathbf{r}_2$ and the distance between them, d. Find an expression for the semi-major axis of the minimum energy ellipse, which will contain both position vectors, as a function of r_1, r_2, d.

Note: The remaining exercises are well suited for computer use, although a few iterations can be made by hand.

5.8 An Earth satellite is observed at two times t_1 and t_2 to have the following positions:

$$\mathbf{r}_1 = \mathbf{I} \text{ DU}$$

$$\mathbf{r}_2 = \mathbf{I} + \mathbf{J} + \mathbf{K} \text{ DU}$$

find p, **e** and **h** for $t_2 - t_1 = 1.0922$ TU. Use the p-iteration technique with $p = 2$ as a starting value.
(Ans. $\mathbf{h} = 1.009\,(-\mathbf{J} + \mathbf{K})$).

5.9 Repeat problem 5.8 using the universal variable.

5.10 An Earth satellite's positions at two times t_1 and t_2 are measured by radar to be

$$\mathbf{r}_1 = \mathbf{I} \text{ DU}$$

$$\mathbf{r}_1 = \mathbf{I} + \frac{1}{8}\mathbf{J} + \frac{1}{8}\mathbf{K} \text{ DU}$$

and $t_2 - t_1 = \frac{1}{8}$ TU

Find the velocity $\mathbf{v}_1$ at t_1 using the f and g series.
(Ans. $\mathbf{v}_1 = 0.061858026\mathbf{I} + 1.00256\,(\mathbf{J} + \mathbf{K})$).

5.11 For the following data sets for $\mathbf{r}_1$, $\mathbf{r}_2$, $t_2 - t_1$ determine $\mathbf{v}_2$ using

a. The universal variable method.
b. The p-iteration technique.
c. The original Gauss method.

Compare the accuracy and speed of convergence.

a. $\mathbf{r}_1 = 0.5\mathbf{I} + 0.6\mathbf{J} + 0.7\mathbf{K}$ DU $t_2 - t_1 = 20$ TU
$\mathbf{r}_2 = -\mathbf{J}$ DU Use "long way" trajectory.
(Answer: $\mathbf{v}_2 = 0.66986992\mathbf{I} + 0.48048471\mathbf{J} + 0.93781789\mathbf{K}$ DU/TU)

b. $\mathbf{r}_1 = 1.2\mathbf{I}$ DU $t_2 - t_1 = 10$ TU
$\mathbf{r}_2 = 2\mathbf{J}$ DU Use " short way" trajectory.

c. $\mathbf{r}_1 = \mathbf{I}$ $t_2 - t_1 = 0.0001$ TU
$\mathbf{r}_2 = \mathbf{J}$ Use "short way" trajectory.

d. $\mathbf{r}_1 = 4\mathbf{I}$ $t_2 - t_1 = 10$ TU
$\mathbf{r}_2 = -2\mathbf{I}$ Use "short way" trajectory.
(Why is this data set insoluable?)

e. $\mathbf{r}_1 = 2\mathbf{I}$ $t_2 - t_1 = 20$ TU
$\mathbf{r}_2 = -2\mathbf{I} - 0.2\mathbf{J}$ Use "long way" trajectory.

LIST OF REFERENCES

1. Bell, Eric Temple. "The Prince of Mathematicians," in *The World of Mathematics*. Vol 1. New York, Simon and Schuster, 1956.

2. Battin, Richard H. *Astronautical Guidance*. New York, NY, McGraw-Hill Book Company, 1964.

3. Escobal, Pedro Ramon. *Methods of Orbit Determination*. New York, NY, John Wiley & Sons, Inc, 1965.

4. Gauss, Carl Friedrich. *Theory of the Motion of the Heavenly Bodies Revolving about the Sun in Conic Sections*, a translation of *Theoria Motus* (1857) by Charles H. Davis. New York, NY, Dover Publications, Inc, 1963.

5. Moulton, Forest Ray. *An Introduction to Celestial Mechanics*. New York, NY, the MacMillan Company, 1914.

6. Herrick, S. and A. Liu. *Two Body Orbit Determination From Two Positions and Time of Flight*. Appendix A, Aeronutronic Pub No C-365, 1959.

CHAPTER 6

BALLISTIC MISSILE TRAJECTORIES

> 'Tis a principle of war that when you can use the lightning 'tis better than cannon.
>
> –Napoleon I

6.1 HISTORICAL BACKGROUND

While the purist might insist that the history of ballistic missiles stretches back to the first use of crude rockets in warfare by the Chinese, long-range ballistic missiles, which concern us in this chapter, have a very short history.

The impetus for developing the long-range rocket as a weapon of war was, ironically, the Treaty of Versailles which ended World War I. It forbade the Germans to develop long-range artillery. As a result the German High Command was more receptive to suggestions for rocket development than were military commands in other countries. The result is well known. Efforts, begun in 1932 under the direction of then Captain Walter Dornberger, culminated in the first successful launch of an A-4 ballistic missile (commonly known as the V-2) on 3 October 1942. During 1943 and 1944 over 280 test missiles were fired from Peenemünde. The first two operational missiles were fired against Paris on 6 September 1944 and an attack on London followed 2 days later. By the end of the war in May 1945 over 3,000 V-2's had been fired in anger.[1]

General Dornberger summed up the use of the long-range missile in World War II as "too late."[2] He might have added that it was not a particularly effective weapon as used. It had a dispersion at the target of 10 miles over a range of 200 miles and carried a warhead of about one ton of the high explosive Amatol.[3] Nevertheless, it was more than just an extension of long-range artillery—it was the first ballistic missile as we know it today.

After the war there was a mad scramble to "capture" the German scientists of Peenemünde who were responsible for this technological miracle. The United States Army obtained the services of most of the key scientists and technicians including Dornberger and Wernher von Braun. The Soviets were able to assemble at Khimki a staff of about 80 men under the former propulsion expert Werner Baum. They were assigned the task of designing a rocket motor with a thrust of 260,000 pounds and later one of 530,000 pounds thrust.[3]

At an intelligence briefing at Wright-Patterson Air Force Base in August 1952, these disquieting facts were revealed by Dornberger who had interviewed many of his former colleagues on a return trip to Germany. Even more disquieting should have been the report that the Soviets had built a separate factory building adjacent to that occupied by the German workers. No German was permitted to enter this separate building.

The experts displayed no particular sense of immediacy. After all, the atomic bomb was still too heavy to be carried by a rocket. According to Dr. Darol Froman of the Los Alamos Scientific Laboratory the key question in the 1950's was "When could the AEC come up with a warhead light enough to make missiles practical?"

The ballistic missile program in this country was still essentially in abeyance when a limited contract for the ICBM called "Atlas" was awarded. But in November 1952 at Eniwetok the thermonuclear "Mike" shot ended all doubts and paved the way for the "Shrimp" shot of March 1954 which revolutionized the program.[4]

Accordingly, in June 1953, Trevor Gardner, Assistant Secretary of the Air Force for Research and Development, convened a special group of the nation's leading scientists known as the Teapot Committee. The group was led by the late Professor John von Neumann of the Institute for Advanced Studies at Princeton. They met in a vacant church in Inglewood, California, and the result of their study was a recommendation to the Air Force that the ballistic missile program be reactivated with top priority.

In October 1953 a study contract was placed with the Ramo-Wooldridge Corporation and by May 1954 the new ICBM program had highest Air Force priority. In July 1954, Brigadier General

Bernard A. Schriever was given the monumental task of directing the accelerated ICBM program and began handpicking a staff of military assistants. When he reported to the West Coast he had with him a nucleus of four officers–among them Lieutenant Colonel Benjamin P. Blasingame, later to become the first Head of the Department of Astronautics at the United States Air Force Academy.

Within a year of its beginning in a converted parochial school building in Inglewood the program had passed from top Air Force priority to top national priority. From 2 main contractors at the beginning, the program had, by mid-1959, 30 main contractors and more than 80,000 people participating directly.

Progress was rapid. After three unsuccessful attempts, the first successful flight of a Series A Atlas took place on 17 December 1957. Only 4 months after the Soviet Union had announced that it had an intercontinental ballistic missile, Atlas was a reality.

While the history of the ballistic missile is both interesting and significant, a knowledge of how and why it works is indispensable for understanding its employment as a weapon.

The trajectory of a missile differs from a satellite orbit in only one respect–it intersects the surface of the Earth. Otherwise, it follows a conic orbit during the free-flight portion of its trajectory and we can analyze its behavior according to principles which you already know.

Ballistic missile targeting is just a special application of the Gauss problem which we treated rigorously in Chapter 5. In this chapter we will present a somewhat simplified scalar analysis of the problem so that you may gain some fresh insight into the nature of ballistic trajectories. To compute precise missile trajectories requires the full complexity of perturbation theory. In this chapter we are concerned mainly with concepts.

6.2 THE GENERAL BALLISTIC MISSILE PROBLEM

A ballistic missile trajectory is composed of three parts–the *powered flight* portion which lasts from launch to thrust cutoff or burnout, the *free-flight* portion which constitutes most of the trajectory, and the *re-entry* portion which begins at some ill-defined point where atmospheric drag becomes a significant force in determining the missile's path and lasts until impact.

Since energy is continuously being added to the missile during powered flight, we cannot use 2-body mechanics to determine its path from launch to burnout. The path of the missile during this critical part of the flight is determined by the guidance and navigation system. This is the topic of an entire course and will not be covered here.

During free-flight the trajectory is part of a conic orbit–almost always an ellipse–which we can analyze using the principles learned in Chapter 1.

Re-entry involves the dissipation of energy by friction with the atmosphere. It will not be discussed in this text.

We will begin by assuming that the Earth does not rotate and that the altitude at which re-entry starts is the same as the burnout altitude. This latter assumption insures that the *free-flight trajectory is symmetrical* and will allow us to derive a fairly simple expression for the free-flight range of a missile in terms of its burnout conditions.

We will then answer a more practical question–"given r_{bo}, v_{bo}, and a desired free-flight range, what flight-path angle at burnout is required?"

Following a discussion of maximum range trajectories, we will determine the time-of-flight for the free-flight portion of the trajectory.

6.2.1 Geometry of the Trajectory. Since you are already familiar with the terminology of orbital mechanics, such terms as "height at burnout," "height of apogee," "flight-path angle at burnout," etc., need not be redefined. There are, however, a few new and unfamiliar terms which you must learn before we embark on any derivations. Figure 6.2.1 defines these new quantities.

6.2.2 The Nondimensional Parameter, Q. We will find it very convenient to define a nondimensional parameter called Q such that

$$\boxed{Q \equiv \frac{v^2 r}{\mu}.} \tag{6.2-1}$$

Q can be evaluated at any point in an orbit and may be thought of as the squared ratio of the speed of the satellite to circular satellite speed at that point. Since $v_{cs} = \sqrt{\mu/r}$,

$$Q = \left(\frac{v}{v_{cs}}\right)^2 = \frac{v^2 r}{\mu}. \tag{6.2-2}$$

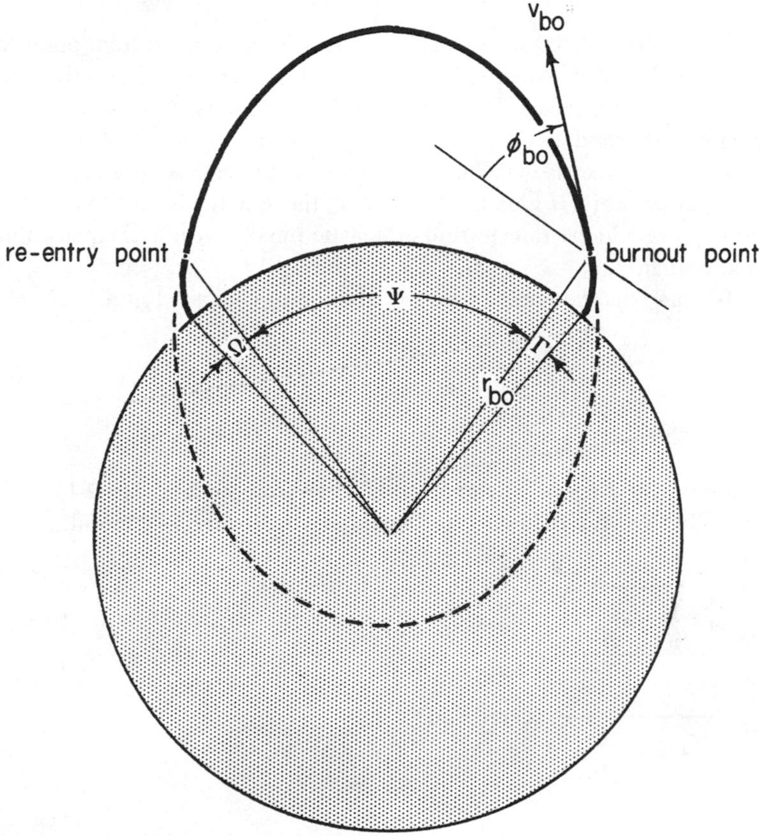

Γ – powered flight range angle

Ψ – free-flight range angle

Ω – re-entry range angle

Λ – total range angle

R_p – ground range of powered flight

R_{ff} – ground range of free-flight

R_{re} – ground range of re-entry

R_t – total ground range

$$\Lambda = \Gamma + \Psi + \Omega$$

$$R_t = R_p + R_{ff} + R_{re}$$

Figure 6.2-1 Geometry of the ballistic missile trajectory

The value of Q is not constant for a satellite but varies from point to point in the orbit. When Q is equal to 1, the satellite has exactly local circular satellite speed. This condition $(Q = 1)$ exists at every point in a circular orbit and at the end of the minor axis of *every* elliptical orbit.

If $Q = 2$ it means that the satellite has exactly escape speed and is on a parabolic orbit. If Q is greater than 2, the satellite is on a hyperbolic orbit. It would be rare to find a ballistic missile with a Q equal to or greater than 2.

We can take the familiar energy equation of Chapter 1,

$$\mathcal{E} = \frac{v^2}{2} - \frac{\mu}{r} = -\frac{\mu}{2a}$$

and substitute for v^2 the expression $\mu Q/r$ from equation (6.2-1). This yields both of the following relationships which will prove useful:

$$\boxed{a = \frac{r}{2-Q}} \tag{6.2-3}$$

or

$$\boxed{Q = 2 - \frac{r}{a}.} \tag{6.2-4}$$

6.2.3 The Free-Flight Range Equation. Since the free-flight trajectory of a missile is a conic section, the general equation of a conic can be applied to the burnout point.

$$r_{bo} = \frac{p}{1 + e \cos \nu_{bo}} \tag{6.2-5}$$

Solving for $\cos \nu_{bo}$, we get

$$\cos \nu_{bo} = \frac{p - r_{bo}}{e r_{bo}} \tag{6.2-6}$$

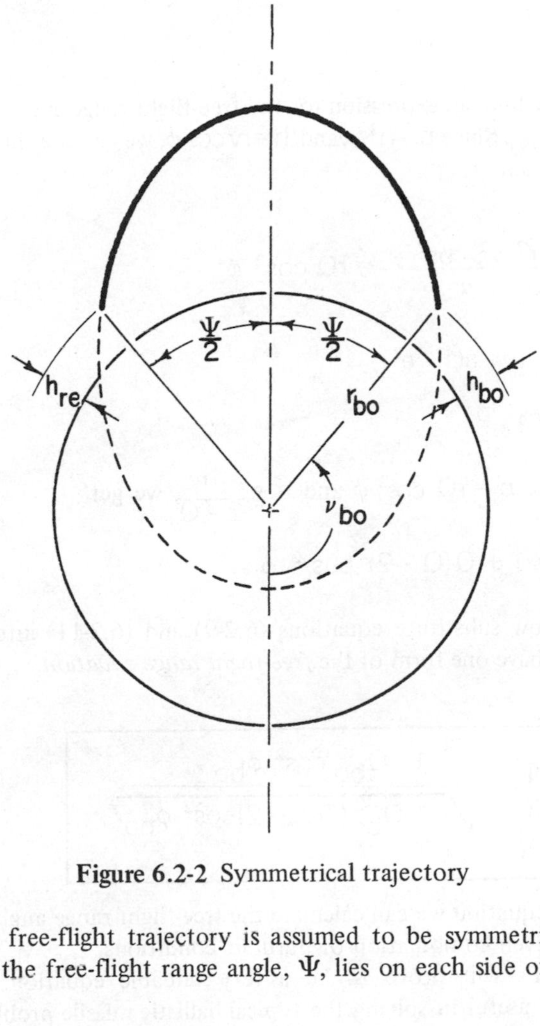

Figure 6.2-2 Symmetrical trajectory

Since the free-flight trajectory is assumed to be symmetrical ($h_{bo} = h_{re}$), half the free-flight range angle, Ψ, lies on each side of the major axis, and

$$\cos\frac{\Psi}{2} = -\cos\nu_{bo}\,. \tag{6.2-7}$$

Equation (6.2-6) above can, therefore, be written as

$$\cos\frac{\Psi}{2}=\frac{r_{bo}-p}{er_{bo}}\,. \tag{6.2-8}$$

We now have an expression for the free-flight range angle in terms of p, e, and r_{bo}. Since $p=h^2/\mu$ and $h=rv\cos\phi$, we can use the definition of Q to obtain

$$p=\frac{r^2\,v^2\,\cos^2\phi}{\mu}=rQ\cos^2\phi\,. \tag{6.2-9}$$

Now, since $p=a(1-e^2)$,

$$e^2=1-\frac{p}{a}\,. \tag{6.2-10}$$

Substituting $p=rQ\cos^2\phi$ and $a=\frac{r}{2-Q}$, we get

$$e^2=1+Q(Q-2)\cos^2\phi \tag{6.2-11}$$

If we now substitute equations (6.2-9) and (6.2-11) into equation (6.2-8) we have one form of the *free-flight range equation:*

$$\boxed{\cos\frac{\Psi}{2}=\frac{1-Q_{bo}\cos^2\phi_{bo}}{\sqrt{1+Q_{bo}(Q_{bo}-2)\cos^2\phi_{bo}}}\,.} \tag{6.2-12}$$

From this equation we can calculate the free-flight range angle resulting from any given combination of burnout conditions, r_{bo}, v_{bo} and ϕ_{bo}.

While this will prove to be a very valuable equation, it is not particularly useful in solving the typical ballistic missile problem which can be stated thus: Given a particular launch point and target, the total range angle, Λ, can be calculated as we shall see later in this chapter. If we know how far the missile will travel during powered flight and re-entry, the required free-flight range angle, Ψ, also becomes known. If we now specify r_{bo} and v_{bo} for the missile, what should the flight-path angle, ϕ_{bo}, be in order that the missile will hit the target?

In other words, it would be nice to have an equation for ϕ_{bo} in terms of r_{bo}, v_{bo} and Ψ.

We could, with a little algebra, choose the straightforward way of solving the range equation for $\cos \phi_{bo}$. Instead, we will derive an expression for ϕ_{bo} geometrically because it demonstrates some rather interesting geometrical properties of the ellipse.

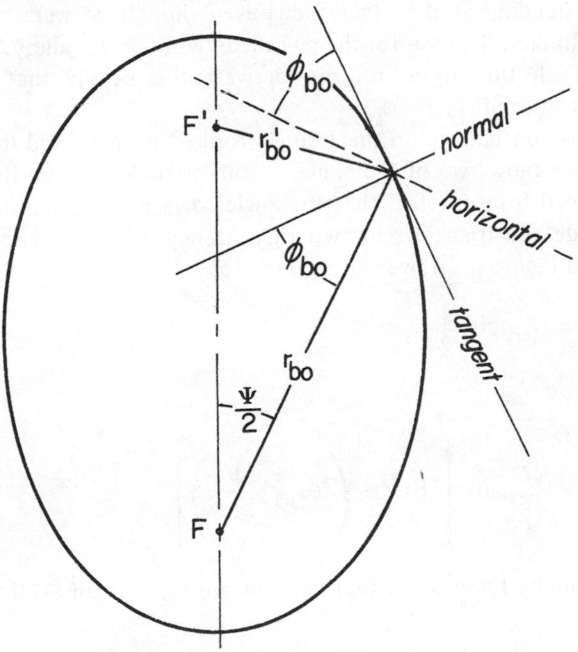

Figure 6.2-3 Ellipse geometry

6.2.4 The Flight-Path Angle Equation. In Figure 6.2-3 we have drawn the local horizontal at the burnout point and also the tangent and normal at the burnout point. The line from the burnout point to the secondary focus, F′, is called r'_{bo}.

The angle between the local horizontal and the tangent (direction of v_{bo}) is the flight-path angle, ϕ_{bo}. Since r_{bo} is perpendicular to the local horizontal, and the normal is perpendicular to the tangent, the angle between r_{bo} and the normal is also ϕ_{bo}.

Now, it can be proven (although we won't do it) that the angle between r_{bo} and r'_{bo} is bisected by the normal. This fact gives rise to many interesting applications for the ellipse. It means, for example, that, if the ellipse represented the surface of a mirror, light emanating from one focus would be reflected to the other focus since the angle of reflection equals the angle of incidence. If the ceiling of a room were made in the shape of an ellipsoid, a person standing at a particular point in the room corresponding to one focus could be heard clearly by a person standing at the other focus even though he were whispering. This is, in fact, the basis for the so-called "whispering gallery."

What all this means for our derivation is simply that the angle between r_{bo} and r'_{bo} is $2\phi_{bo}$.

Let us concentrate on the triangle formed by F, F' and the burnout point. We know two of the angles in this triangle and the third can be determined from the fact that the angles of a triangle sum to 180°. If we divide the triangle into two right triangles by the dashed line, d, shown in Figure 6.2-4, we can express d as

$$d = r_{bo} \sin \frac{\Psi}{2} \tag{6.2-13}$$

and also as

$$d = r'_{bo} \sin \left[180^o - \left(2\phi_{bo} + \frac{\Psi}{2}\right)\right] \tag{6.2-14}$$

Combining these two equations and noting that $\sin(180^o - x) = \sin x$, we get

$$\sin\left(2\phi_{bo} + \frac{\Psi}{2}\right) = \frac{r_{bo}}{r'_{bo}} \sin \frac{\Psi}{2}. \tag{6.2-15}$$

Since $r_{bo} = a(2 - Q_{bo})$ from equation (6.2-3) and $r'_{bo} + r_{bo} = 2a$,

$$\boxed{\sin\left(2\phi_{bo} + \frac{\Psi}{2}\right) = \frac{2 - Q_{bo}}{Q_{bo}} \sin \frac{\Psi}{2}.} \tag{6.2-16}$$

This is called the *flight-path angle equation* and it points out some interesting and important facts about ballistic missile trajectories.

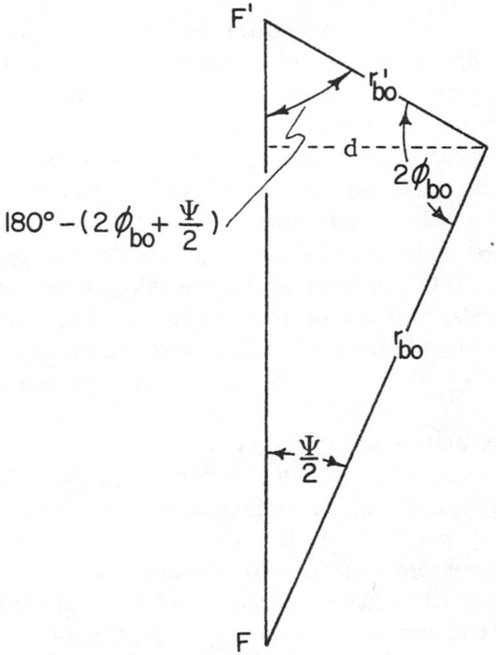

Figure 6.2-4 Ellipse geometry

Suppose we want a missile to travel a free-flight range of 90° and it has $Q_{bo} = .9$. Substituting these values into equation (6.2-16) gives us

$$\sin(2\phi_{bo} + 45^{\circ}) = \frac{2 - .9}{.9} \sin 45^{\circ} = .866$$

But there are *two angles* whose sine equals .866, so

$$2\phi_{bo} + 45^{\circ} = 60^{\circ} \text{ and } 120^{\circ}$$

and

$$\phi_{bo} = 7.5^{\circ} \text{ and } 37.5^{\circ}$$

There are two trajectories to the target which result from the same values of r_{bo} and v_{bo}. The trajectory corresponding to the larger value of flight-path angle is called the *high trajectory*; the trajectory associated with the smaller flight-path angle is the *low trajectory*.

The fact that there are two trajectories to the target should not surprise you since even very short-range ballistic trajectories exhibit this property. A familiar illustration of this result is the behavior of water discharged from a garden hose. With constant water pressure and nozzle setting, the speed of the water leaving the nozzle is fixed. If a target well within the maximum range of the hose is selected, the target can be hit by a flat or lofted trajectory.

The nature of the high and low trajectory depends primarily on the value of Q_{bo}. If Q_{bo} is less than 1 there will be a limit to how large Ψ may be in order that the value of the right side of equation (6.2-16) does not exceed 1. This implies that *there is a maximum range for a missile with* Q_{bo} *less than 1*. This maximum range will always be less than 180° for Q_{bo} less than 1. Provided that Ψ is attainable, there will be *both a high and a low trajectory to the target.*

If Q_{bo} is exactly 1, one of the trajectories to the target will be the circular orbit connecting the burnout and re-entry points. This would not be a very practical missile trajectory, but it does represent the borderline case where ranges of 180° and more are just attainable.

If Q_{bo} is greater than 1, equation (6.2-16) will always yield one positive and one negative value for ϕ_{bo}, regardless of range. A negative ϕ_{bo} is not practical since the trajectory would penetrate the earth, so *only the high trajectory can be realized for* Q_{bo} *greater than 1.*

The real significance of Q_{bo} greater than 1 is that ranges in excess of 180° are possible. An illustration of such a trajectory would be a missile directed at the North American continent from Asia via the south pole. While such a trajectory would avoid detection by our northern radar "fences," it would be costly in terms of payload delivered and accuracy attainable. Nevertheless, the shock value of such a surprise attack in terms of what it might do towards creating chaos among our defensive forces should not be overlooked by military planners.

Since both the high and low trajectories result from the same r_{bo} and v_{bo}, they both have the same energy. Because $a = -\mu/2\mathcal{E}$, the major axis of the high and low trajectories are the same length.

Table 6.2-1 shows which trajectories are possible for various combinations of Q_{bo} and Ψ. Figure 6.2-5 should be helpful in visualizing each case.

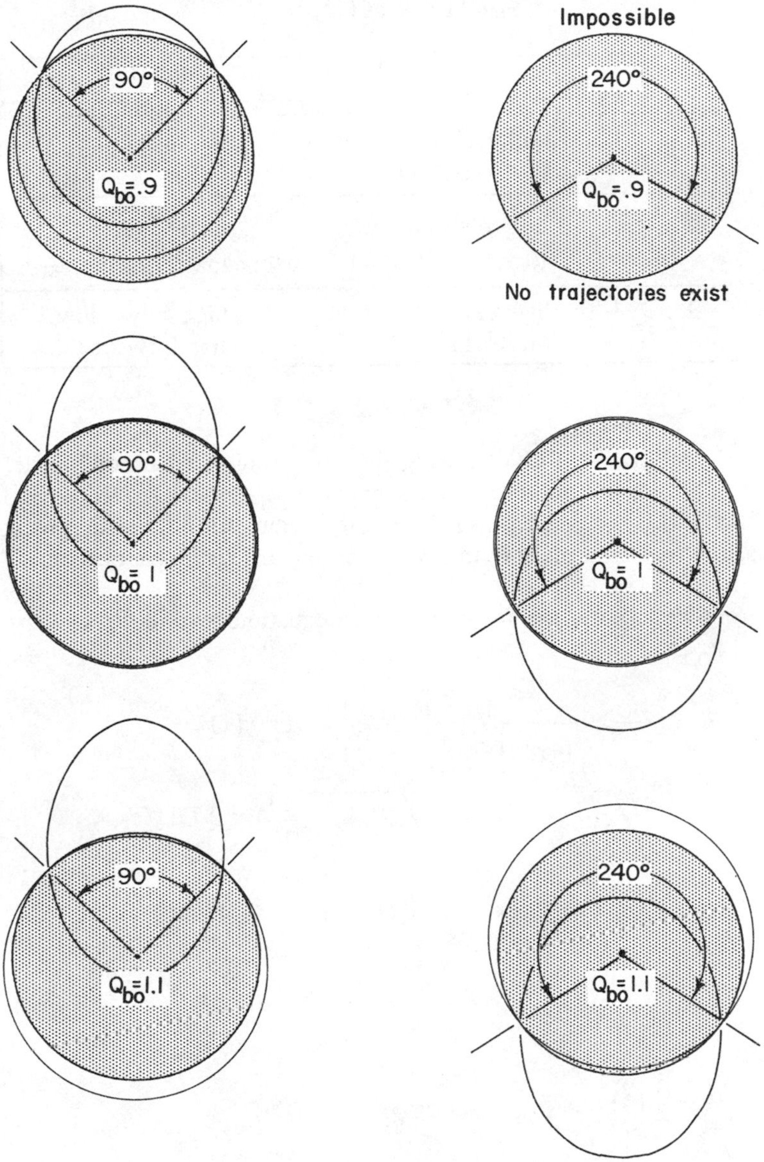

Figure 6.2-5 Example ballistic missile trajectories

Significance of Q_{bo}

	$\Psi < 180^\circ$	$\Psi > 180^\circ$
$Q_{bo} < 1$	both high and low if $\Psi <$ max. range	impossible
$Q_{bo} = 1$	both high and low ($\phi_{bo} = 0^\circ$ for low)	high has $\phi_{bo} = 0^\circ$ one low traj. skims earth
$Q_{bo} > 1$	high only – low traj. hits earth	high only – low traj. hits earth

Table 6.2-1

EXAMPLE PROBLEM. During the test firing of a ballistic missile, the following measurements were made: h_{bo} = 1/5 DU, v_{bo} = 2/3 DU/TU, h_{apogee} = 0.5 DU. Assuming a symmetrical trajectory, what was the free-flight range of the missile during this test in nautical miles?

Before we can use the free-flight range equation to find Ψ, we must find ϕ_{bo} and Q_{bo}

$$\mathcal{E} = \frac{v_{bo}^2}{2} - \frac{\mu}{r_{bo}} = \frac{4}{18} - \frac{1}{1.2} = -\frac{11}{18} \text{ DU}^2/\text{TU}^2$$

$$v_a = \sqrt{2\left(\frac{\mu}{r_a} + \mathcal{E}\right)} = \sqrt{2\left(\frac{1}{1.5} - \frac{11}{18}\right)} = \frac{1}{3} \text{ DU/TU}$$

$$h = r_a v_a = 1.5\left(\frac{1}{3}\right) = \frac{1}{2} \text{ DU}^2/\text{TU}$$

$$h = r_{bo} v_{bo} \cos\phi_{bo}$$

$$\frac{1}{2} = \frac{6}{5}\left(\frac{2}{3}\right)\cos\phi_{bo}, \quad \therefore \cos\phi_{bo} = 0.625$$

$$Q_{bo} = \frac{v_{bo}^2\, r_{bo}}{\mu} = 0.5333$$

Using equation (6.2-12), $\cos \frac{\Psi}{2} = + \frac{19}{20}$.

$$\Psi = 36^{o}24' \text{ or } 36.4^{o}$$

$$R_{ff} = (36.4 \text{Deg.})(60 \frac{\text{nmi}}{\text{Deg}}) = \underline{\underline{2{,}184 \text{ n mi}}}$$

EXAMPLE PROBLEM. A missile's coordinates at burnout are: 30°N, 60°E. Re-entry is planned for 30°S, 60°W. Burnout velocity and altitude are 1.0817 DU/TU and .025 DU respectively. Ψ is less than 180°.

What must the flight-path angle be at burnout?

Before we can use the flight-path angle equation to find ϕ_{bo}, we must find Q_{bo} and Ψ.

$$Q_{bo} = \frac{(1.0817)^2 (1.025)}{1} = 1.2$$

From spherical trigonometry,

$$\cos \Psi = \cos 60^{o} \cos 120^{o} + \sin 60^{o} \sin 120^{o} \cos 120^{o} = -.625$$

$$\therefore \Psi = 128^{o}41'$$

From the flight-path angle equation,

$$\sin(2\phi_{bo} + \frac{128^{o}41'}{2}) = \frac{2-1.2}{1.2} \sin(\frac{128^{o}41'}{2}) = .6$$

$$\therefore 2\phi_{bo} + 64^{o}20.5' = 143^{o}04'$$

or $\phi_{bo} = \underline{\underline{39.36^{o}}}$

6.2.5 The Maximum Range Trajectory. Suppose we plot the free-flight range angle, Ψ, versus the flight-path angle, ϕ_{bo}, for a fixed value of Q_{bo} less than 1. We get a curve like that shown in Figure 6.2-6. As the flight-path angle is varied from 0° to 90° the range first increases then reaches a maximum and decreases to zero again. Notice

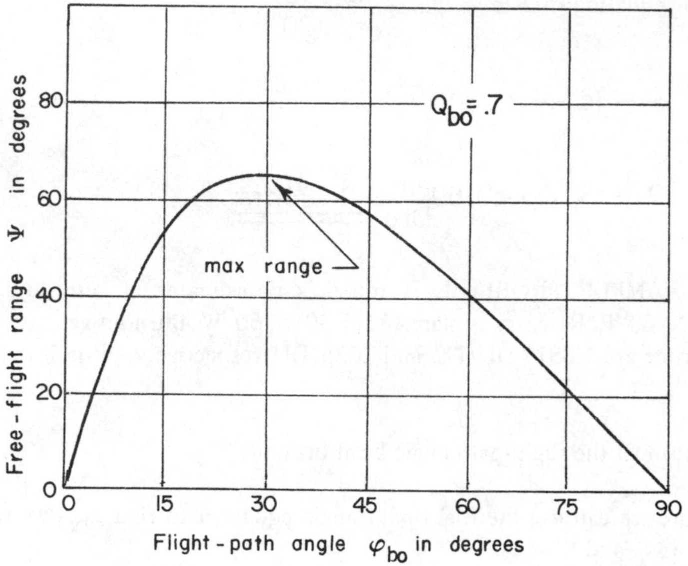

Figure 6.2-6 Range versus ϕ_{bo}

that for every range except the maximum there are two values of ϕ_{bo} corresponding to a high and a low trajectory. *At maximum range there is only one path to the target.*

There are at least two ways that we could derive expressions for the maximum range condition. One way is to derive an expression for $\partial\Psi/\partial\phi$ and set it equal to zero. A simpler method is to see under what conditions the flight-path angle equation yields a single solution.

If the right side of equation (6.2-16) equals exactly 1, we get only a single answer for ϕ_{bo}. This must, then, be the maximum range condition.

$$\sin\left(2\phi_{bo} + \frac{\Psi}{2}\right) = \frac{2 - Q_{bo}}{Q_{bo}} \sin\frac{\Psi}{2} = 1 \tag{6.2-17}$$

from which $2\phi_{bo} + \dfrac{\Psi}{2} = 90^\circ$

and

$$\boxed{\phi_{bo} = \frac{1}{4}(180^\circ - \Psi)} \tag{6.2-18}$$

for *maximum range conditions only.*

We can easily find the maximum range angle attainable with a given Q_{bo}. From equation (6.2-17),

$$\sin \frac{\Psi}{2} = \frac{Q_{bo}}{2 - Q_{bo}} \tag{6.2-19}$$

for *maximum range conditions.*

If we solve this equation for Q_{bo}, we get

$$Q_{bo} = \frac{2 \sin(\Psi/2)}{1 + \sin(\Psi/2)} \tag{6.2-20}$$

for *maximum range conditions.* This latter form of the equation is useful for determining the lowest value of Q_{bo} that will attain a given range angle.

6.2.6 Time of Free-Flight. The time-of-flight methods developed in Chapter 4 are applicable to the free-flight portion of a ballistic missile trajectory, but, due to the symmetry of the case where $h_{re} = h_{bo}$, the equations are considerably simplified. From the symmetry of Figure 6.2-7 you can see that the time-of-flight from burnout to re-entry is just twice the time-of-flight from burnout (point 1) to apogee (point 2).

By inspection, the eccentric anomaly of point 2 is π radians or 180°. The value of E_1 can be computed from equation (4.2-8), noting that $\nu_1 = 180^\circ - \Psi/2$

$$\cos E_1 = \frac{e - \cos \frac{\Psi}{2}}{1 - e \cos \frac{\Psi}{2}} \tag{6.2-21}$$

If we now substitute into equation (4.2-9) on page 186 we get the time of free-flight

$$t_{ff} = 2 \sqrt{\frac{a^3}{\mu}} \, (\pi - E_1 + e \sin E_1) \tag{6.2-22}$$

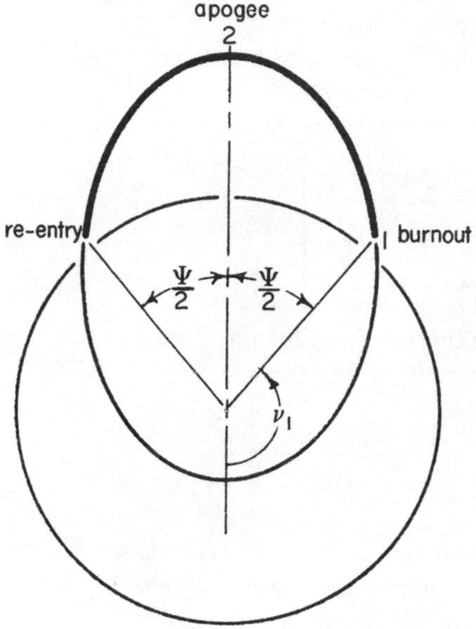

Figure 6.2-7 Time of free-flight

The semi-major axis, a, and the eccentricity, e, can be obtained from equations (6.2-3) and (6.2-11).

Figure 6.2-9 is an excellent chart for making rapid time-of-flight calculations for the ballistic missile. In fact, since five variables have been plotted on the figure, most ballistic missile problems can be solved completely using just this chart.

The free-flight time is read from the chart as the ratio $t_{ff}/\mathbb{P}_{CS}$, where $\mathbb{P}_{CS}$ is the period of a fictitious circular satellite orbiting at the burnout altitude. Values for $\mathbb{P}_{CS}$ may be calculated from

$$\mathbb{P}_{CS} = 2\pi\sqrt{\frac{r_{bo}^3}{\mu}} \qquad (6.2\text{-}23)$$

or they may be read directly from Figure 6.2-8.

EXAMPLE PROBLEM. A ballistic missile was observed to have a burnout speed and altitude of 24,300 ft/sec and 258 nm respectively. What must be the maximum free-flight range capability of this missile?

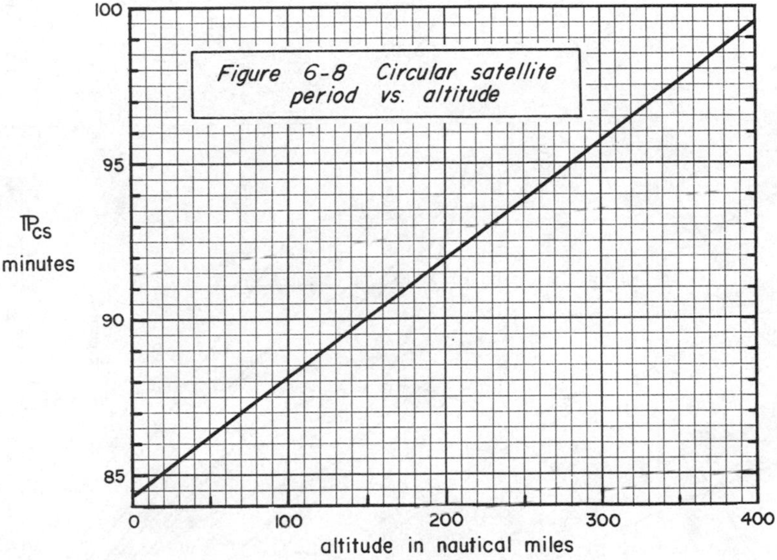

Figure 6.2-8 Circular satellite period vs altitude

In canonical units

$$Q_{bo} = \left(\frac{2.43\times10^4}{2.59\times10^4}\right)^2 \left(1+\frac{258}{3444}\right)$$

$$= (0.884)(1.075) = 0.95$$

From Figure 6.2-9 it is rapidly found that

$$\Psi_{max} = 129^\circ$$

and $R_{ff} = (1)\left(\frac{129}{57.3}\right) = \underline{\underline{2.25\text{ DU}}} = \underline{\underline{7{,}750\text{ nm}}}$

EXAMPLE PROBLEM. It is desired to maximize the payload of a new ballistic missile for a free-flight range of 8,000 nm. The design burnout altitude has been fixed at 344 nm. What should be the design burnout speed?

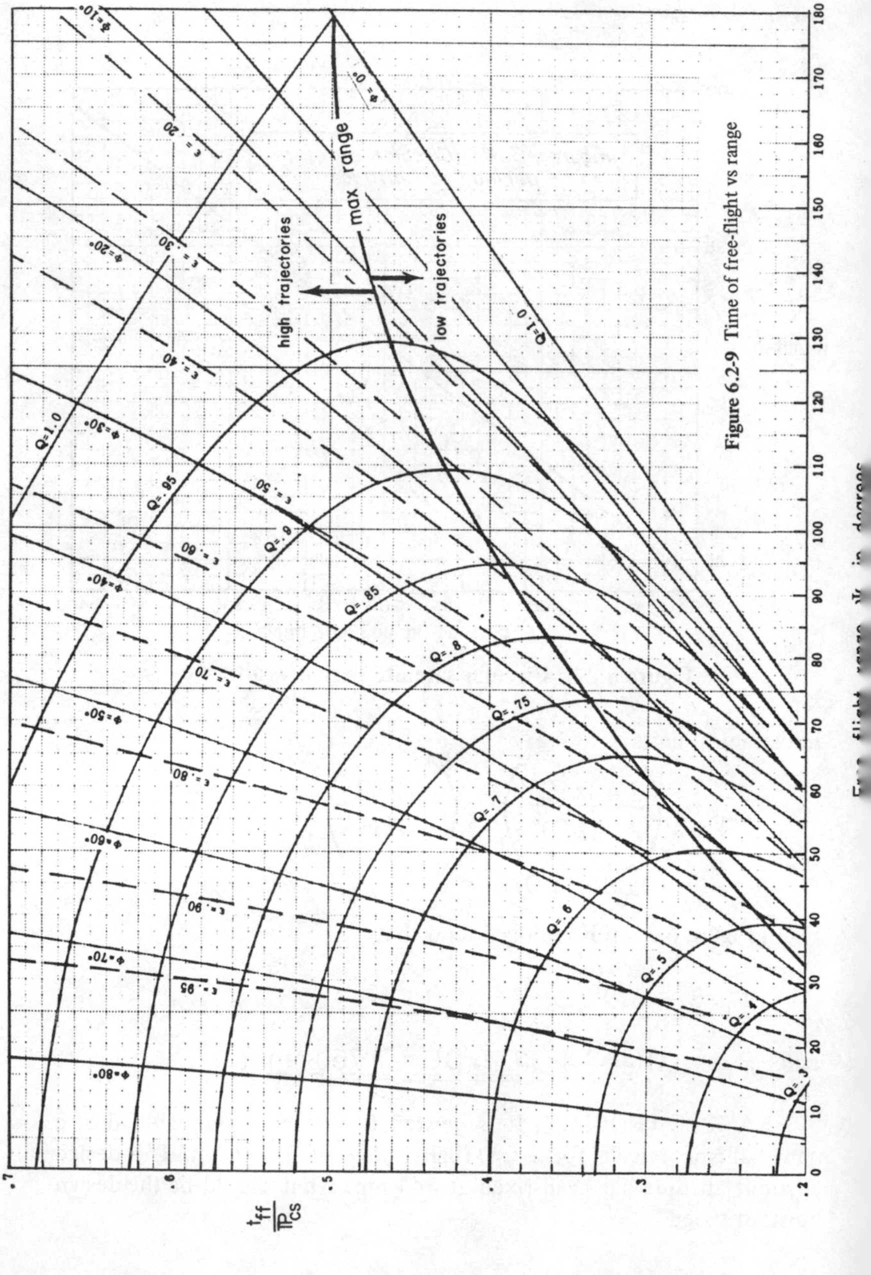

Figure 6.2-9 Time of free-flight vs range

For a given amount of propellant, range may be sacrificed to increase payload and vice-versa. For a fixed burnout altitude, payload may be maximized by minimizing the burnout speed (minimum Q_{bo}).

$$\Psi_{max} = \frac{8{,}000\,\text{nm}}{3{,}444\,\text{nm}} = 2.32 \text{ rads} = 133.3^\circ$$

From equation (6.2-20)

$$Q_{bo\,min} = \frac{2 \sin 66.7^\circ}{1+\sin 66.7^\circ} = 0.957$$

From equation (6.2-1)

$$v_{bo} = \sqrt{\frac{.957}{1.1}} = 0.933 \text{ DU/TU} = \underline{\underline{24{,}200 \text{ ft/sec}}}$$

6.3 EFFECT OF LAUNCHING ERRORS ON RANGE

Variations in the speed, position, and launch direction of the missile at thrust cutoff will produce errors at the impact point. These errors are of two types–errors in the intended plane which cause either a long or a short hit, and out-of-plane errors which cause the missile to hit to the right or left of the target. For brevity, we will refer to errors in the intended plane as "down-range" errors, and out-of-plane errors as "cross-range" errors.

There are two possible sources of cross-range error and these will be treated first.

6.3.1 Effect of a Lateral Displacement of the Burnout Point. If the thrust cutoff point is displaced by an amount, ΔX, perpendicular to the intended plane of the trajectory and all other conditions are nominal, the cross-range error, ΔC, at impact can be determined from spherical trigonometry. In Figure 6.3-1 we show the ground traces of the intended and actual trajectories. For purposes of this example, suppose the intended burnout point is on the equator and the launch azimuth is due north along a meridian toward the intended target at A. The actual burnout point occurs at a point on the equator a distance, ΔX to the east but with the correct launch azimuth of due north. As a result the missile flies up the wrong meridian, impacting at B.

The arc length ΔC represents the cross-range error. It is customary in spherical trigonometry to measure arc length in terms of the angle subtended at the center of the sphere so that both ΔX and ΔC may be thought of as angles.

Applying the law of cosines for spherical trigonometry to triangle OAB in Figure 6.3-1 and noting that the small angle at O is the same as ΔX, we get

$$\cos \Delta C = \sin^2 \Psi + \cos^2 \Psi \cos \Delta X. \tag{6.3-1}$$

Since both ΔX and ΔC will be very small angles, we can use the small angle approximation, $\cos x \approx 1 - \frac{x^2}{2}$ to simplify equation (6.3-1) to

$$\boxed{\Delta C \approx \Delta X \cos \Psi.} \tag{6.3-2}$$

Both ΔX and ΔC are assumed to be expressed as angles in this equation. If they are in radians you can convert them to arc length by multiplying by the radius of the Earth; if they are in degrees, you can use the fact that a 60 nm arc on the surface of the Earth subtends an angle of 1° at the center.

Equation (6.3-2) tells us that cross-range error is zero for a free-flight range of 90° (5,400 nm) regardless of how far the burnout point is displaced out of the intended plane. In Figure 6.3-1, for example, if the intended target had been the north pole, the actual burnout point could occur anywhere on the equator and we would hit the target so long as

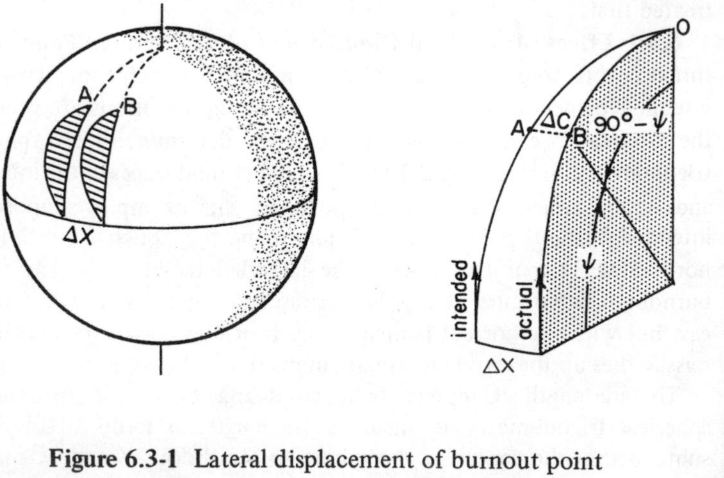

Figure 6.3-1 Lateral displacement of burnout point

launch azimuth and free-flight range, Ψ, are as planned. Since most of our ICBM's are targeted for ranges of approximately 90°, this particular source of cross-range error is not as significant as an error in launch azimuth as we shall see in the next section.

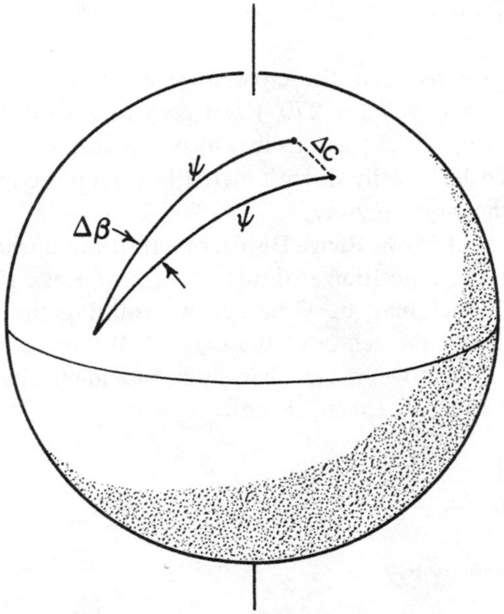

Figure 6.3-2 Azimuth error

6.3.2 Cross-Range Error Due to Incorrect Launch Azimuth. If the actual launch azimuth differs from the intended launch azimuth by an amount, $\Delta\beta$, a cross-range error, ΔC, will result. Figure 6.3-2 illustrates the geometry of an azimuth error. The ground trace of the actual and intended trajectory are shown. Since all launch conditions are assumed to be nominal except launch azimuth, the free-flight range of both the actual and the intended trajectories is Ψ. The third side of the spherical triangle shown in Figure 6.3-2 is ΔC, the cross-range error. As before, we will consider ΔC to be expressed in terms of the angle it subtends at the center of the Earth.

From the law of cosines for spherical triangles we get

$$\cos \Delta C = \cos^2 \Psi + \sin^2 \Psi \cos \Delta\beta \tag{6.3-3}$$

If we assume that both $\Delta\beta$ and ΔC will be very small angles we can use the approximation, $\cos x \approx 1 - \frac{x^2}{2}$, to simplify equation (6.3-3) to

$$\boxed{\Delta C \approx \Delta\beta \sin \Psi} \tag{6.3-4}$$

This time we see that the cross-range error is a maximum for a free-flight range of 90° (or 270°) and goes to zero if Ψ is 180°. In other words, if you have a missile which will travel exactly half way around the Earth, it really doesn't matter in what direction you launch it; it will hit the target anyway.

6.3.3 Effect of Down-Range Displacement of the Burnout Point. An error in down-range position at thrust cutoff produces an equal error at impact. The effect may be visualized by rotating the trajectory in Figure 6.2-1 about the center of the Earth. If the actual burnout point is 1 nm farther down-range than was intended, the missile will overshoot the target by exactly 1 nm.

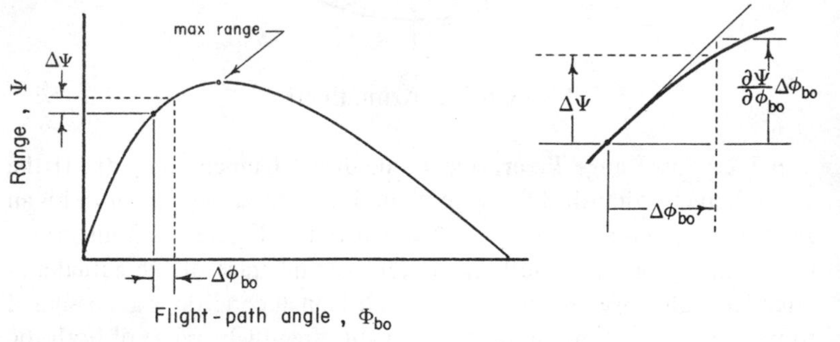

Figure 6.3-3 Effect of flight-path angle errors on range

6.3.4 Errors in Burnout Flight-Path Angle, ϕ_{bo}. In Figure 6.3-3 we show a typical plot of free-flight range versus flight-path angle for a fixed value of r_{bo} and v_{bo}. The intended flight-path angle and intended range, Ψ, are shown by a solid line in the figure. If the actual ϕ_{bo} differs from the intended value by an amount $\Delta\phi_{bo}$, the actual range will be different by an amount $\Delta\Psi$. This $\Delta\Psi$ will represent a down-range error causing the missile to undershoot or overshoot the target.

We could get an approximate value for $\Delta\Psi$ if we knew the slope of the curve (which is $\dfrac{\partial\Psi}{\partial\phi_{bo}}$) at the point corresponding to the intended trajectory. The diagram at the right of Figure 6.3-3 illustrates the fact that

$$\Delta\Psi \approx \frac{\partial\Psi}{\partial\phi_{bo}}\,\Delta\phi_{bo} \tag{6.3-5}$$

which is a good approximation for very small values of $\Delta\phi_{bo}$.

The expression for $\dfrac{\partial\Psi}{\partial\phi_{bo}}$ may be obtained by implicit partial differentiation of the free-flight range equation. But first we will derive an alternate form of the free-flight range equation.

The free-flight range equation was derived as equation (6.2-12) of this chapter. If we call the numerator of this expression α and the denominator β, then

$$\cos\frac{\Psi}{2} = \frac{\alpha}{\beta} \tag{(6.3-6}$$

and $$\cot\frac{\Psi}{2} = \frac{\alpha}{\sqrt{\beta^2 - \alpha^2}}\ . \tag{6.3-7}$$

Substituting for α and β we get

$$\cot\frac{\Psi}{2} = \frac{1 - Q_{bo}\cos^2\phi_{bo}}{Q_{bo}\cos\phi_{bo}\sqrt{1-\cos^2\phi}} \tag{6.3-8}$$

Since $\sqrt{1-\cos^2\phi_{bo}} = \sin\phi_{bo}$,

$$\cot\frac{\Psi}{2} = \frac{1 - Q_{bo}\cos^2\phi_{bo}}{Q_{bo}\cos\phi_{bo}\sin\phi_{bo}} \tag{6.3-9}$$

But since $\cos x \sin x = \frac{1}{2}\sin 2x$, we can simplify further to obtain

$$\cot\frac{\Psi}{2} = \frac{2}{Q_{bo}}\csc 2\phi_{bo} - \cot\phi_{bo}\ . \tag{6.3-10}$$

We now have another form of the free-flight range equation which is much simpler to differentiate. Let us first express equation (6.3-10) in terms of r_{bo}, v_{bo} and ϕ_{bo}.

$$\cot\frac{\Psi}{2} = \frac{2\mu}{v_{bo}^2 r_{bo}} \csc 2\phi_{bo} - \cot\phi_{bo} . \qquad (6.3\text{-}11)$$

We now proceed to differentiate (6.3-11) implicitly with respect to ϕ_{bo}, considering r_{bo} and v_{bo} as constants.

$$-\frac{1}{2}\csc^2\frac{\Psi}{2}\frac{\partial\Psi}{\partial\phi_{bo}} = \qquad (6.3\text{-}12)$$

$$\frac{2\mu}{v_{bo}^2 r_{bo}}(-2\cot 2\phi_{bo}\csc 2\phi_{bo}) + \csc^2\phi_{bo} .$$

Substituting from equation (6.3-11) we get

$$-\frac{1}{2}\csc^2\frac{\Psi}{2}\frac{\partial\Psi}{\partial\phi_{bo}} = -2\cot 2\phi_{bo}\left(\cot\frac{\Psi}{2} + \cot\phi_{bo}\right)$$

$$+ \csc^2\phi_{bo}$$

$$= \quad 2\left(1 - \cot 2\phi_{bo}\cot\frac{\Psi}{2}\right) .$$

Solving for $\dfrac{\partial\Psi}{\partial\phi_{bo}}$,

$$\frac{\partial\Psi}{\partial\phi_{bo}} = 4\left(\cot 2\phi_{bo}\sin\frac{\Psi}{2}\cos\frac{\Psi}{2} - \sin^2\frac{\Psi}{2}\right)$$

$$= 4\left(\frac{1}{2}\cot 2\phi_{bo}\sin\Psi - \sin^2\frac{\Psi}{2}\right)$$

$$= 2(\sin\Psi\cot 2\phi_{bo} + \cos\Psi - 1)$$

$$= 2\frac{\sin\Psi\cos 2\phi_{bo} + \cos\Psi\sin 2\phi_{bo}}{\sin 2\phi_{bo}} - 2$$

which finally reduces to

$$\frac{\partial \Psi}{\partial \phi_{bo}} = \frac{2 \sin(\Psi + 2\phi_{bo})}{\sin 2\phi_{bo}} - 2 \tag{6.3-13}$$

This partial derivative, when used in the manner described above, is called an *influence coefficient* since it influences the size of the range error resulting from a particular burnout error.

While we need equation (6.3-13) to evaluate the magnitude of the flight-path angle influence coefficient, the general effect of errors in flight-path angle at burnout are apparent from Figure 6.3-3.

The maximum range condition separates the low trajectories from the high trajectories. For all low trajectories the slope of the curve $\frac{\partial \Psi}{\partial \phi_{bo}}$ is positive which means that too high a flight-path angle (a positive $\Delta\phi_{bo}$) will cause a positive $\Delta\Psi$ (overshoot); a flight-path angle which is lower than intended (a negative $\Delta\phi_{bo}$) will cause a negative $\Delta\Psi$ (undershoot).

Just the opposite effect occurs for all high trajectories where $\frac{\partial \Psi}{\partial \phi_{bo}}$ is always negative. Too high a ϕ_{bo} on the high trajectory will cause the missile to fall short and too low a ϕ_{bo} will cause an overshoot. If this seems strange, remember that water from a garden hose behaves in exactly the same way.

Figure 6.3-3 also reveals that the high trajectory is less sensitive than the low trajectory to flight-path angle errors. For a typical ICBM fired over a 5,000 nm range, an error of 1 minute (1/60 of a degree) causes a miss of about 1 nm on the high trajectory and nearly 3 nm on the low trajectory.

The maximum range trajectory is the least sensitive to flight-path angle errors. Since $\frac{\partial \Psi}{\partial \phi_{bo}} = 0$ for the maximum range case, equation (6.3-5) tells us that $\Delta\Psi$ will be approximately zero for small values of $\Delta\phi_{bo}$. In fact, the actual range error on a 3,600 nm ICBM flight due to a 1 minute error in ϕ_{bo} is only 4 feet!

6.3.5 Down-Range Errors Caused by Incorrect Burnout Height. We can use exactly the same approach to errors in burnout height as we used in the previous section. A plot of range versus burnout radius, however, is not particularly interesting since it reveals just what we might suspect–if burnout occurs higher than intended, the missile overshoots; if burnout occurs too low, the missile falls short of the target in every case.

Following the arguments in the last section we can say that

$$\Delta\Psi \approx \frac{\partial \Psi}{\partial r_{bo}} \Delta r_{bo} \qquad (6.3\text{-}14)$$

for small values of Δr_{bo}.

The partial derivative with respect to r_{bo} is much simpler. Again, differentiating the range equation (6.3-11) implicitly,

$$-\frac{1}{2} \csc^2 \frac{\Psi}{2} \frac{\partial \Psi}{\partial r_{bo}} = \frac{-2\mu}{v_{bo}^2 r_{bo}^2} \csc 2\phi_{bo} \,. \qquad (6.3\text{-}15)$$

Solving for $\dfrac{\partial \Psi}{\partial r_{bo}}$ we get

$$\frac{\partial \Psi}{\partial r_{bo}} = \frac{4\mu}{v_{bo}^2 r_{bo}^2} \frac{\sin^2 \frac{\Psi}{2}}{\sin 2\phi_{bo}} \qquad (6.3\text{-}16)$$

A burnout error of 1 nm in height on a 5,000 nm range trajectory will cause a miss of about 2 nm on the high trajectory and about 5 nm on the low trajectory.

6.3.6 Down-Range Errors Caused by Incorrect Speed at Burnout. Speed at burnout affects range in just the way we would expect–too fast and the missile overshoots; too slow and the missile falls short. The magnitude of the error is

$$\Delta\Psi \approx \frac{\partial \Psi}{\partial v_{bo}} \Delta v_{bo} \qquad (6.3\text{-}17)$$

where $\partial\Psi/\partial v_{bo}$ is given by implicit differentiation of equation (6.3-11) as

$$\boxed{\frac{\partial \Psi}{\partial v_{bo}} = \frac{8\mu}{v_{bo}^3 r_{bo}} \frac{\sin^2 \frac{\Psi}{2}}{\sin 2\phi_{bo}}} \ . \qquad (6.3\text{-}18)$$

A rough rule-of-thumb is that an error of 1 ft/sec will cause a miss of about 1 nm over typical ICBM range.

An analysis of equation (6.3-18) would reveal that, like the other two influence coefficients, $\partial\Psi/\partial v_{bo}$ is larger on the low trajectory than on the high. Most ICBM's are programmed for the high trajectory for the simple reason that is revealed here—the guidance requirements are less stringent and the accuracy is better.

EXAMPLE PROBLEM. A ballistic missile has the following nominal burnout conditions:

$$v_{bo} = .905 \text{ DU/TU}, \ r_{bo} = 1.1 \text{ DU}, \ \phi_{bo} = 30^\circ$$

The following errors exist at burnout:

$$\Delta v_{bo} = -5 \times 10^{-5} \text{ DU/TU}, \ \Delta r_{bo} = 5 \times 10^{-4} \text{ DU}$$

$$\Delta\phi_{bo} = -10^{-4} \text{ radians.}$$

How far will the missile miss the target? What will be the direction of the miss relative to the trajectory plane?

We will find total down-range error. There is no cross-range error. We will need Q_{bo} (.9 for this case) and Ψ ($\Psi \approx 100^\circ$ from Figure 6.2-9). Total down-range error:

$$\Delta\Psi_{TOT} = \frac{\partial\Psi}{\partial r_{bo}}\Delta r_{bo} + \frac{\partial\Psi}{\partial v_{bo}}\Delta v_{bo} + \frac{\partial\Psi}{\partial\phi_{bo}}\Delta\phi_{bo}$$

$$\frac{\partial\Psi}{\partial r_{bo}} = \frac{4(\sin 50^\circ)^2}{(.905)^2(1.1)^2 \sin 60^\circ} = 2.735 \frac{\text{rad}}{\text{DU}}$$

$$\frac{\partial\Psi}{\partial v_{bo}} = \frac{2r_{bo}}{v_{bo}}\left[\frac{\partial\Psi}{\partial r_{bo}}\right] = \frac{2(1.1)2.735}{.905} = 6.649 \frac{\text{rad}}{\text{DU/TU}}$$

$$\frac{\partial\Psi}{\partial\phi_{bo}} = \frac{2\sin 160^{\circ}}{\sin 60^{\circ}} - 2 = -1.21$$

$$\Delta\Psi_{TOT} = 2.735(5\times10^{-4}) + 6.649(-5\times10^{-5}) - 1.21\,(-10^{-4})$$

$$= 11.56\times10^{-4} \text{ rad}$$

$$\Delta R_{ff} = (11.56\times10^{-4})(3444) \approx \underline{\underline{4 \text{ n.mi. overshoot}}}$$

6.4 THE EFFECT OF EARTH ROTATION

Up to this point the Earth has been considered as nonrotating. In the two sections which follow we will see what effect earth rotation has on the problem of sending a ballistic missile from one fixed point on the Earth to another.

The Earth rotates once on its axis in 23 hrs 56 min producing a surface velocity at the equator of 1,524 ft/sec. The rotation is from west to east.

The free-flight portion of a ballistic missile trajectory is inertial in character. That is, it remains fixed in the XYZ inertial frame while the Earth runs under it. Relative to this inertial XYZ frame, both the launch point and the target are in motion.

We will compensate for motion of the launch site by recognizing that the "true velocity" of the missile at burnout is the velocity relative to the launch site (which could be measured by radar) plus the initial eastward velocity of the launch site due to earth rotation.

We will compensate for motion of the target by "leading it" slightly. That is, we will send our missile on a trajectory that passes through the point in space where the target will be when our missile arrives. If we know the time-of-flight of the missile, we can compute how much the Earth (and target) will turn in that time and can aim for a point the proper distance to the east of the target.

6.4.1 Compensating for the Initial Velocity of the Missile Due to Earth Rotation. We describe the speed and direction of a missile at burnout in terms of its speed, v, its flight-path angle, ϕ, and its azimuth angle, β. If measurements of these three quantities are made from the surface of the rotating Earth (by radar, for example), then all three

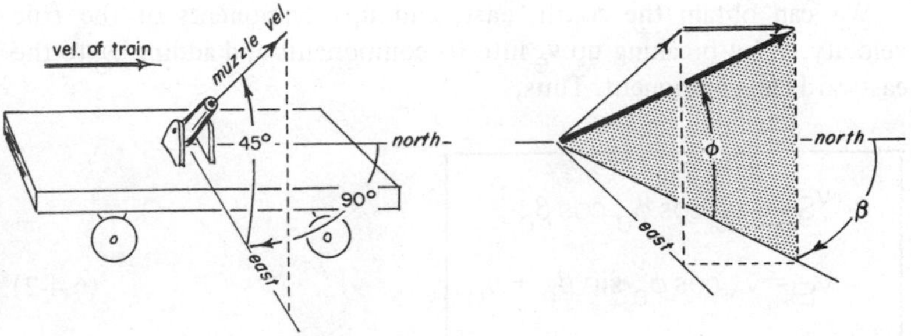

Figure 6.4-1 "True" velocity and direction

measurements are erroneous in that they do not indicate the "true" speed and direction of the missile. A simple example should clarify this point.

Suppose we set up a cannon on a train which is moving along a straight section of track in a northerly direction. If we point the cannon due east (perpendicular to the track) and upward at a 45° angle and then fire it, an observer on the moving train would say that the projectile had a flight-path angle, ϕ, of 45°, an azimuth, β, of 90° (east), and a speed of, say 1,000 ft/sec.

An observer at rest would not agree. He would correctly see that the "true" velocity of the projectile includes the velocity of the train relative to the "fixed frame." The vector diagram at the right of Figure 6.4-1 illustrates the true situation and shows that the speed is actually somewhat greater than 1,000 ft/sec, the flight-path angle slightly less than 45°, and the azimuth considerably less than 90°.

Like the observer on the moving train, we make our measurements in a moving reference frame. This frame is called the topocentric-horizon system. Hereafter, we will refer to measurements of burnout direction in this frame as ϕ_e and β_e.

Since a point on the equator has a speed of 1,524 ft/sec in the eastward direction, we can express the speed of any launch point on the surface of the earth as

$$v_o = 1524 \cos L_o \text{ (ft/sec).} \tag{6.4-1}$$

where L_o is the latitude of the launch site. The subscript "o" is used to indicate that this is the initial eastward velocity of the missile even while it is on the launch pad.

We can obtain the south, east, and up components of the true velocity, **v**, by breaking up $\mathbf{v}_e$ into its components and adding v_o to the eastward (E) component. Thus,

$$v_S = -v_e \cos\phi_e \cos\beta_e$$

$$v_E = v_e \cos\phi_e \sin\beta_e + v_o \qquad (6.4\text{-}2)$$

$$v_Z = v_e \sin\phi_e$$

The true speed, flight-path angle, and azimuth can then be found from

$$v = \sqrt{v_S^2 + v_E^2 + v_Z^2} \qquad (6.4\text{-}3)$$

$$\sin\phi = \frac{v_Z}{v} \qquad (6.4\text{-}4)$$

$$\tan\beta = \frac{-v_E}{v_S}. \qquad (6.4\text{-}5)$$

The inverse problem of determining v_e, ϕ_e and β_e if you are given v, ϕ and β can be handled in a similar manner; first, break up **v** into its components, then subtract v_o from the eastward component to obtain the components of $\mathbf{v}_e$. Once you have the components of $\mathbf{v}_e$, finding v_e, ϕ_e and β_e is easy.

One word of caution: you will have to determine in which quadrant the azimuth, β, lies. If you can draw even a crude sketch and visualize the geometry of the problem, this will not be difficult.

It is, of course, **v** at burnout which determines the missile's trajectory. The rocket booster only has to add $\mathbf{v}_e$ to the initial velocity $\mathbf{v}_o$. If the desired launch velocity is eastward the rocket will not have to provide as much speed as it would for a westward launch.

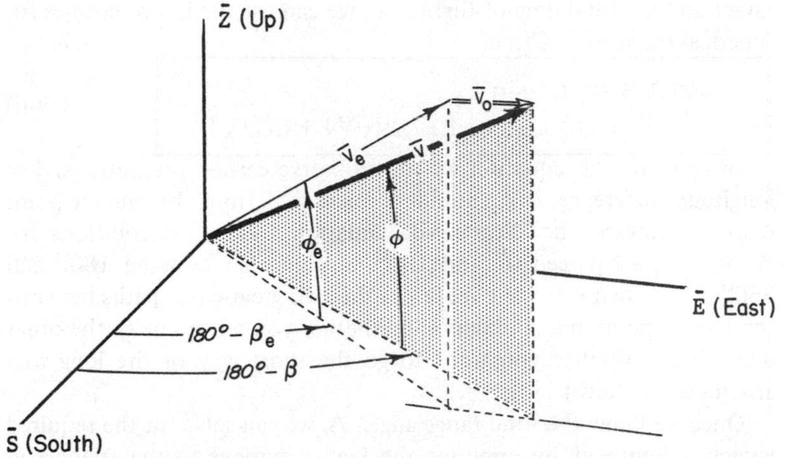

Figure 6.4-2 "True" speed and direction at burnout

6.4.2 Compensating for Movement of the Target Due to Earth Rotation. In Figure 6.4-3 we show the Earth at the instant a missile is launched. The trajectory goes from the launch point at A to an "aiming point" at B. This aiming point does not coincide with the target at the time the missile is launched. Rather, it is a point at the same latitude as the target but east of it an amount equal to the number of degrees the Earth will turn during the total time the missile is in flight. Hopefully, when the missile arrives at point B, the target will be there also.

The latitude and longitude coordinates of the launch point are L_o and N_o, respectively, so the arc length OA in Figure 6.4-3 is just $90^\circ - L_o$. If the coordinates of the target are (L_t, N_t) then the latitude and longitude of the aiming point should be L_t and $N_t + \omega_\oplus t_\Lambda$, respectively. The term, $\omega_\oplus t_\Lambda$, represents the number of degrees the Earth turns during the time t_Λ. The angular rate, $\omega_\oplus$, at which the earth turns is approximately 15°/hr.

Arc length OB is simply $90^\circ - L_t$, and the third side of the spherical triangle (the dashed line) is the ground trace of the missile trajectory which subtends the angle Λ. The included angle at A is the launch azimuth, β. The angle formed at O is just the difference in longitude between the launch point and the aiming point, $\Delta N + \omega_{\oplus} t_{\Lambda}$, where ΔN is the difference in longitude between launch point and target.

If we assume that we know the coordinates of the launch point and target and the total time-of-flight, t_{Λ}, we can use the law of cosines for spherical triangles to obtain

$$\cos \Lambda = \sin L_o \sin L_t + \cos L_o \cos L_t \cos (\Delta N + \omega_{\oplus} t_{\Lambda}) . \tag{6.4-6}$$

In applying this equation we must observe certain precautions. The longitude difference, ΔN, should be measured from the launch point *eastward* toward the target. The equation yields two solutions for Λ–one angle between 0° and 180°, and another between 180° and 360°. These two solutions represent the two great-circle paths between the launch point and aiming point. Whether you select one or the other depends on whether you want to go the short way or the long way around to the target.

Once we know the total range angle, Λ, we can solve for the required launch azimuth, β, by applying the law of cosines to the triangle in Figure 6.4-3 again–this time considering β as the included angle:

$$\sin L_t = \sin L_o \cos \Lambda + \cos L_o \sin \Lambda \cos \beta . \tag{6.4-7}$$

Solving for $\cos \beta$, we get

$$\cos \beta = \frac{\sin L_t - \sin L_o \cos \Lambda}{\cos L_o \sin \Lambda} \tag{6.4-8}$$

Again, this equation yields two solutions for β–one between 0° and 180° and the other between 180° and 360°. A simple rule exists for determining which value of β is correct: If you are going the "short way" to the target, and if $\Delta N + \omega_{\oplus} t_{\Lambda}$ lies between 0° and 180° then so does β. If you are firing the missile the "long way" around, then a value of $\Delta N + \omega_{\oplus} t_{\Lambda}$ between 0° and 180° requires that β lie between 180° and 360°.

It is worth going back for a moment to look at the equations for

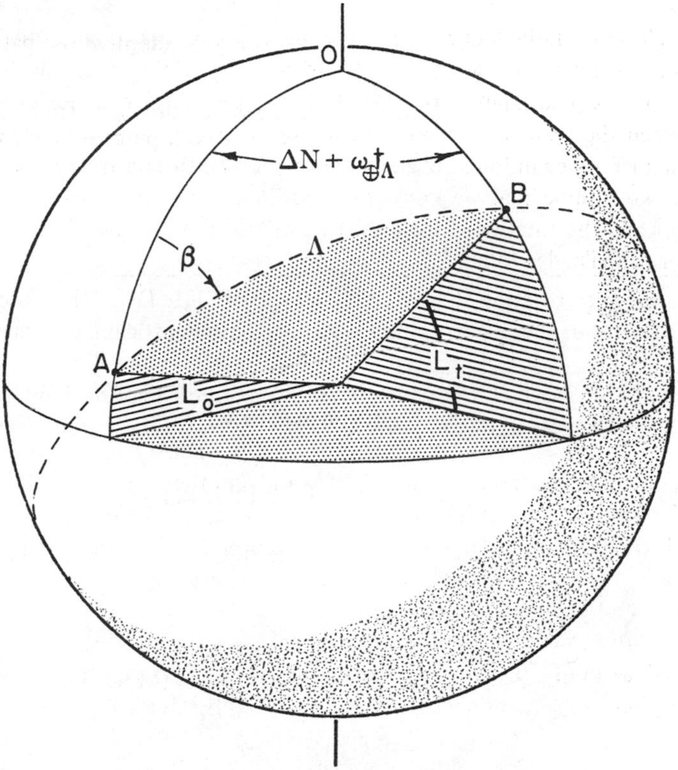

Figure 6.4-3 Launch site and "aiming point" at the instant of launch

total range, Λ, and launch azimuth, β. In order to compute β we must know the coordinates of the launch point and target as well as the range, Λ. In order to compute Λ we must know the time-of-flight, t_Λ. But, how can we know the time-of-flight if we do not know the range being flown?

The situation is not entirely desperate. In actual practice you would begin by "guessing" a reasonable time for t_Λ. This value would be used to compute an initial estimate for Λ which in turn would allow you to get a first estimate of t_{ff}. By adding the times of powered flight and re-entry (which also depend somewhat on Λ) to t_{ff} you get a value of t_Λ which you can use as your second "guess." The whole process is then repeated until the computed value of t_Λ agrees with the estimated value. Needless to say, the digital computer is more suited for this type of trial and error computation than the average student whose patience is limited.

EXAMPLE PROBLEM. A ballistic missile launched from 29°N, 79.3°W burns out at 30°N, 80°W after developing an increase in velocity of .9 DU/TU. Its elevation and azimuth relative to the Earth are:

$$\phi_e = 30^\circ \quad \beta_e = 330^\circ$$

Assuming a rotating Earth and a r_{bo} of 1.1 DU, what are the coordinates of the re-entry point? Assume a symmetrical trajectory.

The inertial speed, flight-path angle and azimuth may be found using equations (6.4-2) through (6.4-5):

$$v_S = -.9\cos30^\circ\cos330^\circ = -.675 \text{ DU/TU}$$

$$v_E = .9\cos30^\circ\sin330^\circ + .0588\cos29^\circ = -.339 \text{ DU/TU}$$

$$v_Z = .9\sin30^\circ = .45 \text{ DU/TU}$$

$$v = \sqrt{(-.675)^2 + (-.339)^2 + (.45)^2} = .879 \text{ DU/TU}$$

$$\sin\phi_{bo} = \frac{v_Z}{v} = \frac{.45}{.879} = .51195$$

$$\phi_{bo} = 30.8^\circ$$

$$\tan\beta = \frac{-v_E}{v_S} = -\frac{(-.339)}{-.675} = -.5022$$

$$\beta = 333.33^\circ$$

$$Q_{bo} = \frac{(.879)^2(1.1)}{1} = .85$$

From Figure 6.2-9, using $Q_{bo} = .85$ and $\phi_{bo} = 31^\circ$,

$$\Psi = 90^\circ$$

Using the Law of Cosines from spherical trigonometry,

$$\sin L_{re} = \sin L_{bo} \cos\Psi + \cos L_{bo} \sin\Psi \cos(360^{\circ} - \beta)$$

$$= \sin 30^{\circ} \cos 90^{\circ} + \cos 30^{\circ} \sin 90^{\circ} \cos(26.67^{\circ})$$

$$= .7739$$

$$L_{re} = \underline{\underline{50^{\circ}42'N}} \text{ (re-entry latitude)}$$

Using the Law of Cosines again, we have

$$\cos\Psi = \sin L_{re} \sin L_{bo} + \cos L_{re} \cos L_{bo} \cos(\Delta N + \omega_{\oplus} t_{ff})$$

$$\cos 90^{\circ} = \sin 50^{\circ}42' \sin 30^{\circ}$$

$$+ \cos 50^{\circ}42' \cos 30^{\circ} \cos(\Delta N + \omega_{\oplus} t_{ff})$$

$$\cos(\Delta N + \omega_{\oplus} t_{ff}) = -0.7055$$

$$\Delta N + \omega_{\oplus} t_{ff} = 135^{\circ} 8'$$

From Figure 6.2-9, using $Q_{bo} = .85$ and $\phi_{bo} = 30.78^{\circ}$,

$$\frac{t_{ff}}{\mathbb{P}_{cs}} \approx .46$$

$$\mathbb{P}_{cs} = 2\pi \sqrt{\frac{r_{bo}^{3}}{\mu}} = 2\pi \sqrt{1.1^{3}} = 7.25 \text{ TU}$$

$$\therefore t_{ff} = 3.334 \text{ TU}$$

$$\Delta N = 135^{\circ} - \omega_{\oplus}(3.334) = 135^{\circ} - 3.3709 \frac{\text{DEG}}{\text{TU}} (3.334 \text{TU})$$

$$= 123.77^{\circ}$$

$$N_{re} = N_{bo} + \Delta N = -80^{\circ} - 123.77^{\circ} = -203.77^{\circ}$$

$$N_{re} = \underline{\underline{156.23^{\circ}E}} \text{ (re-entry longitude)}$$

EXERCISES

6.1 The following measurements were obtained during the testing of an ICBM:

$$v_{bo} = .926 \text{ DU/TU}, \quad r_{bo} = 1.05 \text{ DU}$$

$$\phi_{bo} = 10^o, \quad R_p = 60 \text{ n. mi.}$$

$$R_{re} = 300 \text{ n. mi.}$$

What is R_t?

6.2 A ballistic missile is launched from a submarine in the Atlantic (30°N, 75°W) on an azimuth of 135°. Burnout speed relative to the submarine is 16,000 ft/sec and at an angle of 30° to the local horizontal. Assume the submarine lies motionless in the water during the firing. What is the true speed of the missile relative to the center of the rotating Earth?
(Ans. v = 16,840.6 ft/sec)

6.3 For a ballistic missile having:

$$r_{bo} = 1.1 \text{ DU}$$

$$v_{bo} = 0.5 \text{ DU/TU}$$

What will be the maximum range ϕ_{bo}?

6.4 What values of Q_{bo} may be used in equation (6.2-19)? Why?

6.5 A ballistic missile's burnout point is at the end of the semi-minor axis of an ellipse. Assuming burnout altitude equals re-entry altitude, and a spherical Earth, what will the value of Q be at re-entry?

6.6 What is the minimum velocity required for a ballistic missile to travel a distance measured on the surface of the Earth of 5,040 n mi? Neglect atmosphere and assume r_{bo} = 1 DU.

6.7 In general, how many possible trajectories are there for a given range and Q_{bo}? ($Q_{bo}<1$) What is the exception to this rule?

6.8 An enterprising young engineer was able to increase a certain rocket's Q_{bo} from 0.98 to 1.02. Using the equations

$$\sin\frac{\Psi}{2}=\frac{Q_{bo}}{2\text{-}Q_{bo}}$$

$$\phi_{bo}=\frac{1}{4}(180^{\circ}-\Psi)$$

he was unable to obtain a new maximum range ϕ_{bo} for $Q_{bo}=1.02$. Why?

6.9 A ballistic missile is capable of achieving a burnout velocity of .83 DU/TU at an altitude of 1.06 DU. What is the maximum free-flight range of this missile in nautical miles? Assume a symmetrical trajectory. Do not use charts.
(Ans. $R_{ff}=4{,}212$ nm)

6.10 During a test flight, an ICBM is observed to have the following position and velocity at burnout:

$$\mathbf{r}_{bo}=\mathbf{I}-3/4\mathbf{J}\ \text{DU}$$

$$\mathbf{v}_{bo}=1/5\mathbf{J}+\sqrt{3/5}\mathbf{K}\ \text{DU/TU}$$

What is the maximum range capability of this missile in nautical miles?
(Ans. $R_{ff}=5{,}000$ nm)

6.11 A rocket testing facility located at 30°N, 100°W launches a missile to impact at a latitude of 70°S. A lateral displacement, ΔX, in the launch causes the rocket to burn out east of the intended burnout point. In what direction will the error at impact be?

6.12 Assuming that the maximum allowable cross-range error at the impact point of a ballistic missile is 1.0 n mi where the free flight range of the ballistic missile is 5,400 n mi, how large can Δx and $\Delta\beta$ be?

6.13 A malfunction causes the flight-path angle of a ballistic missile to be greater than nominal. How will this affect the missile's free-flight range? Consider three separate cases:

a. ϕ_{bo} was less than ϕ_{bo} for maximum range.
b. ϕ_{bo} was greater than ϕ_{bo} for maximum range.
c. ϕ_{bo} was equal to ϕ_{bo} for maximum range.

6.14 A ballistic missile has been launched with the following burnout errors:

$$\Delta r = 0.6888 \text{ n mi}$$

$$\Delta v = -25.936 \text{ fps}$$

$$\Delta\phi = 0.25^\circ$$

where the influence coefficients have been calculated to be:

$$\frac{\partial\Psi}{\partial\phi} = 1.5$$

$$\frac{\partial\Psi}{\partial r} = 3.0 \frac{1}{DU_\oplus}$$

$$\frac{\partial\Psi}{\partial v} = 5.0\ TU_\oplus/DU_\oplus$$

a. Determine the error at the impact point in n mi.
(Ans. 7.346 n mi long)
b. Is this a high, low, or maximum range trajectory? Why?

6.15 In general will a given $\Delta\phi_{bo}$ cause a larger error in a high or low trajectory? Why?

6.16 Assuming r_{bo} = 1.0 DU for a ballistic missile, what is the minimum burnout velocity required to achieve a free-flight range of 1,800 nautical miles?
(Ans. v_{bo} = 0.642 DU/TU)

6.17 A ballistic missile whose maximum free-flight range is 3,600 n mi is to be launched from the equator on the Greenwich meridian toward a target located at 45°N, 30°E, using a minimum time-of-flight trajectory. What should the flight-path angle be at burnout? Neglect the atmosphere and assume $Q_{bo} = Q_{bo}$ maximum.

6.18 Show that for maximum range: $Q_{bo} = 1 - e^2$ where e is the eccentricity.

6.19 An ICBM is to be flight-tested over a total range, R_t of 4,700 n mi using a high trajectory. Burnout will occur 45 n mi down-range at a Q of .8 and an altitude of 150 n mi. Re-entry range is calculated at 155 n mi. What will be the time of free-flight?
(Ans. $t_{ff} = 40.4$ min)

6.20 A ballistic missile which burned out at 45°N, 150°E, at an altitude of 2.092574×10^6 ft will re-enter at 45°N, 120°W, at the same altitude, using a "backdoor" trajectory ($\Psi > 180°$). If the velocity at burnout was 28,530 ft/sec, what was the flight-path angle at burnout?

6.21 A ballistic missile's trajectory is a portion of an ellipse whose apogee is 1.5 DU and whose perigee is .5 DU. Assuming burnout occurred at sea level on a spherical earth, what is the free-flight range expressed in nautical miles? Assume a symmetrical trajectory. Do not use charts.

6.22 The range error equation could be written as:

$$\Delta\Psi = \frac{\partial\Psi}{\partial Q_{bo}} \Delta Q_{bo} + \frac{\partial\Psi}{\partial\phi_{bo}} \Delta\phi_{bo}$$

Derive and expression for $\dfrac{\partial\Psi}{\partial Q_{bo}}$ in terms of Q_{bo}, Ψ, ϕ_{bo}

and analyze the result, i.e., determine whether the influence coefficient is always positive or negative and if so what it means.

6.23 A ballistic missile has the following nominal burnout conditions:

v_{bo}=23,500 ft/sec=0.905 $DU_\oplus/TU_\oplus$

r_{bo}=22.99(10)^6 ft =1.1 $DU_\oplus$

ϕ_{bo}=30°

The following errors exist at burnout:

Δv_{bo}= −1.3 ft/sec = -5(10)$^{-5}$ $DU_\oplus/TU_\oplus$

Δr_{bo}= +1.72 nm = +5(10)$^{-4}$ $DU_\oplus$

$\Delta\phi_{bo}$= −0.0057^o = -10^{-4} radians

a. How far will the missile miss the target?
(Ans. 4.02 nm)
b. Is the shot long or short?
c. Is this a high, low, or maximum range trajectory?

6.24 A rocket booster is programmed for a true velocity relative to the center of the Earth at burnout of

$$\mathbf{v}_{bo} = -1045.92\mathbf{S} - 10608.66\mathbf{E} + 20784.6\mathbf{Z} \text{ (ft/sec)}$$

What must the speed, elevation, and azimuth relative to the launch site be at burnout? Launch site coordinates are 28°N, 120°W. Do *not* assume a nonrotating Earth.
(Partial answer: ϕ_e = 60°)

***6.25** An ICBM is to be flight tested. It is desired that the missile display the following nominal parameters at burnout:

h = 2.0926 x 10^5 ft

v_e = 20.48966 x 10^4 ft/sec

β_e = 315.2^o

ϕ_e = 44.5^o

All angles and velocities are measured relative to the launch site located at 30°N, 120°W.

During the actual firing, the following errors were measured: $\Delta\phi_e = 30'$, $\Delta\beta_e = -12'$, down-range displacement of burnout point = .73 n mi.

What are the coordinates of the missile's re-entry point? Assume a symmetrical trajectory and do not assume a nonrotating Earth.
(Ans. $L_t = 54°31'$, $N_t = 179°23'W$)

***6.26** An ICBM located at 60°N, 160°E is programmed against a target located on the equator at 115°W using a minimum velocity trajectory. What should the time of flight be? Assume a spherical, rotating, atmosphereless Earth, and $r_{bo} \approx r_{\oplus}$.
(Ans. $t_{ff} = 32.57$ min)

***6.27** A requirement exists for a ballistic missile with a total range of 7,400 nautical miles where:

$R_p = 140$ n mi

$R_{RE} = 60$ n mi

$r_{bo} = 21.8\,(10)^6$ feet

a. Assuming a symmetric orbit, what is the minimum burnout velocity required to reach a target at this range?
b. What is the required ϕ_{bo}?
(Ans. $\phi_{bo} = 15°$)
c. What will be the time of free flight (t_{ff})?
d. In order to overshoot the impact point would you increase or decrease the elevation angle? Explain.

***6.28** A ballistic missile burns out at an altitude of 172.1967 n mi with a $Q = 1$. The maximum altitude achieved during the ensuing flight is 1,618.649 n mi. What was the free-flight range, in nautical miles? Assume a symmetrical trajectory.

***6.29** Derive the flight-path angle equation (6.2-16) from the free-flight range equation (6.2-12). (Hint: See section 6.3.4)

***6.30** A ballistic missile is targeted with the following parameters: $Q_{bo} = 4/3$, $R_{ff} = 10{,}800$ n mi, altitude at bo = .02 DU. What will be the time of free-flight? Do not use charts.

***6.31** The approximation:

$$\Delta\Psi \approx \frac{\partial\Psi}{\partial\phi_{bo}}\Delta\phi_{bo} + \frac{\partial\Psi}{\partial r_{bo}}\Delta r_{bo} + \frac{\partial\Psi}{\partial v_{bo}}\Delta v_{bo}$$

is only useful for small values of the in-plane errors $\Delta\phi_{bo}$, Δr_{bo} and Δv_{bo}. How could this equation be modified to accomodate larger errors? (Hint: Use a Taylor series expansion, and truncate all terms third degree and higher).

LIST OF REFERENCES

1. Burgess, Eric. *Long-Range Ballistic Missiles*. New York, NY, The MacMillan Company, 1962.

2. Dornberger, Walter. *V-2*. New York, NY, The Viking Press, 1955.

3. Schwiebert, Ernest G. *A History of the U.S. Air Force Ballistic Missiles*. New York, NY, Frederick A. Praeger, Publishers, 1965.

4. Gantz, Lt Col Kenneth, ed. *The United States Air Force Report on the Ballistic Missile*. Garden City, NY, Doubleday & Company, Inc, 1958.

5. Wheelon, Albert D. "Free Flight," Chapter 1 of *An Introduction to Ballistic Missiles*. Vol 2. Los Angeles, Calif, Space Technology Laboratories, Inc, 1960.

CHAPTER 7

LUNAR TRAJECTORIES

> Which is more useful, the Sun or the Moon? . . . The Moon is the more useful, since it gives us light during the night, when it is dark, whereas the Sun shines only in the daytime, when it is light anyway.
>
> –fictitious philosopher
> created by George Gamow[1]

7.1 HISTORICAL BACKGROUND

According to the British astronomer Sir Richard A. Proctor, "Altogether the most important circumstance in what may be called the history of the Moon is the part which she has played in assisting the progress of modern exact astronomy. It is not saying too much to assert that if the Earth had no satellite the law of gravitation would never have been discovered."[2]

The first complete explanation of the irregularities in the motion of the Moon was given by Newton in Book I of the *Principia* where he states:

> For the Moon, though principally attracted by the Earth, and moving round it, does, together with the Earth, move round the Sun once a year, and is, according as she is nearer or farther from the Sun, drawn by him more or less than the center of the Earth, about which she moves; whence arise several irregularities in her motion, of all which, the

> Author in this book, with no less subtility than industry, has given a full account.[3]

Newton nevertheless regarded the Lunar Theory as very difficult and confided to Halley in despair that it "made his head ache and kept him awake so often that he would think of it no more."

In the 18th century Lunar Theory was developed analytically by Euler, Clairaut, D'Alembert, Lagrange and Laplace. Much of the work was motivated by the offer of substantial cash prizes by the English Government and numerous scientific societies to anyone who could produce accurate lunar tables for the use of navigators in determining their position at sea.

A more exact theory based on new concepts and developed by new mathematical methods was published by G. W. Hill in 1878 and was finally brought to perfection by the research of E. W. Brown.

In the first part of this chapter we will describe the Earth-Moon system together with some of the irregularities of the Moon's motion which have occupied astronomers since Newton's time. In the second part we will look at the problem of launching a vehicle from the Earth to the Moon.

The motion of near-Earth satellites or ballistic missiles may be described by 2-body orbital mechanics where the Earth is the single point of attraction. Even interplanetary trajectories, which are the subject of the next chapter, may be characterized by motion which is predominantly shaped by the presence of a single center of attraction, in this case the Sun. What distinguishes these situations from the problem of lunar trajectories is that the vast majority of the flight time is spent in the gravitational environment of a single body.

A significant feature of lunar trajectories is not merely the presence of two centers of attraction, but the relative sizes of the Earth and Moon. Although the mass of the Moon is only about 1/80th the mass of the Earth, this ratio is far larger than any other binary system in our solar system. Thus the Earth-Moon system is a rather singular event, not merely because we find our abode on the Earth, but because it comes close to being a double planet.

7.2 THE EARTH-MOON SYSTEM

The notion that the Moon revolves about the Earth is somewhat erroneous; it is more precise to say that both the Earth and the Moon revolve about their common center of mass. The mean distance between the center of the Earth and the center of the Moon is 384,400 km[4] and the mass of the Moon is 1/81.30[4] of the mass of the Earth. This puts the center of the system 4,671 km from the center of the Earth or about 3/4 of the way from the center to the surface.

Describing the motion of the Earth-Moon system is a fairly complex business and begins by noting that the center of mass revolves around the Sun once per year (by definition). The Earth and Moon both revolve about their common center of mass once in 27.3 days. As a result, the longitude of an object such as the Sun or a nearby planet exhibits fluctuations with a period of 27.3 days arising from the fact that we observe it from the Earth and not from the center of mass of the Earth-Moon system. These periodic fluctuations in longitude were, in fact, the most reliable source for determining the Moon's mass until Ranger 5 flew within 450 miles of the Moon in October 1962.

The orbital period of the Moon is not constant but is slowly increasing at the same time the distance between the Earth and Moon is increasing. According to one theory advanced by G. H. Darwin, son of the great biologist Charles Darwin, the Moon was at one time much closer to the Earth than at present. The slow recession of the Moon can be explained by the fact that the tidal bulge in the Earth's oceans raised by the Moon is carried eastward by the Earth's rotation. This shifts the center of gravity of the Earth to the east of the line joining the centers of mass of the Earth and Moon and gives the Moon a small acceleration in the direction of its orbital motion causing it to speed up and slowly spiral outward.

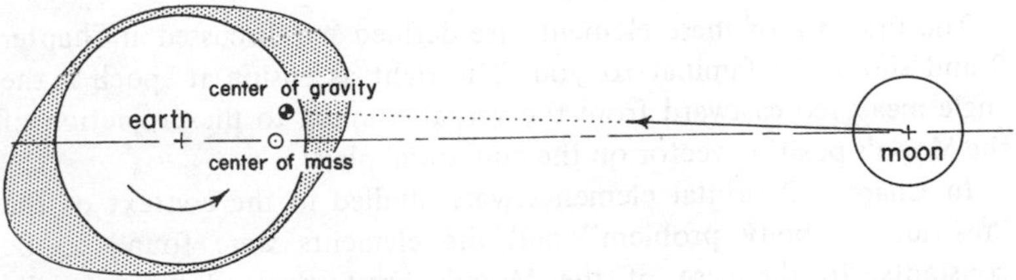

Figure 7.2-1 Acceleration of moon caused by earth's tidal bulge

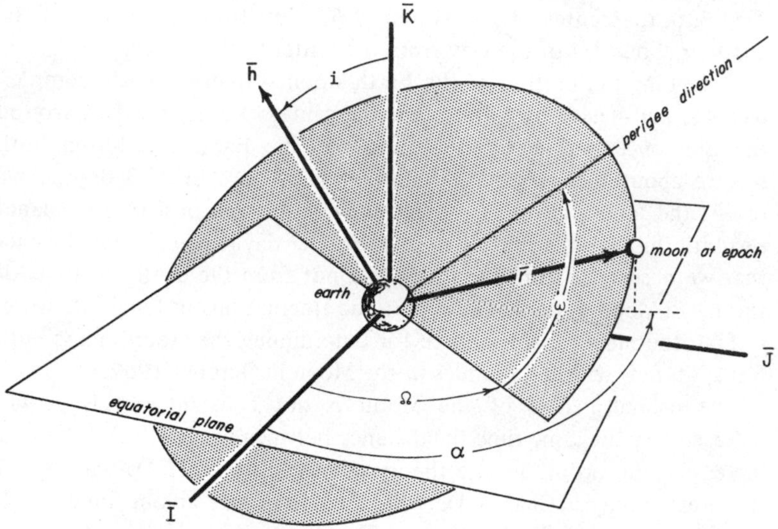

Figure 7.2-2 Lunar orbital elements

7.2.1 Orbital Elements of the Moon. When viewed from the center of the Earth the Moon's orbit can be described by six classical orbital elements, for example:

a–semi-major axis
e–eccentricity
i–inclination
Ω–longitude of the ascending node
ω–argument of perigee
α–right ascension at epoch

The first five of these elements are defined and discussed in Chapter 2 and should be familiar to you. The right ascension at epoch is the angle measured eastward from the vernal equinox to the projection of the Moon's position vector on the equatorial plane.

In Chapter 2 orbital elements were studied in the context of the "restricted 2-body problem" and the elements were found to be constants. In the case of the Moon's orbit, due primarily to the perturbative effect of the Sun, the orbital elements are constantly changing with time; their value at any particular time can be obtained from a lunar ephemeris such as is published in the *American Ephemeris and Nautical Almanac.*

We will mention some of the principal perturbations of the Moon's motion in order to illustrate its complexity:

a. The mean value of the semi-major axis is 384,400 km. The average time for the Moon to make one complete revolution around the Earth relative to the stars is 27.31661 days. Due to solar perturbations the sidereal period may vary by as much as 7 hours.

b. The mean eccentricity of the Moon's orbit is 0.054900489. Small periodic changes in the orbital eccentricity occur at intervals of 31.8 days. This effect, called "evection," was discovered more than 2,000 years ago by Hipparchus.

c. The Moon's orbit is inclined to the ecliptic (plane of the Earth's orbit) by about $5^{\circ}8'$. The line of nodes, which is the intersection of the

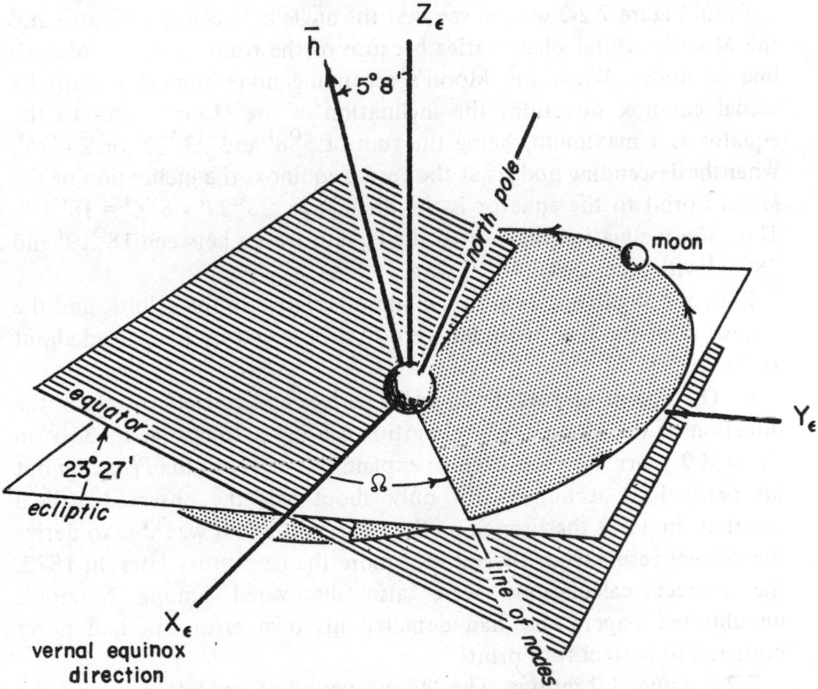

Figure 7.2-3 Rotation of the moon's line of nodes

Moon's orbital plane with the ecliptic, rotates westward, making one complete revolution in 18.6 years.

The node where the Moon crosses the ecliptic from south to north is called the ascending node; the other, where the Moon crosses from north to south, is the descending node. Only when the Moon is at one of these nodal crossing points can eclipses occur, for only then can Sun, Earth and Moon be suitably aligned. The average time for the Moon to go around its orbit from node to (the same) node is 27.21222 days and is called the "draconitic period," in reference to the superstition that a dragon was supposed to swallow the Sun at a total eclipse.

d. The inclination of the Moon's orbit to the ecliptic actually varies between $4^{o}59'$ and $5^{o}18'$; its mean value is $5^{o}8'$. The Earth's equator is inclined to the ecliptic by $23^{o}27'$, and except for the slow precession of the Earth's axis of rotation with a period of 26,000 years, the equatorial plane is relatively stationary.

From Figure 7.2-3 we can see that the angle between the equator and the Moon's orbital plane varies because of the rotation of the Moon's line of nodes. When the Moon's ascending node coincides with the vernal equinox direction, the inclination of the Moon's orbit to the equator is a maximum, being the sum of $5^{o}8'$ and $23^{o}27'$ or $28^{o}35'$. When the descending node is at the vernal equinox, the inclination of the Moon's orbit to the equator is the difference, $23^{o}27' - 5^{o}8' = 18^{o}19'$. Thus, the inclination relative to the equator varies between $18^{o}19'$ and $28^{o}35'$ with a period of 18.6 years.

Both the slight variation in inclination relative to the ecliptic and the regression of the line of nodes were first observed by Flamsteed about 1670.

e. The line of apsides (line joining perigee and apogee) rotates in the direction of the Moon's orbital motion causing ω to change by 360° in about 8.9 years. Newton tried to explain this effect in the *Principia* but his predictions accounted for only about half the observed apsidal rotation. In 1749 the French mathematician Clairaut was able to derive the correct result from theory, but more than a century later, in 1872, the correct calculations were also discovered among Newton's unpublished papers: he had detected his own error but had never bothered to correct it in print!

7.2.2 Lunar Librations. The Moon's period of revolution around the Earth is exactly equal to its period of rotation on its axis, so it always keeps the same face turned toward the Earth. If the Moon's orbit were circular and if its axis of rotation were perpendicular to its orbit, we would see exactly half of its surface.

Actually we see, at one time or another, about 59 percent of the lunar surface because of a phenomenon known as "lunar libration." The libration or "rocking motion" of the Moon is due to two causes. The geometrical libration in latitude occurs because the Moon's equator is inclined 6.5° to the plane of its orbit. At one time during the month the Moon's north pole is tipped toward the Earth and half a month later the south pole is tipped toward us allowing us to see slightly beyond each pole in turn.

The geometrical libration in longitude is due to the eccentricity of the orbit. The rotation of the Moon on its axis is uniform but its angular velocity around its orbit is not since it moves faster near perigee and slower near apogee. This permits us to see about 7.75° around each limb of the moon.

In addition to the apparent rocking motion described above there is an actual rocking called "physical libration" caused by the attraction of the Earth on the long diameter of the Moon's triaxial ellipsoid figure.

7.3 SIMPLE EARTH-MOON TRAJECTORIES

The computation of a precision lunar trajectory can only be done by numerical integration of the equations of motion, taking into account the oblate shape of the Earth, solar perturbations, solar pressure, and the terminal attraction of the Moon, among other things. Because of the complex motions of the Moon, actually mission planning places heavy reliance on a lunar ephemeris, which is a tabular listing of the Moon's position at regular intervals of chronological time. As a result, lunar missions are planned on an hour-by-hour, day-by-day, month-by-month basis.

The general procedure is to assume the initial conditions, $\mathbf{r}_O$ and $\mathbf{v}_O$, at the injection point and then use a Runge-Kutta or similar numerical method to determine the subsequent trajectory. Depending on how well we select the initial $\mathbf{r}_O$ and $\mathbf{v}_O$, the trajectory may hit the Moon or miss it entirely. The idea is to adjust the injection conditions by trial-and-error until a suitable lunar impact occurs.

Even on a high-speed digital computer this procedure can take hours of computation time for a single launch date. If we have to explore a large number of different launch dates and a variety of injection conditions, the computer time required could become prohibitive, and some approximate analytical method is needed to narrow down the choice of launch time and injection conditions.

It is not important that the analytical method be precise, but only that it retain the predominant features of the actual problem. With this thought in mind, we will look at a few simple Earth-Moon trajectories in order to gain some insight into the problem of selecting optimum launch dates and approximate injection conditions.

7.3.1 Some Simplifying Assumptions. In order to study the basic dynamics of lunar trajectories, we will assume that the Moon's orbit is circular with a radius of 384,400 km. Since the mean eccentricity of the actual orbit is only about .0549, this will not introduce significant errors. We will also assume that we can neglect the terminal attraction of the Moon and simply look at some trajectories that intersect the Moon's orbit.

In the analysis which follows we will also assume that the lunar trajectory is coplanar with the Moon's orbit. In a precision trajectory calculation the launch time is selected so that this is approximately true in order to minimize the Δv required for the mission since plane changes are expensive in terms of velocity.

7.3.2 Time-of-Flight Versus Injection Speed. With the assumption stated above we can proceed to investigate the effect of injection speed on the time-of-flight of a lunar probe. We can compute the energy and angular momentum of the trajectory from

$$\mathcal{E} = \frac{v_0^2}{2} - \frac{\mu}{r_0}$$

$$h = r_0 v_0 \cos \phi_0 .$$

The parameter, semi-major axis, and eccentricity are then obtained from

$$p = \frac{h^2}{\mu}$$

$$a = \frac{-\mu}{2\mathcal{E}}$$

$$e = \sqrt{1 - p/a} .$$

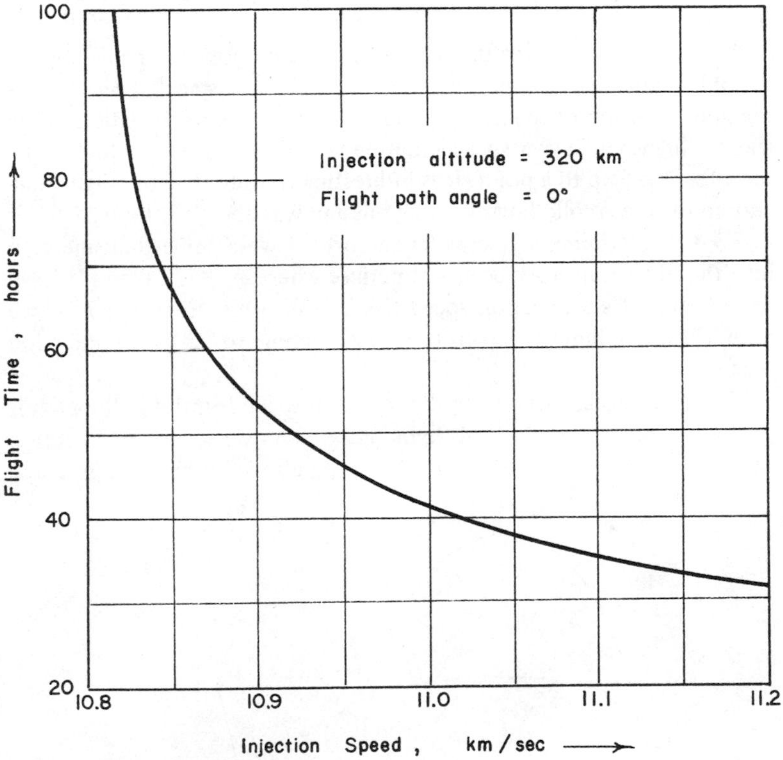

Figure 7.3-1 Lunar flight time vs injection speed

Solving the polar equation of a conic for true anomaly, we get

$$\cos \nu = \frac{p - r}{er} .$$

If we let $r = r_0$ in this expression, we can solve for ν_0; if we let r equal the radius of the Moon's orbit, we can find the true anomaly upon arrival at the Moon's orbit.

We now have enough information to determine the time-of-flight from Earth to Moon for any set of injection conditions using the equations presented in Chapter 4.

In Figure 7.3-1 we have plotted time-of-flight vs injection speed for an injection altitude of .05 Earth radii (320 km) and a flight-path angle of 0°. Actually, the curve is nearly independent of flight-path angle at injection.

We see from this curve that a significant reduction in time-of-flight is possible with only modest increases in injection speed. For manned missions life-support requirements increase with mission duration so the slightly higher injection speed required to achieve a shorter flight time pays for itself up to a point. It is interesting to note that the flight time chosen for the Apollo lunar landing mission was about 72 hours.

7.3.3 The Minimum Energy Trajectory. If we assume that injection into the lunar trajectory occurs at perigee where $\phi_o = 0^\circ$, then it is easy to see what effect injection speed has on the orbit. In Figure 7.3-2 we have shown a family of orbits corresponding to different injection speeds.

For the limiting case where the injection speed is infinite, the path is a straight line with a time-of-flight of zero. As we lower the injection speed the orbit goes from hyperbolic, to parabolic, to elliptical in shape

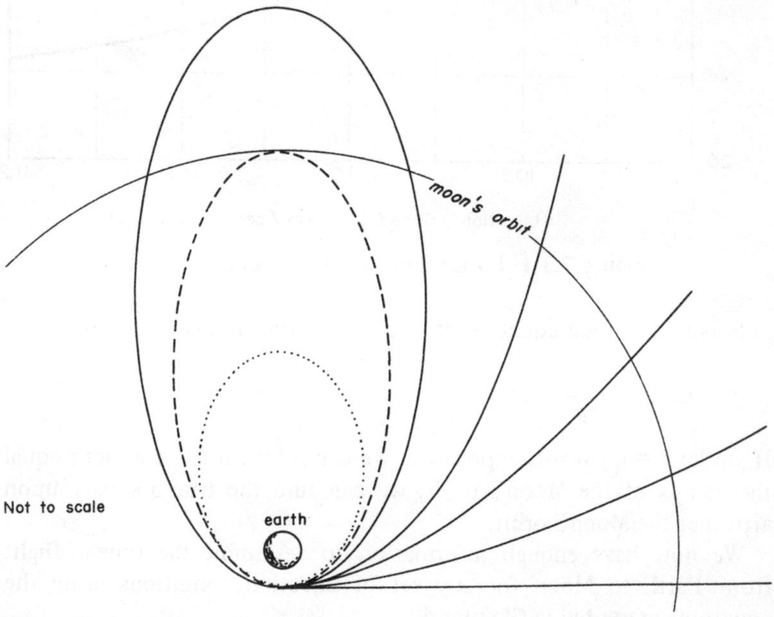

Figure 7.3-2 Effect of injection speed on trajectory shape

and the time-of-flight increases. Eventually, if we keep reducing the injection speed, we will arrive at the dashed orbit in Figure 7.3-2 whose apogee just barely reaches to the distance of the Moon. Assuming an injection altitude of .05 Earth radii (320 km), this *minimum injection speed* is 10.82 km/sec. If we give our lunar probe less than this speed it will fail to reach the moon's orbit like the dotted path in Figure 7.3-2.

The time-of-flight for this minimum energy lunar trajectory is 7,172 minutes or about 120 hours. Because all other trajectories that reach to the Moon's orbit have shorter flight times, this represents an approximate *maximum time-of-flight* for a lunar mission; if we try to take longer by going slower, we will never reach the Moon at all.

The eccentricity of the minimum energy trajectory is 0.966 and represents the *minimum eccentricity* for an elliptical orbit that reaches as far as the Moon from our assumed injection point.

Assuming no lunar gravity, the speed upon arrival at the Moon's orbit for the minimum energy trajectory is 0.188 km/sec. This represents the slowest approach speed possible for our example. Since the Moon's orbital speed is about 1 km/sec, the Moon would literally run into our probe from behind, resulting in an impact on the "leading edge" or eastern limb of the Moon. Probes which have a higher arrival speed would tend to impact somewhere on the side of the Moon facing us.

From Figure 7.3-2 we see that, as the injection speed is decreased, the geocentric angle swept out by the lunar probe from injection to lunar intercept increases from 0° to 180° for the minimum energy case. In general the sweep angle, which we can call ψ, is a function of injection speed for a fixed injection altitude and flight-path angle; an increase in sweep angle (up to 180°) corresponds to a decrease in injection speed. If we are trying to minimize injection speed, we should try to select an orbit which has a sweep angle of nearly 180°. We will make use of this general principal later in this chapter.

7.3.4 Miss Distance at the Moon Caused by Injection Errors. In trying for a direct hit on the Moon we would time our launch so that the probe crosses the Moon's orbit just at the instant the Moon is at the intercept point. Using our simplified model of the Earth-Moon system and neglecting lunar gravity, we can get some idea of how much we will miss the center of the Moon if, due to errors in guidance or other factors, the injection conditions are not exactly as specified.

In such a case, both the sweep angle and the time-of-flight will differ from their nominal values and the trajectory of the probe will cross the

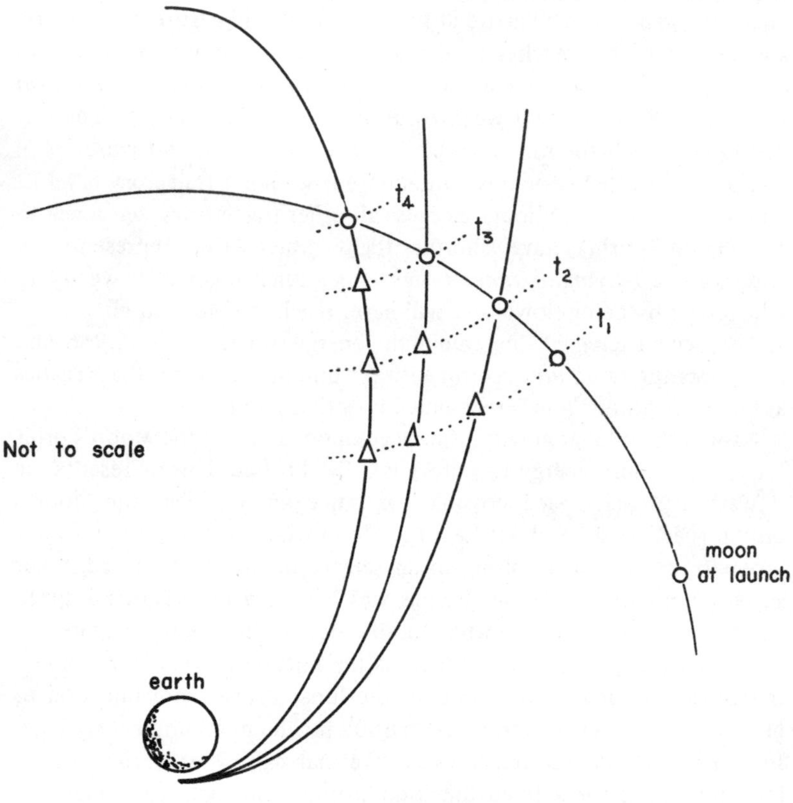

Figure 7.3-3 Effects of launch error may tend to cancel

Moon's orbit at a different point and time than predicted. In the case of a direct (eastward) launch, the effects of such injection errors tend to cancel. This may be seen from Figure 7.3-3.

If, for example, the initial velocity is too high, the geocentric sweep angle will be smaller than predicted; that is, the probe will cross the Moon's orbit west of the predicted point by some amount $\Delta\psi$. But, the time-of-flight will be shorter by an amount Δt, so the Moon will be west of the predicted intercept point by an amount $\omega_m \Delta t$, where ω_m is the angular velocity of the Moon in its orbit. Neglecting lunar gravity, the angular miss distance along the Moon's orbit is the difference between $\Delta\psi$ and $\omega_m \Delta t$.

It is possible to show that for an injection speed of about 11.0 km/sec, which results in a sweep angle of about 160°, the effects of errors in injection altitude and speed exactly cancel and, for all practical purposes, the miss distance at the Moon is a function only of errors in the flight-path angle. For this condition an error of 1° in flight-path angle produces a miss of about 1,300 km.[5] The effect of lunar gravity is to reduce this miss distance.

7.4 THE PATCHED-CONIC APPROXIMATION

While it is acceptable to neglect lunar gravity if we only are interested in determining approximate injection conditions that result in a lunar impact, it is necessary to account for the terminal attraction of the Moon if we want to predict the lunar arrival conditions more exactly.

We can take lunar gravity into account and still use 2-body orbital mechanics by the simple expedient of considering the probe to be under the gravitational influence of the Earth alone until it enters the gravitational "sphere of influence" of the Moon and then assuming that it moves only under the gravitational influence of the Moon. In effect, we pick a point in the vicinity of the Moon where we "turn off the Earth and turn on the Moon." Before we show how this is done it should be clear to you that what we are about to do is an approximation. The transition from geocentric motion to selenocentric motion is a gradual process which takes place over a finite arc of the trajectory where both Earth and Moon affect the path equally. There is, however, evidence to show that this simple strategy of patching two conics together at the edge of the Moon's sphere of influence is a sufficiently good approximation for preliminary mission analysis.[5] *However, the patched-conic analysis is not good for the calculation of the return trajectory to the Earth because of errors in the encounter with the Moon's sphere of influence. It will also be in error for the same reason in calculating the perilune altitude and lunar trajectory orientation. It is good primarily for outbound delta-v calculations.*

For the analysis which follows we will adopt the definition of sphere of influence suggested by Laplace. This is a sphere centered at the Moon and having a radius, R_S, given by the expression

$$R_S = D \left(\frac{m_{moon}}{m_{\oplus}}\right)^{2/5} \qquad (7.4\text{-}1)$$

where D is the distance from the Earth to the Moon. A full derivation of equation (7.4-1) can be found in Battin.[6] To give at least an elementary idea of its origin, consider the equation of motion viewed from an Earth frame,

$$\mathbf{a} = \begin{pmatrix}\text{central}\\ \text{acceleration}\\ \text{due to earth}\end{pmatrix} + \begin{pmatrix}\text{perturbing}\\ \text{acceleration}\\ \text{due to moon}\end{pmatrix}$$

$$= \mathbf{a}_e + \mathbf{a}_{pm} \,.$$

Then, the equation of motion viewed from the Moon is

$$\mathbf{A} = \begin{pmatrix}\text{central}\\ \text{acceleration}\\ \text{due to moon}\end{pmatrix} + \begin{pmatrix}\text{perturbing}\\ \text{acceleration}\\ \text{due to earth}\end{pmatrix}$$

$$= \mathbf{A}_m + \mathbf{A}_{pe} \,.$$

The sphere of influence is the approximation resulting from equating the ratios

$$\frac{a_{pm}}{a_e} = \frac{A_{pe}}{A_m} \,.$$

Equation (7.4-1) yields the value

$$Rs = 66{,}300 \text{ km}$$

or about 1/6 the distance from the Moon to the Earth.

7.4.1 The Geocentric Departure Orbit. Figure 7.4-1 shows the geometry of the geocentric departure orbit. The four quantities that completely specify the geocentric phase are

$$r_0,\ v_0,\ \phi_0,\ \gamma_0$$

where γ_0 is called the "phase angle at departure."

The difficulty with selecting these four quantities as the independent variables is that the determination of the point at which the geocentric trajectory crosses the lunar sphere of influence involves an iterative procedure in which time-of-flight must be computed during each iteration. This difficulty may be by-passed by selecting three initial conditions and one arrival condition as the independent variables. A particularly convenient set is

$$r_0, v_0, \phi_0, \lambda_1$$

where the angle λ_1 specifies the point at which the geocentric trajectory crosses the lunar sphere of influence.

Given these four quantities we can compute the remaining arrival conditions, r_1, v_1, ϕ_1 and γ_1. We will assume that the geocentric trajectory is direct and that lunar arrival occurs prior to apogee of the geocentric orbit. The energy and angular momentum of the orbit can be determined from

$$\mathcal{E} = \frac{v_0^2}{2} - \frac{\mu}{r_0} \tag{7.4-2}$$

$$h = r_0 v_0 \cos \phi_0 \tag{7.4-3}$$

From the law of cosines, the radius, r_1, at lunar arrival is

$$r_1 = \sqrt{D^2 + R_s^2 - 2DR_s \cos \lambda_1}\ . \tag{7.4-4}$$

The speed and flight path angle at arrival follow from conservation of energy and momentum:

$$v_1 = \sqrt{2(\mathcal{E} + \mu/r_1)} \tag{7.4-5}$$

$$\cos \phi_1 = \frac{h}{r_1 v_1} \tag{7.4-6}$$

where ϕ_1 is known to lie between 0° and 90° since arrival occurs prior to apogee.

Finally, from geometry

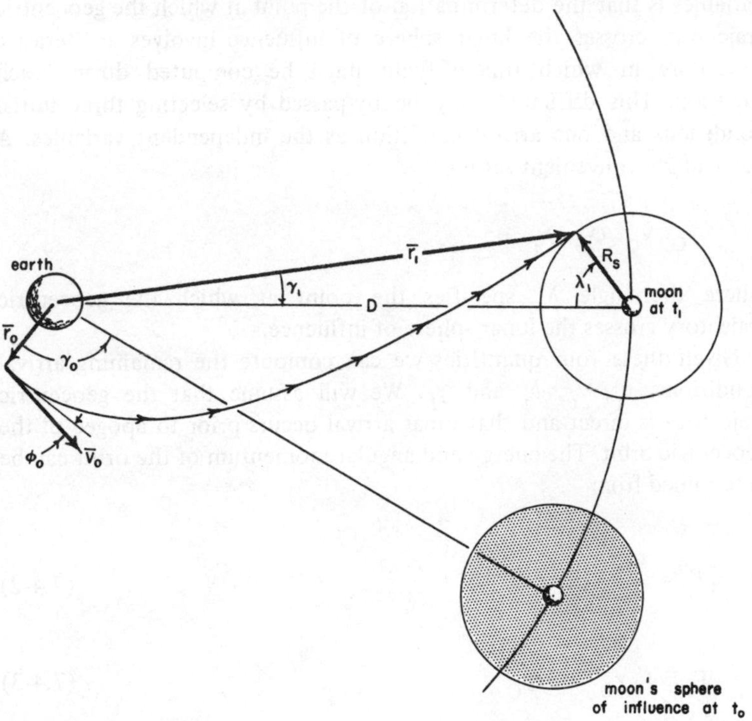

Figure 7.4-1 Geocentric transfer to the lunar sphere of influence

$$\sin \gamma_1 = \frac{R_s}{r_1} \sin \lambda_1 \ . \tag{7.4-7}$$

The time-of-flight, $t_1 - t_0$, from injection to arrival at the lunar sphere of influence can be computed once ν_0 and ν_1 are determined. Before the true anomalies can be found we must determine p, a and e of the geocentric trajectory from

$$p = \frac{h^2}{\mu} \tag{7.4-8}$$

$$a = \frac{-\mu}{2\mathcal{E}} \tag{7.4-9}$$

$$e = \sqrt{1 - p/a} \ . \tag{7.4-10}$$

Then ν_0 and ν_1 follow from the polar equation of a conic:

$$\cos \nu_0 = \frac{p - r_0}{r_0 e} \tag{7.4-11}$$

$$\cos \nu_1 = \frac{p - r_1}{r_1 e} \; . \tag{7.4-12}$$

Next, we can determine the eccentric anomalies, E_0 and E_1 from

$$\cos E_0 = \frac{e + \cos \nu_0}{1 + e \cos \nu_0} \tag{7.4-13}$$

$$\cos E_1 = \frac{e + \cos \nu_1}{1 + e \cos \nu_1} \tag{7.4-14}$$

Finally, time-of-flight is obtained from

$$t_1 - t_0 = \sqrt{\frac{a^3}{\mu}} \, [(E_1 - e \sin E_1) - (E_0 - e \sin E_0)] \tag{7.4-15}$$

The Moon moves through an angle $\omega_m (t_1 - t_0)$ between injection and arrival at the lunar sphere of influence, where ω_m is the angular velocity of the Moon in its orbit. Based on our simplified model of the Earth-Moon system,

$$\omega_m = 2.649 \times 10^{-6} \text{ rad/sec}$$

$$= 2.137 \times 10^{-3} \text{ rad/TU}_\oplus$$

The phase angle at departure, γ_0, is then determined from

$$\gamma_0 = \nu_1 - \nu_0 - \gamma_1 - \omega_m (t_1 - t_0) \tag{7.4-16}$$

Actually, the time-of-flight and phase angle at departure need not be computed until we have verified that the values r_0, v_0, ϕ_0 and λ_1, which we chose arbitrarily, result in a satisfactory lunar approach trajectory or impact. We will do this in the sections that follow. If the lunar trajectory is not satisfactory, then we will adjust the values of r_o, v_o, ϕ_o and λ_1 until it is. Only after this trial-and-error procedure is complete should we perform the computations embodied in equations (7.4-8) through (7.4-16) above.

Before proceeding to a discussion of the patch condition, a few remarks are in order concerning equation (7.4-5). The energy in the geocentric departure orbit is completely determined by r_o and v_o. The geocentric radius, r_1, at arrival is completely determined by λ_1. It may happen that the trajectory is not sufficiently energetic to reach the specified point on the lunar sphere of influence as determined by λ_1. If so, the quantity under the radical sign in equation (7.4-5) will be negative and the whole computational process fails.

7.4.2 Conditions at the Patch Point. We are now ready to determine the trajectory inside the Moon's sphere of influence where only lunar gravity is assumed to act on the spacecraft. Since we must now consider the Moon as the central body, it is necessary to find the speed and direction of the spacecraft *relative to the center of the Moon*. In Figure 7.4-2 the geometry of the situation at arrival is shown in detail.

If we let the subscript 2 indicate the initial conditions relative to the Moon's center, then the selenocentric radius, r_2, is

$$r_2 = R_s \tag{7.4-17}$$

The velocity of the spacecraft relative to the center of the Moon is

$$\mathbf{v}_2 = \mathbf{v}_1 - \mathbf{v}_m$$

where $\mathbf{v}_m$ is velocity of the Moon relative to the center of the Earth. The orbital speed of the Moon for our simplified Earth-Moon model is

$$v_m = 1.018 \text{ km/sec.} \tag{7.4-18}$$

The selenocentric arrival speed, v_2, may be obtained by applying the law of cosines to the vector triangle in Figure 7.4-2:

$$v_2 = \sqrt{v_1^2 + v_m^2 - 2v_1 v_m \cos(\phi_1 - \gamma_1)}\,. \tag{7.4-19}$$

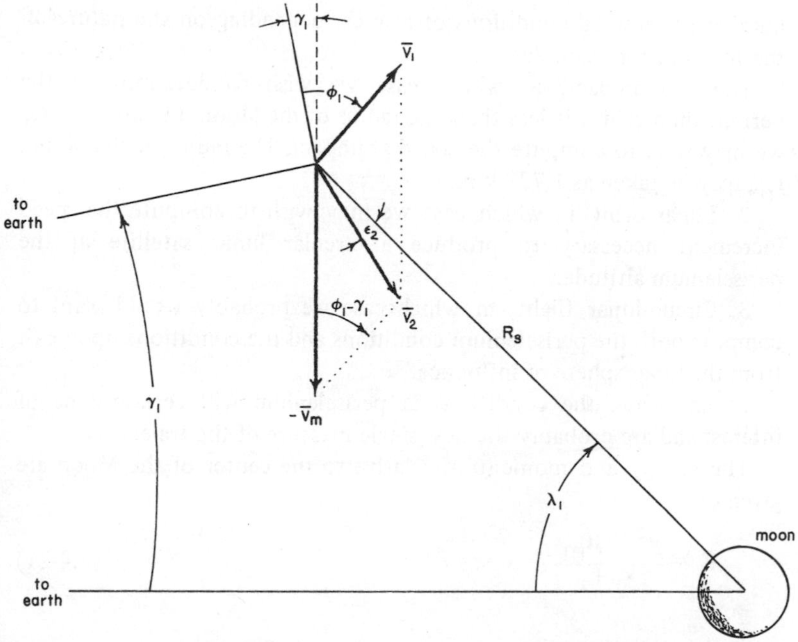

Figure 7.4-2 The patch condition

The angle ϵ_2 defines the direction of the initial selenocentric velocity relative to the Moon's center. Equating the components of $\mathbf{v}_2$ perpendicular to $\mathbf{r}_2$ yields

$$v_2 \sin \epsilon_2 = v_m \cos \lambda_1 - v_1 \cos(\lambda_1 + \gamma_1 - \phi_1)$$

from which

$$\epsilon_2 = \sin^{-1}\left[\frac{v_m}{v_2}\cos\lambda_1 - \frac{v_1}{v_2}\cos(\lambda_1 + \gamma_1 - \phi_1)\right]. \quad (7.4\text{-}20)$$

It is obvious that for a dead-center hit on the Moon ϵ_2 must be exactly zero.

7.4.3 The Selenocentric Arrival Orbit. The selenocentric initial conditions r_2, v_2 and ϵ_2 are now known and so we can compute terminal conditions at other points on the trajectory. There are a number of terminal conditions of interest depending on the nature of the mission, for example:

1. Lunar impact, in which case we wish to determine if the periselenium radius is less than the radius of the Moon. In case $r_p < r_m$ we may wish to compute the speed at impact. The radius of the Moon, r_m, may be taken as 1,738 km.

2. Lunar orbit, in which case we may wish to compute the speed increment necessary to produce a circular lunar satellite at the periselenium altitude.

3. Circumlunar flight, in which case we probably would want to compute both the periselenium conditions and the conditions upon exit from the lunar sphere of influence.

In any case, the conditions at periselenium will certainly be of interest and are probably the best single measure of the trajectory.

The energy and momentum relative to the center of the Moon are given by

$$\mathcal{E} = \frac{v_2^2}{2} - \frac{\mu_m}{r_2} \tag{7.4-21}$$

$$h = r_2\, v_2 \sin \epsilon_2 \tag{7.4-22}$$

where μ_m is the gravitational parameter of the Moon. Since the mass of the Moon is 1/81.3 of the Earth's mass, μ_m may be determined from

$$\mu_m = \frac{\mu_\oplus}{81.3}$$

$$\mu_m = 4.90287 \times 10^3 \ \text{km}^3/\text{sec}^2 \ .$$

The parameter and eccentricity of the selenocentric orbit can be computed from

$$p = \frac{h^2}{\mu_m} \tag{7.4-23}$$

$$e = \sqrt{1 + 2\mathcal{E} h^2/\mu_m^2} \ . \tag{7.4-24}$$

The conditions at periselenium are then obtained from

$$r_p = \frac{p}{1 + e} \tag{7.4-25}$$

$$v_p = \sqrt{2(\mathcal{E} + \mu_m/r_p)} \quad . \tag{7.4-26}$$

If the periselenium conditions are not satisfactory, either the injection conditions, r_o, v_o, ϕ_o or the angle λ_1 should be adjusted by trial-and-error until the trajectory is acceptable.

EXAMPLE PROBLEM. A lunar probe is sent to the Moon on a trajectory with the following injection conditions:

$$r_o = 1.05 \text{ DU}$$

$$v_o = 1.372 \text{ DU/TU}$$

$$\phi_o = 0^o$$

Upon arrival at the Moon's sphere of influence, $\lambda_1 = 30^o$.

Calculate the initial phase angle γ_0 and the altitude at closest approach to the Moon for the probe.

The given information in canonical units based on the Earth is:

$$r_o = 1.05 \text{ DU}$$

$$v_o = 1.372 \text{ DU/TU}$$

$$\phi_o = 0^o$$

$\lambda_1 = 30^o$ (arrival condition at the moon's SOI)

From equation (7.4-2)

$$\mathcal{E} = -0.011 \text{ DU}^2/\text{TU}^2$$

From equation (7.4-3)

$$h = 1.44 \text{ DU}^2/\text{TU}$$

We know that

$$D = 384{,}400\text{km} = 60.27 \text{ DU}$$

$$R_s = 66{,}300\text{km} = 10.395 \text{ DU}$$

Using equation (7.4-4)

$$r_1 = \sqrt{D^2 + R_s^2 - 2DR_s \cos\lambda_1} = 51.53 \text{ DU}$$

From equation (7.4-5)

$$v_1 = \sqrt{2(\mathcal{E} + \frac{\mu}{r_1})} = 0.1296 \text{ DU/TU}$$

and from equation (7.4-6)

$$\phi_1 = \cos^{-1}\left(\frac{h}{r_1 v_1}\right) = 77.6^\circ$$

($\phi_1 < 90^\circ$ since arrival at the Moon's SOI occurs prior to apogee)

From equation (7.4-7) we get the phase angle of the Moon at arrival,

$$\gamma_1 = \sin^{-1}\left(\frac{R_s}{r_1}\sin\lambda_1\right) = 5.78^\circ = 0.1 \text{ rad.}$$

In order to calculate the time of flight to the Moon's SOI, we need the parameters p, a, e, E_0, E_1 for the geocentric trajectory. Using equations developed earlier we obtain:

$$p = 2.074 \text{ DU}, \quad a = 45.45 \text{ DU}, \quad e = 0.977$$

$\nu_o = 0^\circ$ since $\phi_o = 0^\circ$ (the probe burns out at perigee)

$\nu_1 = 169.2^\circ = 2.95$ rad.

$E_0 = 0^\circ$ since $\nu_0 = 0^\circ$

$E_1 = 97.4^\circ = 1.7$ rad.

$\sin E_1 = 0.992$

Using equation (7.4-15) we get the TOF:

$$t_1 - t_0 = \sqrt{\frac{a^3}{\mu}}\,[(E_1 - e \sin E_1) - (E_0 - e \sin E_0)]$$

$$= 306.45[0.731] = 223.98 \text{ TU}$$

$$= 50.198 \text{ hours.}$$

From equation (7.4-16) we get the phase angle at departure, γ_0.

$$\gamma_0 = \nu_1 - \nu_0 - \gamma_1 - \omega_m(t_1 - t_0) = 2.371 \text{ rad} = 135.87^\circ$$

where $\omega_m = 2.137(10^{-3})$ rad/TU.

At the Moon's SOI it is necessary to convert v_1 and R_s to units based on the Moon as the gravitational center of attraction. Using km and km/sec we get:

$v_1 = 0.1296$ DU/TU $= 1.024$ km/sec

$R_s = 10.395$ DU $= 66{,}300$ km

and $\mu_m = \dfrac{\mu_\oplus}{81.3} = 4.903(10^3)\text{ km}^3/\text{sec}^2$

$v_m = 1.018$ km/sec.

Using equation (7.4-19)

$v_2 = 1.198$ km/sec.

From equation (7.4-20)

$$\epsilon_2 = 5.68^\circ$$

Using equations (7.4-21) through (7.4-26) we can determine the minimum distance of approach to the Moon.

$$h = 7.871(10^3)\ \text{km}^2/\text{sec}$$

$$\mathcal{E} = \frac{v_2^2}{2} - \frac{\mu_m}{R_s} = 0.6437 \text{ (Why is } \mathcal{E} > 0?)$$

$$p = \frac{h^2}{\mu_m} = 1.264(10^4)\text{km}$$

$$e = \sqrt{1 + \frac{2\mathcal{E}h^2}{\mu_m^2}} = 2.078$$

$$r_p = 4.105(10^3)\text{km}$$

$$h_p = r_p - r_m = \underline{2.367(10^3)\text{km}}$$

This is the minimum distance of approach.

7.5 NONCOPLANAR LUNAR TRAJECTORIES

The preceding analysis has been based on the assumption that the lunar trajectory lies in the plane of the Moon's orbit. The inclination of the Moon's orbit varies between about 18.2° and 28.5° over a period of 18.6 years. Since it is impossible to launch an Earth satellite into an orbit whose inclination is less than the latitude of the launch site (see section 2.14-2), a coplanar trajectory originating from Cape Kennedy, whose latitude is 28.5°, is possible only when the inclination of the Moon's orbit is at its maximum value. This occured in the early part of 1969 and will occur again in 1978.

Launches which occur at times other than these must necessarily result in noncoplanar trajectories. In the sections which follow we will examine noncoplanar trajectories and develop a method for selecting acceptable launch dates and injection conditions.

7.5.1 Some Typical Constraints on Lunar Trajectories. If there are no restrictions on the launch conditions of a spacecraft or on the conditions at lunar approach, then there are no limitations on the time of the lunar month at which the spacecraft can approach the Moon. Practical considerations, such as launch site location, missile-range safety, accuracy tolerances, and the limited range of attainable injection conditions, impose restrictions on the lunar intercept declination which can be accommodated. It is interesting to examine the limitations on the possible launch times for lunar missions that are imposed by some of these restrictions.

A typical design restriction for lunar missions is the specification of the lighting conditions on the surface of the Moon as determined by the phase of the Moon. For a particular year, the declination of the Moon at a given phase varies between maximum and minimum values which correspond approximately to the mean inclination of the Moon's orbit for that year.

Another typical restriction concerns the permissible directions of launch from a particular site.

In the analysis which follows the launch site is assumed to be Cape Kennedy which has a latitude of 28.5°. The launch azimuth, β_o, must be between 40° and 115° as specified by Eastern Test Range safety requirements.

7.5.2 Determining the Geocentric Sweep Angle. An important parameter in determining acceptable launch times is the total geocentric angle swept out by the spacecraft from launch to lunar intercept. The total sweep angle ψ_t, consists of the free-flight sweep angle, ψ_{ff}, from injection to intercept plus the geocentric angular travel from launch to injection, ψ_c. Depending on the launch technique used, ψ_c may be simply the burning arc of the booster for a direct-ascent launch or it may be the burning arc plus the angular distance traveled during a coasting period prior to injection.

While the angle ψ_c may be selected arbitrarily, the free-flight sweep angle, ψ_{ff}, is determined by the injection conditions r_o, v_o and ϕ_o. The angle ψ_{ff} is just the difference in true anomaly between injection and lunar intercept and may be computed from the equations in section 7.3.2. In Figure 7.5-2 we have plotted the free-flight sweep angle for

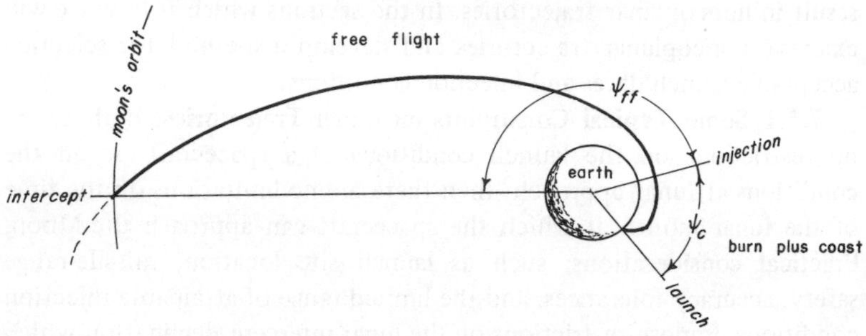

Figure 7.5-1 Geocentric sweep angle

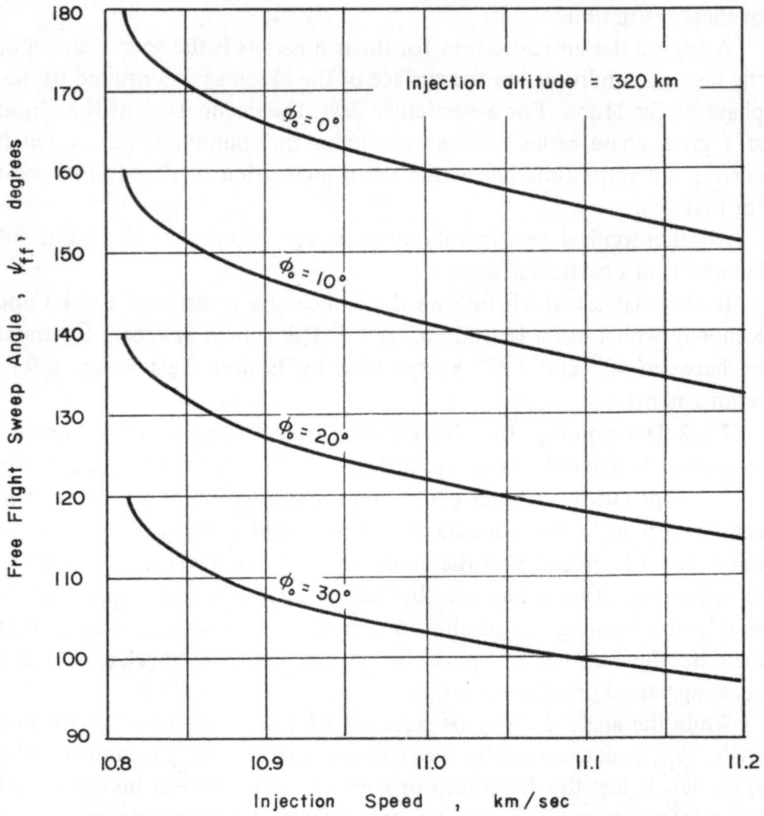

Figure 7.5-2 Free flight sweep angle vs injection speed

several values of v_o and ϕ_o and a fixed injection altitude of 320 km or about 200 miles. Lunar intercept is assumed to occur at a distance of 384,400 km.

By selecting the injection conditions, r_o, v_o and ϕ_o we can determine ψ_{ff}. If we arbitrarily select ψ_c, we can obtain the total sweep angle from

$$\psi_t = \psi_{ff} + \psi_c \tag{7.5-1}$$

Since the latitude or declination of the launch site is known, we may determine the declination of the spacecraft after it has traversed an arc ψ_t if we know the launch azimuth, β_o. This is essentially a problem in spherical trigonometry and is illustrated in Figure 7.5-3.

From the law of cosines for spherical triangles, we obtain

$$\sin \delta_1 = \sin \delta_o \cos \psi_t + \cos \delta_o \sin \psi_t \cos \beta_o \ . \tag{7.5-2}$$

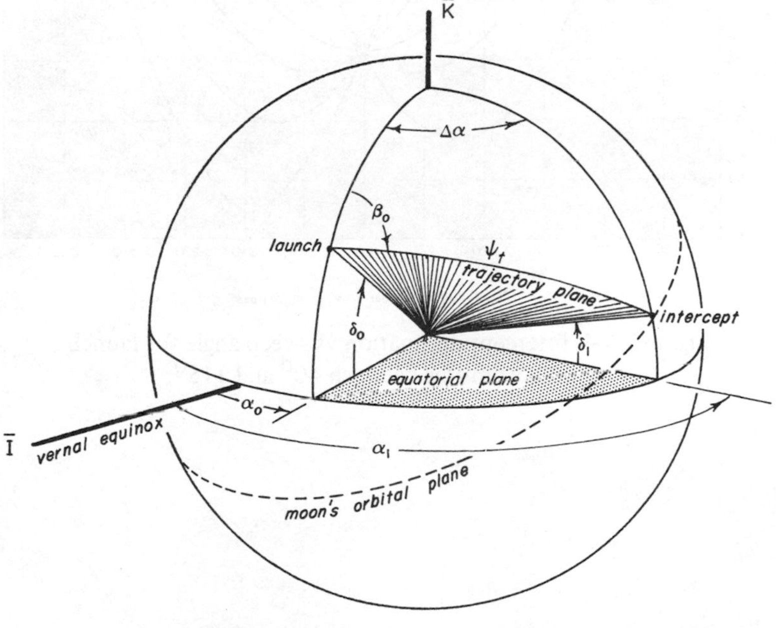

Figure 7.5-3 Lunar interception angular relationships

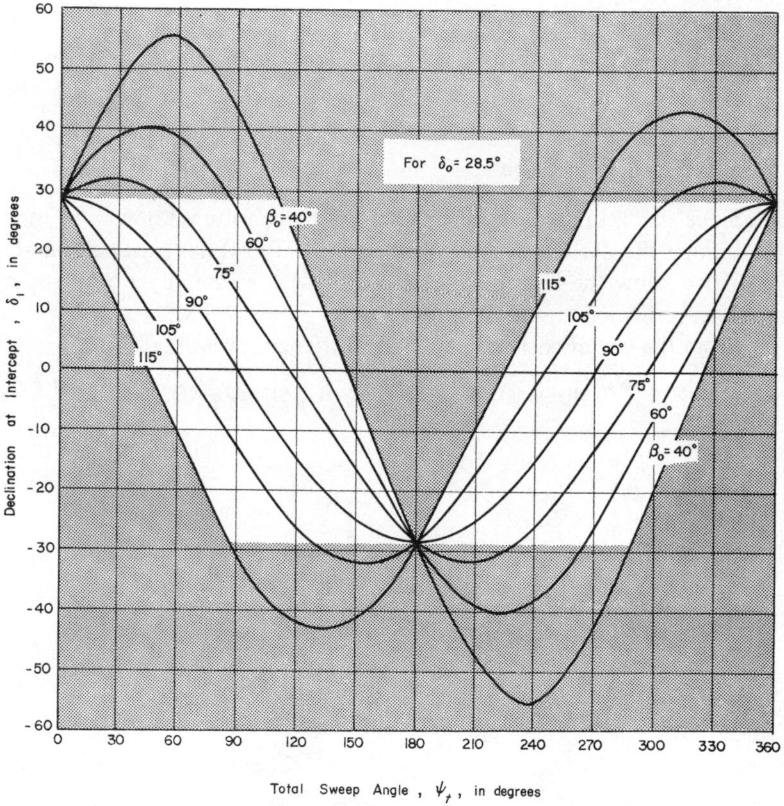

Figure 7.5-4 Intercept declination vs sweep angle for launch azimuths between 40° and 115°

We have plotted values of δ_1 obtained from equation (7.5-2) versus total sweep angle for launch azimuths between 40° and 115° in Figure 7.5-4. Since the Moon's declination at intercept is limited between $+28.5^o$ and -28.5^o (during 1969) and because of the range safety restrictions on launch azimuth, those portions of the graph that are shaded represent impossible launch conditions.

There are several interesting features of Figure 7.5-4 worth noticing. Perhaps the most interesting is the fact that a sweep angle of 180° is possible only if we intercept the Moon at its maximum southern declination of -28.5^o. For this condition any launch azimuth is correct.

From Figure 7.5-2 we see that free-flight sweep angles of less than about 120° are impossible to achieve without going to very high injection speeds or large flight path angles, both of which are undesirable. As a result, if we are interested in minimizing injection speed, we must intercept the Moon when it is near its maximum southern declination for direct-ascent launches or launches where the coasting period is small enough to keep ψ_t less than 180°.

However, if we add a coasting arc, ψ_c, large enough to make ψ_t greater than 180°, we can intercept the Moon at any point along its orbit.

It should be obvious from equation (7.5-2) and Figure 7.5-4 that launch azimuth and lunar declination at intercept are not independent; once ψ_t is known selecting either launch azimuth or declination at intercept determines the other uniquely. If lunar declination at intercept is specified, then launch azimuth may be read directly from Figure 7.5-4 or computed more accurately from

$$\cos \beta_o = \frac{\sin \delta_1 - \sin \delta_o \cos \psi_t}{\cos \delta_o \sin \psi_t} . \tag{7.5-3}$$

7.5.3 Selecting an Acceptable Launch Date. Once the lunar declination at intercept is determined the next step is to search through a lunar ephemeris to find a time when the Moon is at the correct declination. If lighting conditions are important, we must find a time when both declination and phase are simultaneously correct. The *American Ephemeris and Nautical Almanac (AE)* lists the right ascension and declination of the Moon for every hour as well as the phases of the Moon. Figure 7.5-5 shows a typical page portion from a lunar ephemeris.

MOON, 1969
FOR EACH HOUR OF EPHEMERIS TIME

Hour	Apparent Right Ascension		Apparent Declination	
	May 5			
h	h m s	s	° ′ ″	″
0	17 26 39.167		−28 19 53.11	
		167.190		−144.78
1	17 29 26.357		28 22 17.89	
		167.267		131.97
2	17 32 13.624		28 24 29.86	
		167.333		119.14
3	17 35 00.957		28 26 29.00	
		167.385		106.31
4	17 37 48.342		28 28 15.31	
		167.424		93.45
5	17 40 35.766		28 29 48.76	
		167.449		80.58
6	17 43 23.215		28 31 09.34	
		167.461		67.71
7	17 46 10.676		28 32 17.05	
		167.460		54.85
8	17 48 58.136		28 33 11.90	
		167.445		41.97
9	17 51 45.581		28 33 53.87	
		167.417		29.10
10	17 54 32.998		28 34 22.97	
		167.376		16.24
11	17 57 20.374		28 34 39.21	
		167.321		− 3.38
12	18 00 07.695		28 34 42.59	
		167.254		+ 9.44
13	18 02 54.949		28 34 33.15	
		167.172		22.27
14	18 05 42.121		28 34 10.88	
		167.078		35.07
15	18 08 29.199		28 33 35.81	
		166.970		47.84
16	18 11 16.169		28 32 47.97	
		166.851		60.58
17	18 14 03.020		28 31 47.39	
		166.717		73.31
18	18 16 49.737		28 30 34.08	
		166.571		85.98
19	18 19 36.308		28 29 08.10	
		166.413		98.63
20	18 22 22.721		28 27 29.47	
		166.242		111.23
21	18 25 08.963		28 25 38.24	
		166.058		123.80
22	18 27 55.021		28 23 34.44	
		165.862		136.31
23	18 30 40.883		−28 21 18.13	
		165.655		+148.78
	May 6			
0	18 33 26.538		−28 18 49.35	
		165.435		+161.19
1	18 36 11.973		28 16 08.16	
		165.203		173.56

Figure 7.5-5 Typical page portion from lunar ephemeris

Suppose we now select a time, t_1, for lunar intercept that meets the declination and lighting constraints. The right ascension of the moon at t_1 should be noted from the *AE*. This is the angle α_1 in Figure 7.5-3.

The difference in right ascension, $\Delta\alpha$, between launch and intercept is fixed by the geometry of Figure 7.5-3. Applying the law of cosines to the spherical triangle in Figure 7.5-3, we get

$$\cos \Delta\alpha = \frac{\cos \psi_t - \sin \delta_o \sin \delta_1}{\cos \delta_o \cos \delta_1} \tag{7.5-4}$$

The next step is to establish the exact time of launch, t_o, and the right ascension of the launch site, α_o, at this launch time. To establish t_o we need to compute the total time from launch to intercept. This consists of the free-flight time from injection to intercept, which may be computed from the injection conditions, plus the time to traverse the burning and coasting arc, ψ_c. Thus, the total time, t_t, is

$$t_t = t_{ff} + t_c \tag{7.5-5}$$

where t_{ff} is the free-flight time and t_c is the burn-plus-coast time.

The launch time, t_o, can now be obtained from

$$t_o = t_1 - t_t. \tag{7.5-6}$$

The right ascension of the launch point, α_o, is the same as the "local sidereal time," θ, at the launch point (see Figure 2.8-4), and may be obtained from equation (2.8-9):

$$\alpha_o = \theta = \theta_g + \lambda_E \tag{7.5-7}$$

where λ_E is the east longitude of the launch site and θ_g is the Greenwich Sidereal Time at t_o. Values of θ_g are tabulated in the *AE* for 0^h UT on every day of the year and may be obtained by interpolation for any hour or by the method outlined in section 2.9.

It would be sheer coincidence if the difference $\alpha_1 - \alpha_o$ were the same as the $\Delta\alpha$ calculated from equation (7.5-4). Since the right ascension of the Moon changes very slowly whereas the right ascension of the launch point changes by nearly 360^0 in a day, it is possible to adjust t_1, slightly and recompute t_o and α_o until $\alpha_1 - \alpha_o$ and the required $\Delta\alpha$ from equation (7.5-4) agree. The lunar declination at intercept will change very slightly as t_1 is adjusted and it may be necessary to go back to equation (7.5-3) and redetermine the launch azimuth, β_o.

We now know the injection conditions, r_o, v_o, ϕ_o, β_o and the exact day and hour of launch and lunar intercept that will satisfy all of the constraints set forth earlier. These launch conditions should provide us with sufficiently accurate initial conditions to begin the computation of a precision lunar trajectory using numerical methods.

EXERCISES

7.1 Calculate the burnout velocity required to transfer a probe between the vicinity of the Earth (assume $r_{bo} = 1$ DU) and the Moon's orbit using a Hohmann transfer. What additional Δv would be required to place the probe in the same orbit as that of the Moon. Neglect the Moon's gravity in both parts.

7.2 For a lunar vehicle which is injected at perigee near the surface of the Earth, determine the eccentricity of the trajectory that just reaches the sphere of influence of the Moon.

7.3 For a value of $\epsilon = +20^o$ determine the maximum value of v_2 which will allow lunar impact.

7.4 The following quantities are with respect to the Moon:

$\lambda_1 = 30^o$	$h_m = 33{,}084 \text{ km}^2/\text{sec}$
$\epsilon_2 = -30^o$	$p_m = 224{,}000 \text{ km}$
$v_1 = 1 \text{ km/sec}$	$e_m = 6.3$
$\mathcal{E}_m = 0.426 \text{ km}^2/\text{sec}^2$	

Determine λ of perilune.

7.5 It is desired that a lunar vehicle have its perilune 262 km above the lunar surface with direct motion about the Moon.

a. At the sphere of influence $v_2 = 500$ m/sec. Calculate the initial value of ϵ_2 which must exist to satisfy the conditions.

b. Determine the Δv necessary to place the probe in a circular orbit in which perilune is an altitude of 262 km.

7.6 A space probe was sent to investigate the planet Mars. On the way it crosses the Moon's orbit. The burnout conditions are:

$r_{bo} = 7972$ km $\qquad \phi_{bo} = 0^\circ$

$v_{bo} = 1600$ m/sec

For $v_{moon} = 1024$ m/sec and $D = 384{,}400$ km:

a. What is the speed of the probe as it is at the distance D from the Earth?

b. What is the elevation angle of the probe at D?

c. What is the angle through which the Moon will have moved 30 hours after launch of the probe?

7.7 A lunar vehicle arrives at the sphere of influence of the Moon with $\lambda_1 = 0^\circ$. The speed of the vehicle relative to the Earth is 200 m/sec and the flight-path angle relative to the Earth is 80°. The vehicle is in direct motion relative to the Earth. Find v_2 relative to the Moon and ϵ_2. Is the vehicle in retrograde or direct motion relative to the Moon?

7.8 Given a lunar declination at intercept of 15° and a total geocentric sweep angle of 150° find the launch azimuth. Is this an acceptable launch azimuth?

7.9 A sounding rocket is fired vertically from the surface of the Earth so that its maximum altitude occurs at the distance of the Moon's orbit. Determine the velocity of the sounding rocket at apogee relative to the Moon (neglecting Moon gravity) if the Moon were nearby.

7.10 Discuss the change in energy with respect to the Earth due to passing by the Moon in front or behind the Moon's orbit (retrograde or direct orbit with respect to the Moon). Which would be best for a lunar landing?

7.11 Design a computer program which will solve the basic patched conic lunar transfer problem. Specifically, given that you wish to arrive at perilune at a specified altitude, select r_o, v_o, ϕ_o to meet this requirement. One approach is to select a reasonable r_o, v_o, ϕ_o and iterate on λ_1 until the lunar orbit conditions are met. A suggested set of data is $h_{bo} = 200$ km, $\phi_o = 0^o$ and vary the v_o to show the effect of insufficient energy to reach the Moon. Choose a reasonable value of maximum available velocity (of which v_o is a part) and consider having enough velocity remaining at perilune to inject the probe into a circular lunar orbit at perilune altitude.

***7.12** A "free-return" lunar trajectory is one which passes around the Moon in such a manner that it will return to the Earth with no additional power to change its orbit. Determine such an orbit using the patched conic method. Find the v_{bo}, ϕ_{bo} and γ_o such that the return trajectory will have $h_{perigee} = 50$ n mi and thus insure re-entry. Use $h_{bo} = 100$ n mi. Attempt to have a perilune altitude of about 100 n mi. (Hint: Use a computer and don't expect an exact solution. There are several orbits that may meet the criteria. You will have to extend the iterative technique of this chapter.)

***7.13** To show that the calculations in problem 7.12 cannot really be used in practice, vary v_{bo}, r_{bo} and ϕ_{bo} by some small amount (.1 percent) to find the sensitivity of perilune distance and return trajectory perigee to the burnout parameters. Then make some estimates as to how inaccurate the patched-conic approximation is and discuss how great the real-world or actual errors would be (use the sensitivity calculations as a basis for the discussion).

LIST OF REFERENCES

1. Gamow, George. *The Moon*. London and New York, Abelard-Schuman, 1959.

2. Fisher, Clyde. *The Story of the Moon*. Garden City, NY, Doubleday, Doran and Company, Inc, 1943.

3. Newton, Isaac. *Principia*. Motte's translation revised by Cajori. Berkeley and Los Angeles, University of California Press, 1962.

4. Baker, Robert M. L. and Maud W. Makemson. *An Introduction to Astrodynamics*. Second Edition. New York, NY, Academic Press, 1967.

5. "Inertial Guidance for Lunar Probes." ARMA Document DR 220-16, November 1959.

6. Battin, Richard H. *Astronautical Guidance*. New York, NY, McGraw-Hill Book Company, 1964.

CHAPTER 8

INTERPLANETARY TRAJECTORIES

There are seven windows in the head, two nostrils, two eyes, two ears, and a mouth; so in the heavens there are two favorable stars, two unpropitious, two luminaries, and Mercury alone undecided and indifferent. From which and many other similar phenomena of nature, such as the seven metals, etc., which it were tedious to enumerate, we gather that the number of planets is necessarily seven.

—Francesco Sizzi
(argument against Galileo's discovery of the satellites of Jupiter)[1]

8.1 HISTORICAL BACKGROUND

The word "planet" means "wanderer." That the naked-eye planets wander among the stars was one of the earliest astronomical observations. At first it was not understood. The ancient Greeks gradually saw its significance; indeed, Aristarchus of Samos realized that the planets must revolve around the Sun as the central body of the solar system. But the tide of opinion ebbed, and an Earth-centered system held the field until Copernicus rediscovered the heliocentric system in the 16th century.

Copernicus, who was born in Polish Prussia in 1473, compiled tables of the planetary motions that remained useful until superceded by the more accurate measurements of Tycho Brahe. By 1507 he was convinced that the planets revolved around the Sun and in 1530 he wrote a treatise setting forth his revolutionary concept. It is not well known that this work was dedicated to Pope Paul III and that a cardinal paid for the printing; indeed, during the lifetime of Copernicus his work received the approval of the Church. Not until 1616 was it declared "false, and altogether opposed to Holy Scripture," in spite of the fact that Kepler published his first two laws of planetary motion in 1609.

Such was the atmosphere of the Dark Ages that even the telescopic observations of Galileo in 1610 failed to change the Church's position.

Galileo's data seemed to point decisively to the heliocentric hypothesis: the moons of Jupiter were a solar system in miniature. Galileo's books which set forth cogent and unanswerable astronomical arguments in favor of the Copernican theory were suppressed and Galileo himself, at the age of 70, was forced by the Inquisition to renounce what he knew to be true. After swearing that the Earth was "fixed" at the center of the solar system he is said to have murmured under his breath "it does move, nevertheless."

The publication of Newton's *Principia* in 1687 laid to rest forever the Earth-centered concept of the solar system. With Newton the process of formulating and understanding the motions within the solar system was brought to completion.

8.2 THE SOLAR SYSTEM

The Sun is attended by an enormous number of lesser bodies, the members of the solar system. Most conspicuous are the nine planets–Mercury, Venus, Earth, Mars, Jupiter, Saturn, Uranus, Neptune, and Pluto. Between Mars and Jupiter circulate the minor planets, or asteroids which vary in size from a few hundred miles to a few feet in diameter. In addition, comets, some of which pass near the Sun, are spread much more widely throughout the system.

8.2.1 Bode's Law. The mean distances of the principal planets from the Sun show a simple relationship which is known as Bode's Law after the man who formulated it in 1772. If we write down the series 0, 3, 6, 12 . . . , add 4 to each number and divide by 10, the numbers thus obtained represent the mean distances of the planets from the Sun in Astronomical Units (AU). An Astronomical Unit is the mean distance from the Sun to the Earth. The "law," as may be seen in Table 8.2-1, predicts fairly well the distances of all the planets except Neptune and Pluto.

BODE'S LAW

Planet	Bode's Law Distance	Actual Distance
Mercury	0.4	0.39
Venus	0.7	0.72
Earth	1.0	1.00
Mars	1.6	1.52
Asteroids (average)	2.8	2.65
Jupiter	5.2	5.20
Saturn	10.0	9.52
Uranus	19.6	19.28
Neptune	38.8	30.17
Pluto	77.2	39.76

Table 8.2-1

Whether Bode's Law is an empirical accident or is somehow related to the origin and evolution of the solar system by physical laws is a question which remains unanswered.

8.2.2 Orbital Elements and Physical Constants. Except for Mercury and Pluto, the orbits of the planets are nearly circular and lie nearly in the plane of the ecliptic. Pluto's orbit is so eccentric that the perihelion point lies inside the orbit of Neptune; this suggests that Pluto may be an escaped satellite of Neptune.

The size, shape and orientation of the planetary orbits is described by five classical orbital elements which remain relatively fixed except for slight perturbations caused by the mutual attraction of the planets. The sixth orbital element, which defines the position of the planet in its orbit, changes rapidly with time and may be obtained for any date from the *American Ephemeris and Nautical Almanac.* A complete set of orbital elements for the epoch 1969 June 28.0 is presented in Table 8.2-2.

Table 8.2-3 summarizes some of the important physical characteristics of the planets.

8.3 THE PATCHED-CONIC APPROXIMATION

An interplanetary spacecraft spends most of its flight time moving under the gravitational influence of a single body–the Sun. Only for brief periods, compared with the total mission duration, is its path

ORBITAL ELEMENTS OF THE PLANETS*
FOR THE EPOCH 1969 JUNE 28.0

Planet	Semi-major axis a, [AU]	Orbital eccentricity e	Inclination to ecliptic i	Longitude of ascending node Ω	Longitude of perihelion Π	True longitude at epoch ℓ_0
Mercury	.3871	.2056	$7^0.004$	$47^0.970$	$76^0.981$	$341^0.111$
Venus	.7233	.0068	$3^0.394$	$76^0.405$	$131^0.142$	$326^0.400$
Earth	1.000	.0167	$0^0.000$	undefined	$102^0.416$	$276^0.117$
Mars	1.524	.0934	$1^0.850$	$49^0.322$	$335^0.497$	$265^0.096$
Jupiter	5.203	.0482	$1^0.306$	$100^0.139$	$13^0.684$	$188^0.568$
Saturn	9.519	.0539	$2^0.489$	$113^0.441$	$93^0.828$	$31^0.074$
Uranus	19.28	.0514	$0^0.773$	$73^0.916$	$171^0.513$	$183^0.225$
Neptune	30.17	.0050	$1^0.773$	$131^0.397$	$52^0.275$	$237^0.573$
Pluto	39.76	.2583	$17^0.136$	$109^0.870$	$222^0.894$	$175^0.423$

*From reference 2

Table 8.2-2

PHYSICAL CHARACTERISTICS OF THE SUN AND PLANETS*

Planet	Orbital Period years	Mean distance 10^6 km	Orbital speed km/sec	Mass Earth = 1	μ km^3/sec^2	Equatorial radius km	Inclination of equator to orbit
Sun	–	–	–	333432	1.327×10^{11}	696000	7^0 15'
Mercury	.241	57.9	47.87	.056	2.232×10^4	2487	?
Venus	.615	108.1	35.04	.817	3.257×10^5	6187	32^0
Earth	1.000	149.5	29.79	1.000	3.986×10^5	6378	23^0 27'
Mars	1.881	227.8	24.14	.108	4.305×10^4	3380	23^0 59'
Jupiter	11.86	778	13.06	318.0	1.268×10^8	71370	3^0 04'
Saturn	29.46	1426	9.65	95.2	3.795×10^7	60400	26^0 44'
Uranus	84.01	2868	6.80	14.6	5.820×10^6	23530	97^0 53'
Neptune	164.8	4494	5.49	17.3	6.896×10^6	22320	28^0 48'
Pluto	247.7	5896	4.74	.9?	3.587×10^5?	7016?	?

*From reference 3

Table 8.2-3

shaped by the gravitational field of the departure or arrival planet. The perturbations caused by the other planets while the spacecraft is pursuing its heliocentric course are negligible.

Just as in lunar trajectories, the computation of a precision orbit is a trial-and-error procedure involving numerical integration of the complete equations of motion where all perturbation effects are considered. For preliminary mission analysis and feasibility studies it is sufficient to have an approximate analytical method for determining the total Δv required to accomplish an interplanetary mission. The best method available for such an analysis is called the patched-conic approximation and was introduced in Chapter 7.

The patched-conic method permits us to ignore the gravitational influence of the Sun until the spacecraft is a great distance from the Earth (perhaps a million kilometers). At this point its speed relative to the Earth is very nearly the "hyperbolic excess speed" referred to in Chapter 1. If we now switch to a heliocentric frame of reference, we can determine both the velocity of the spacecraft relative to the Sun and the subsequent heliocentric orbit. The same procedure is followed in reverse upon arrival at the target planet's sphere of influence.

The first step in designing a successful interplanetary trajectory is to select the heliocentric transfer orbit that takes the spacecraft from the sphere of influence of the departure planet to the sphere of influence of the arrival planet.

8.3.1 The Heliocentric Transfer Orbit. For transfers to most of the planets, we may consider that the planetary orbits are both circular and coplanar. In Chapter 3 we discussed the problem of transferring between coplanar circular orbits and found that the most economical method, from the standpoint of Δv required, was the Hohmann transfer. A Hohmann transfer between Earth and Mars is pictured in Figure 8.3-1.

While it is always desirable that the transfer orbit be tangential to the Earth's orbit at departure, it may be preferrable to intercept Mars' orbit prior to apogee, especially if the spacecraft is to return to Earth. The Hohmann transfer, if continued past the destination planet, would not provide a suitable return trajectory. For a one-way trip this is irrelevent; however, for a probe which is to be recovered or for a manned mission, this consideration is important. The outbound trip to Mars on the Hohmann trajectory consumes between 8 and 9 months. If the spacecraft continued its flight it would return to the original point

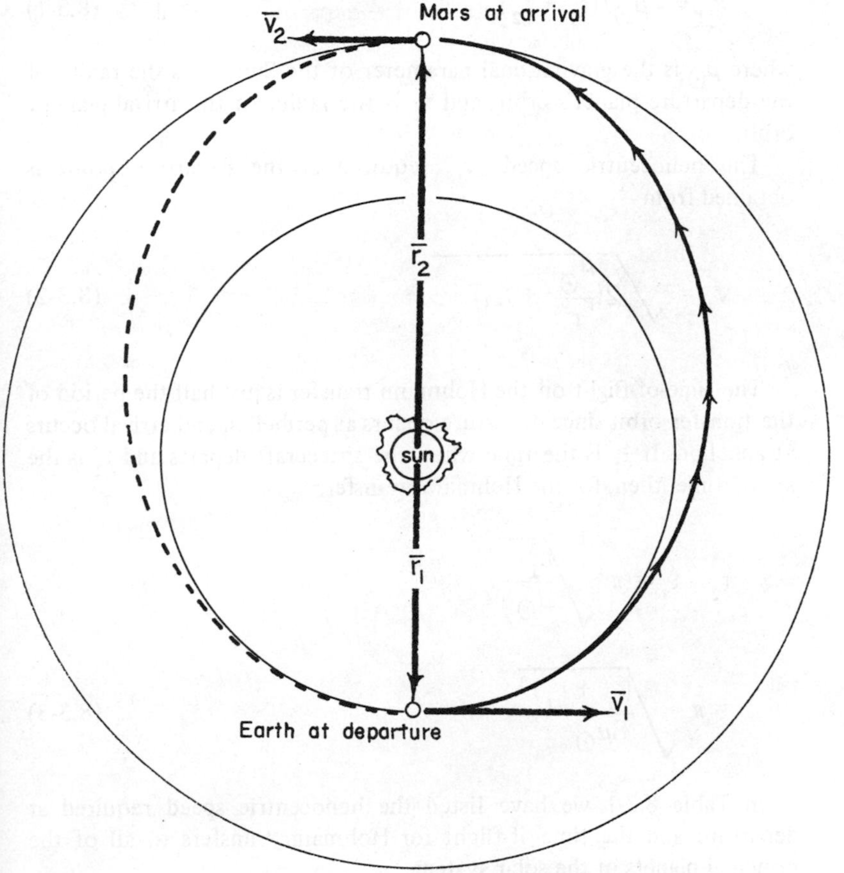

Figure 8.3-1 Hohmann transfer to Mars

of departure only to find the Earth nearly on the opposite side of its orbit. Therefore, either the spacecraft must loiter in the vicinity of Mars for nearly 6 months or the original trajectory must be modified so that the spacecraft will encounter the Earth at the point where it recrosses the earth's orbit.

Nevertheless, the Hohmann transfer provides us with a convenient yardstick for determining the minimum Δv required for an interplanetary mission. The energy of the Hohmann transfer orbit is given by

$$\mathcal{E}_t = -\mu_\odot/(r_1 + r_2) \tag{8.3-1}$$

where $\mu_\odot$ is the gravitational parameter of the Sun, r_1 is the radius of the departure planet's orbit, and r_2 is the radius of the arrival planet's orbit.

The heliocentric speed, v_1, required at the departure point is obtained from

$$v_1 = \sqrt{2(\frac{\mu_\odot}{r_1} + \mathcal{E}_t)} \ . \tag{8.3-2}$$

The time-of-flight on the Hohmann transfer is just half the period of the transfer orbit since departure occurs at perihelion and arrival occurs at aphelion. If t_1 is the time when the spacecraft departs and t_2 is the arrival time, then, for the Hohmann transfer,

$$t_2 - t_1 = \pi \sqrt{\frac{a_t^3}{\mu_\odot}}$$

$$= \pi \sqrt{\frac{(r_1 + r_2)^3}{8\mu_\odot}} \tag{8.3-3}$$

In Table 8.3-1 we have listed the heliocentric speed required at departure and the time-of-flight for Hohmann transfers to all of the principal planets of the solar system.

The orbital speed of the Earth is 1 $AU_\odot/TU_\odot$ or 29.78 km/sec. From Table 8.3-1 it is obvious that transfers to the inner planets require that the spacecraft be launched in the direction opposite the Earth's orbital motion so as to cancel some of the Earth's orbital velocity; transfers to the outer planets require that the spacecraft depart the Earth parallel to the Earth's orbital velocity.

In any case, the difference between the heliocentric speed, v_1, and the Earth's orbital speed represents the speed of the spacecraft *relative to the Earth* at departure which we will call Δv_1.

The method for computing Δv_1 for other than Hohmann transfers is presented in section 3.3.3 and should be reviewed.

HOHMANN TRAJECTORIES FROM EARTH

Planet	Heliocentric Speed at Departure, v_1		Time-of-Flight, $t_2 - t_1$	
	$AU_\odot/TU_\odot$	km/sec	days	years
Mercury	.748	22.28	105.5	–
Venus	.916	27.28	146.1	–
Mars	1.099	32.73	258.9	–
Jupiter	1.295	38.57	–	2.74
Saturn	1.345	40.05	–	6.04
Uranus	1.379	41.07	–	16.16
Neptune	1.391	41.42	–	30.78
Pluto	1.397	41.60	–	46.03

Table 8.3-1

The important thing to remember is that Δv_1 represents the speed of the spacecraft relative to the Earth *upon exit from the Earth's sphere of influence.* In other words, it is the hyperbolic excess speed left over after the spacecraft has escaped the Earth.

EXAMPLE PROBLEM. Calculate the heliocentric departure speed and time of flight for a Hohmann transfer from Earth to Mars. Assume both planets are in circular coplanar orbits. Neglect the phasing requirements due to their relative positions at the time of transfer.

From Table 8.2-2 the radius to Mars from the Sun, $r_♂$, and the radius to Earth from the Sun, $r_\oplus$, are respectively:

$$r_♂ = 1.524 \text{ AU}$$

$$r_\oplus = 1.0 \text{ AU}$$

From equation (8.3-1) the energy of the Hohmann transfer trajectory is:

$$\mathcal{E}_t = -\mu_\odot/(r_1 + r_2) = -\mu_\odot/(r_\oplus + r_♂) = -.396 \text{ AU}_\odot^2/\text{TU}_\odot^2 .$$

From equation (8.3-2) the heliocentric speed, v_1, required at the departure point is:

$$v_1 = \sqrt{2\left(\frac{\mu_\odot}{r_\oplus} + \mathcal{E}_t\right)} = 1.099 \text{AU/TU}_\odot$$
$$= 32.73 \text{Km/sec}.$$

From equation (8.3-3) the time-of-flight from Earth to Mars for the Hohmann transfer is

$$t_2 - t_1 = \pi \sqrt{\frac{a_t^3}{\mu_\odot}} = \pi \sqrt{\frac{(r_\oplus + r_♂)^3}{8\mu_\odot}}$$

$$= 4.4538 \text{ TU}_\odot = .7088 \text{ years} = 258.9 \text{ days}$$

8.3.2 Phase Angle at Departure. If the spacecraft is to encounter the target planet at the time it crosses the planet's orbit then obviously the Earth and the target planet must have the correct angular relationship at departure. The angle between the radius vectors to the departure and arrival planets is called γ_1, the phase angle at departure, and is illustrated in Figure 8.3-2 for a Mars trajectory.

The total sweep angle from departure to arrival is just the difference in true anomaly at the two points, $\nu_2 - \nu_1$, which may be determined from the polar equation of a conic once p and e of the heliocentric transfer orbit have been selected.

$$\cos \nu_2 = \frac{p - r_2}{er_2} \tag{8.3-4}$$

$$\cos \nu_1 = \frac{p - r_1}{er_1} \tag{8.3-5}$$

The time-of-flight may be determined from the Kepler equation which was derived in Chapter 4. The target planet will move through an angle $\omega_t(t_2 - t_1)$ while the spacecraft is in flight, where ω_t is the angular velocity of the target planet. Thus, the correct phase angle at departure is

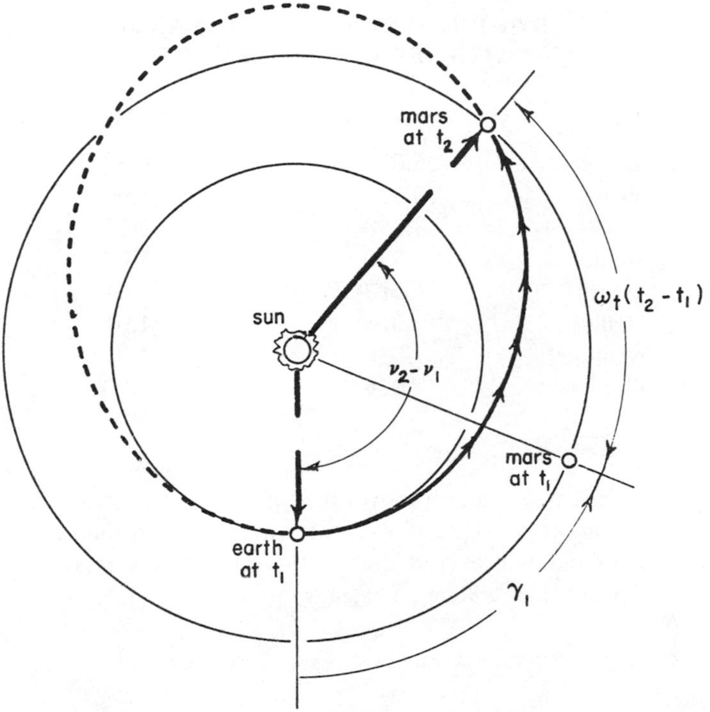

Figure 8.3-2 Phase angle at departure, γ_1

$$\gamma_1 = (\nu_2 - \nu_1) - \omega_t(t_2 - t_1) \,. \tag{8.3-6}$$

The requirement that the phase angle at departure be correct severely limits the times when a launch may take place. The heliocentric longitudes of the planets are tabulated in the *American Ephemeris and Nautical Almanac,* pp 160-176, and these may be used to determine when the phase angle will be correct. The heliocentric longitude of the Earth may be obtained from pp 20-33 of the *AE* by adding 180° to the tabulated geocentric longitudes of the Sun.

If we miss a particular launch opportunity, how long will we have to wait until the correct phase angle repeats itself? The answer to this question depends on the "synodic period" between the Earth and the particular target planet. Synodic period is defined as the time required for any phase angle to repeat itself.

SYNODIC PERIODS OF THE PLANETS WITH RESPECT TO THE EARTH

Planet	ω_t, rad/yr	Synodic Period, yrs
Mercury	26.071	0.32
Venus	10.217	1.60
Mars	3.340	2.13
Jupiter	.530	1.09
Saturn	.213	1.04
Uranus	.075	1.01
Neptune	.038	1.01
Pluto	.025	1.00

Table 8.3-2

In a time, τ_S, the Earth moves through an angle $\omega_\oplus \tau_S$ and the target planet advances by $\omega_t \tau_S$. If τ_S is the synodic period, then the angular advance of one will exceed that of the other by 2π radians and the original phase angle will be repeated, so

$$\omega_\oplus \tau_S - \omega_t \tau_S = \pm 2\pi$$

$$\tau_S = \frac{2\pi}{\left| \omega_\oplus - \omega_t \right|} \qquad (8.3\text{-}7)$$

The synodic periods of all the planets relative to Earth are given in Table 8.3-2.

It is clear that for the two planets nearest the Earth, Mars and Venus, the times between the reoccurence of a particular phase angle is quite long. Thus, if a Mars or Venus launch is postponed, we must either compute a new trajectory or wait a long time for the same launch conditions to occur again.

8.3.3 Escape From the Earth's Sphere of Influence. Once the heliocentric transfer orbit has been selected and Δv_1 determined, we can proceed to establish the injection or launch conditions near the surface of the Earth which will result in the required hyperbolic excess speed. Since the Earth's sphere of influence has a radius of about 10^6 km, we assume that

$$v_\infty \approx \Delta v_1 \; . \qquad (8.3\text{-}8)$$

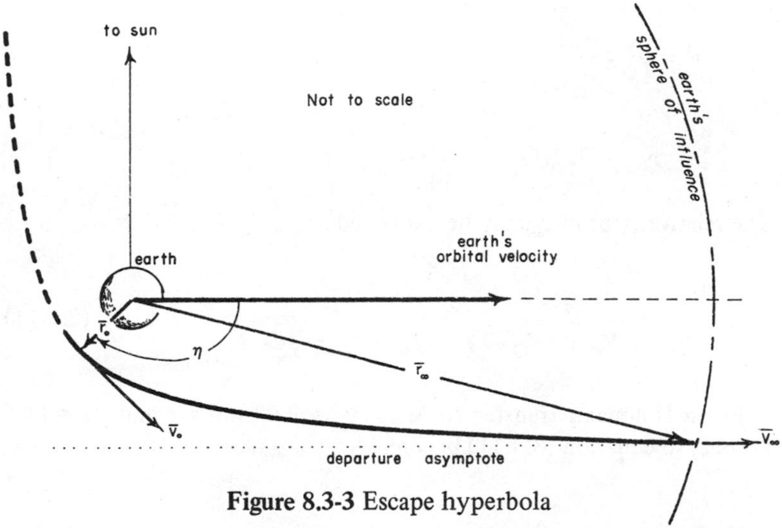

Figure 8.3-3 Escape hyperbola

Since energy is constant along the geocentric escape hyperbola, we may equate $\mathcal{E}$ at injection and $\mathcal{E}$ at the edge of the sphere of influence where $r = r_\infty$.

$$\mathcal{E} = \frac{v_o^2}{2} - \frac{\mu}{r_o} = \frac{v_\infty^2}{2} - \frac{\mu}{r_\infty} \text{ (neglect)} \quad . \tag{8.3-9}$$

Solving for v_o, we get

$$v_o = \sqrt{v_\infty^2 + 2\mu/r_o} \quad . \tag{8.3-10}$$

The hyperbolic excess speed is extremely sensitive to small errors in the injection speed. This may be seen by solving equation (8.3-9) for v_∞^2 and taking the differential of both sides of the resulting equation assuming that r_o is constant:

$$v_\infty^2 = v_o^2 - \frac{2\mu}{r_o}$$

$$2v_\infty dv_\infty = 2v_o dv_o \ .$$

The relative error in v_∞ may be expressed as

$$\frac{dv_\infty}{v_\infty} = \left(\frac{v_o}{v_\infty}\right)^2 \frac{dv_o}{v_o} \ . \qquad (8.3\text{-}11)$$

For a Hohmann transfer to Mars, v_∞ = 2.98 km/sec and v_o = 11.6 km/sec, so equation (8.3-11) becomes

$$\frac{dv_\infty}{v_\infty} = 15.2 \ \frac{dv_o}{v_o}$$

which says that a 1 percent error in injection speed results in a 15.2 percent error in hyperbolic excess speed.

For transfer to one of the outer planets, the hyperbolic excess velocity should be parallel to the Earth's orbital velocity as shown in Figure 8.3-3. Assuming that injection occurs at perigee of the departure hyperbola, the angle, η, between the Earth's orbital velocity vector and the injection radius vector may be determined from the geometry of the hyperbola.

From Figure 8.3-4 we see that the angle, η, is given by

$$\cos \eta = - \frac{a}{c}$$

but since $e = c/a$ for any conic, we may write

$$\cos \eta = -\frac{1}{e} \tag{8.3-12}$$

where the eccentricity, e, is obtained directly from the injection conditions:

$$\mathcal{E} = \frac{v_0^2}{2} - \frac{\mu}{r_0} \tag{8.3-13}$$

$$h = r_0 v_0 \text{ (for injection at perigee)} \tag{8.3-14}$$

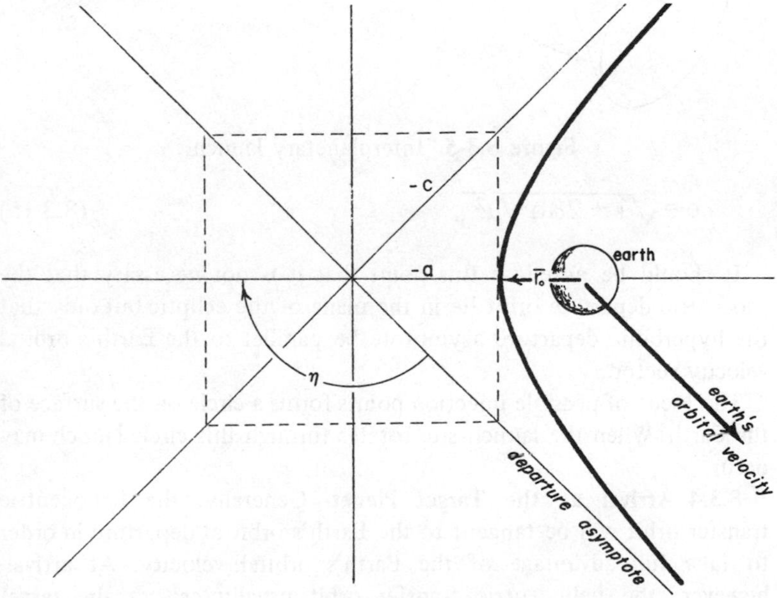

Figure 8.3-4 Geometry of the departure hyperbola

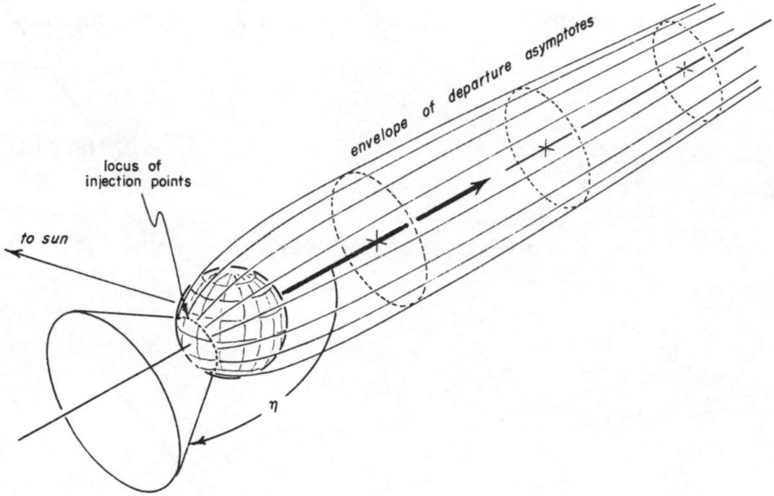

Figure 8.3-5 Interplanetary launch

$$e = \sqrt{1 + 2\mathcal{E}h^2/\mu^2}\ . \tag{8.3-15}$$

It should be noted at this point that it is not necessary that the geocentric departure orbit lie in the plane of the ecliptic but only that the hyperbolic departure asymptote be parallel to the Earth's orbital velocity vector.

The locus of possible injection points forms a circle on the surface of the Earth. When the launch site rotates through this circle launch may occur.

8.3.4 Arrival at the Target Planet. Generally, the heliocentric transfer orbit will be tangent to the Earth's orbit at departure in order to take full advantage of the Earth's orbital velocity. At arrival, however, the heliocentric transfer orbit usually crosses the target planet's orbit at some angle, ϕ_2, as shown in Figure 8.3-6.

If $\mathcal{E}_t$ and h_t are the energy and angular momentum of the heliocentric transfer orbit, then v_2 and ϕ_2 may be determined from

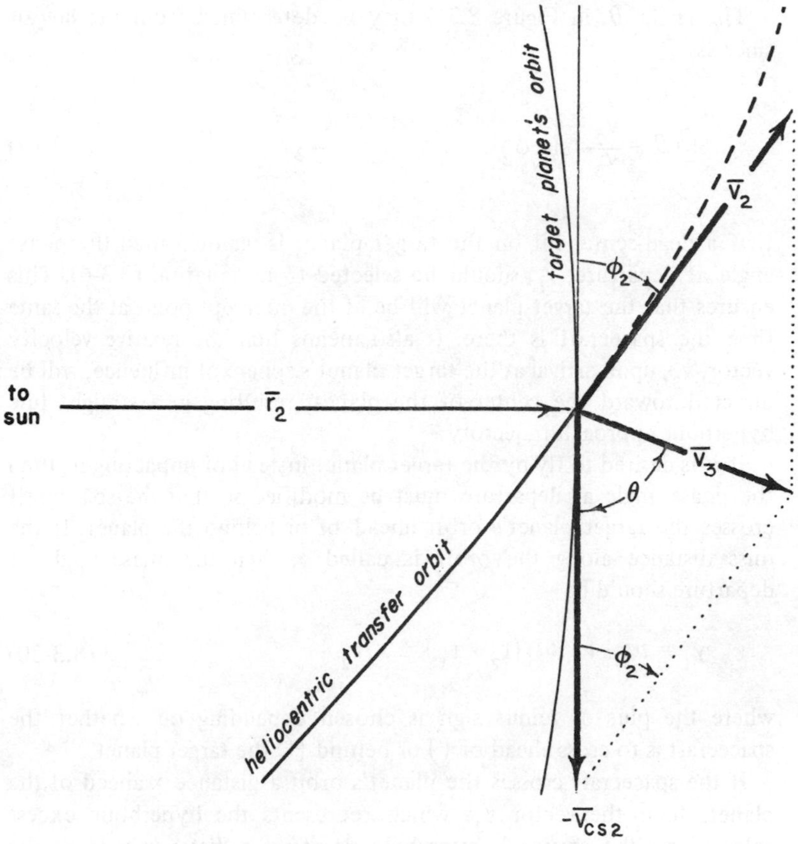

Figure 8.3-6 Relative velocity at arrival

$$v_2 = \sqrt{2(\mu_\odot/r_2 + \mathcal{E}_t)} \tag{8.3-16}$$

$$\cos\phi_2 = h_t/r_2 v_2 \,. \tag{8.3-17}$$

If we call the velocity of the spacecraft *relative to the target planet* $\mathbf{v}_3$, then from the law of cosines,

$$v_3^2 = v_2^2 + v_{cs_2}^2 - 2v_2 v_{cs_2} \cos\phi_2 \tag{8.3-18}$$

where v_{cs2} is the orbital speed of the target planet.

The angle, θ, in Figure 8.3-6 may be determined from the law of sines as

$$\sin \theta = \frac{v_2}{v_3} \sin \phi_2 \ . \tag{8.3-19}$$

If a dead-center hit on the target planet is planned then the phase angle at departure, γ_1, should be selected from equation (8.3-6). This ensures that the target planet will be at the intercept point at the same time the spacecraft is there. It also means that the relative velocity vector, $\mathbf{v}_3$, upon arrival at the target planet's sphere of influence, will be directed toward the center of the planet, resulting in a straight line hyperbolic approach trajectory.

If it is desired to fly by the target planet instead of impacting it, then the phase angle at departure must be modified so that the spacecraft crosses the target planet's orbit ahead of or behind the planet. If the miss distance along the orbit is called x then the phase angle at departure should be

$$\gamma_1 = \nu_2 - \nu_1 - \omega_t (t_2 - t_1) \pm x/r_2 \tag{8.3-20}$$

where the plus or minus sign is chosen depending on whether the spacecraft is to cross ahead of (-) or behind (+) the target planet.

If the spacecraft crosses the planet's orbit a distance x ahead of the planet, then the vector $\mathbf{v}_3$, which represents the hyperbolic excess velocity on the approach hyperbola, is offset a distance y from the center of the target planet as shown in Figure 8.3-7.

From Figure 8.3-7 we see that

$$y = x \sin \theta \tag{8.3-21}$$

Once the offset distance is known, the distance of closest approach or periapsis radius may be computed.

8.3.5 Effective Collision Cross Section. In Figure 8.3-8 the hyperbolic approach trajectory is shown, where $\mathbf{v}_3$ is the hyperbolic excess velocity upon entrance to the target planet's sphere of influence and y is the offset distance.

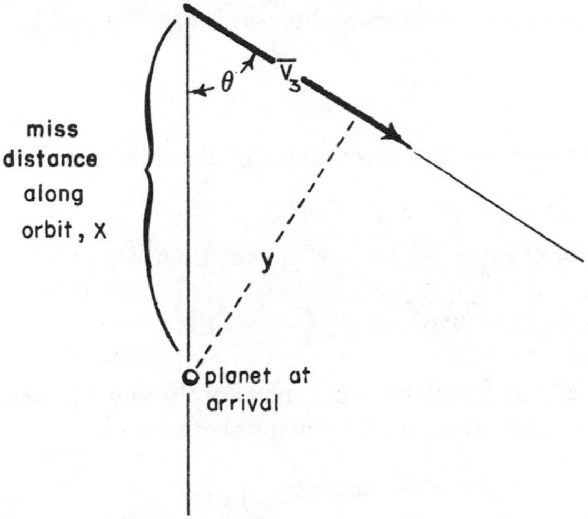

Figure 8.3-7 Offset miss distance at arrival

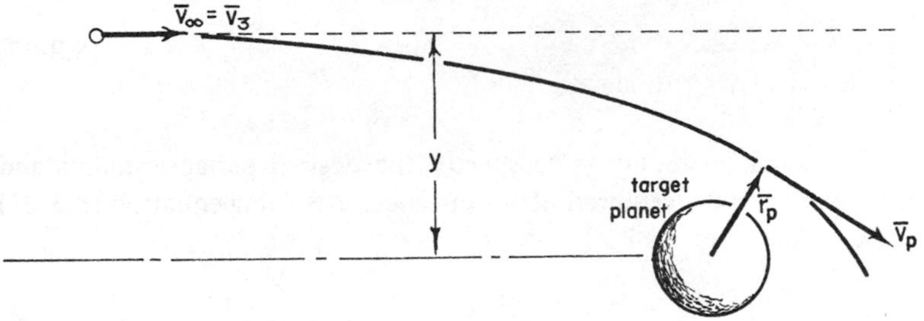

Figure 8.3-8 Hyperbolic approach orbit

From the energy equation we know that the energy in the hyperbolic approach trajectory is

$$\mathcal{E} = \frac{v_3^2}{2} - \cancelto{0}{\frac{\mu_t}{r_\infty}} \quad . \tag{8.3-22}$$

The angular momentum is obtained from

$$h = yv_3 \tag{8.3-23}$$

The parameter and eccentricity of the approach trajectory follow from

$$p = h^2/\mu_t \tag{8.3-24}$$

$$e = \sqrt{1 + 2\mathcal{E}h^2/\mu_t^2} \tag{8.3-25}$$

where μ_t is the gravitational parameter of the target planet.

We may now compute the periapsis radius from

$$r_p = \frac{p}{1+e} \,. \tag{8.3-26}$$

Because angular momentum is conserved, the speed at periapsis is simply

$$v_p = \frac{yv_3}{r_p} \,. \tag{8.3-27}$$

The usual procedure is to specify the desired periapsis radius and then compute the required offset distance, y. Solving equation (8.3-27) for y yields

$$y = \frac{r_p v_p}{v_3} \tag{8.3-28}$$

Equating the energy at periapsis with the energy upon entrance to the sphere of influence, we obtain

$$\mathcal{E} = \frac{v_3^2}{2} = \frac{v_p^2}{2} - \frac{\mu_t}{r_p}$$

and solving for v_p and substituting it into equation (8.3-28) we get

$$y = \frac{r_p}{v_3}\sqrt{v_3^2 + \frac{2\mu_t}{r_p}}\,. \tag{8.3-29}$$

It is interesting to see what offset distance results in a periapsis radius equal to the radius of the target planet, r_t. This particular offset distance is called b, the "impact parameter," since any offset less than this will result in a collision. In this connection it is convenient to assign to the target planet an impact size which is larger than its physical size. This concept is also employed by atomic and nuclear physicists and is called "effective collision cross section." The radius of the effective collision cross section is just the impact parameter, b.

We can determine the impact parameter by setting $r_p = r_t$ in equation (8.3-29).

$$b = \frac{r_t}{v_3}\sqrt{v_3^2 + \frac{2\mu_t}{r_t}}\,. \tag{8.3-30}$$

The effective cross section of the planet represents a rather large target as shown in Figure 8.3-9. If we wish to take advantage of atmospheric braking, then a much smaller target must be considered; only a thin annulus of radius b and width db. The cross section for hitting the atmosphere of the target planet is called the "re-entry corridor" and may be very small indeed.

EXAMPLE PROBLEM. It is desired to send an interplanetary probe to Mars on a Hohmann ellipse around the Sun. The launch vehicle burns out at an altitude of 0.05 DU. Determine the burnout velocity required to accomplish this mission.

The given information is

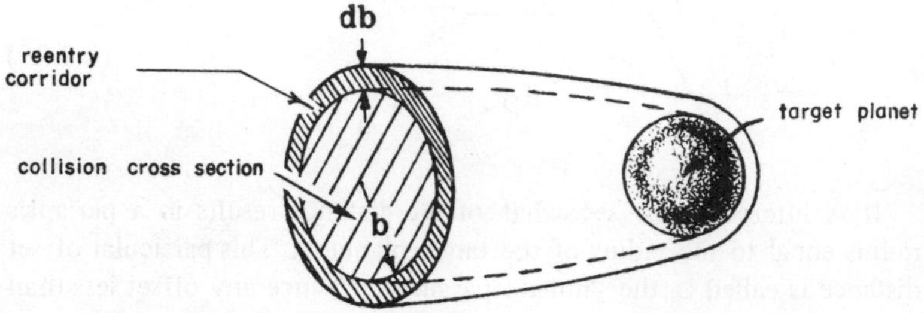

Figure 8.3-9 Effective Cross Section

$$h_{bo} = 0.05 \text{ DU}_{\oplus}$$

Therefore $r_{bo} = r_{\oplus} + h_{bo} = 1.05$ DU

From Table 8.3-1

$v_1 = 1.099$ AU/TU on the Hohmann ellipse at the region of the earth

$$\therefore \Delta v_1 = 1.099 - v_{\oplus} = 1.099 - 1 = 0.099 \text{ AU/TU}$$

$$= 0.373 \text{ DU/TU}$$

At the earth's SOI, $v_\infty = \Delta v_1 = 0.373$ DU/TU and we can write the energy equation

$$\frac{v_\infty^2}{2} - \frac{\mu}{r_\infty} \overset{0}{\nearrow} = \frac{v_{bo}^2}{2} - \frac{\mu}{r_{bo}}$$

Hence

$$v_{bo}^2 = v_\infty^2 + \frac{2\mu}{r_{bo}} = (0.373)^2 + \frac{2(1)}{1.05}$$

$$v_{bo}^2 = .139 + 1.905 = 2.044$$

$$v_{bo} = 1.43 \text{ DU/TU} = \underline{\underline{37{,}076 \text{ ft/sec}}}$$

8.4 NONCOPLANAR INTERPLANETARY TRAJECTORIES

Up to now we have assumed that the planetary orbits all lie in the plane of the ecliptic. From Table 8.2-2 it may be seen that some of the planetary orbits are inclined several degrees to the ecliptic. Fimple[4] has shown that a good procedure to use when the target planet lies above or below the ecliptic plane at intercept is to launch the spacecraft into a transfer orbit which lies in the ecliptic plane and then make a simple plane change during mid-course when the true anomaly change remaining to intercept is 90°. This minimizes the magnitude of the plane change required and is illustrated in Figure 8.4-1.

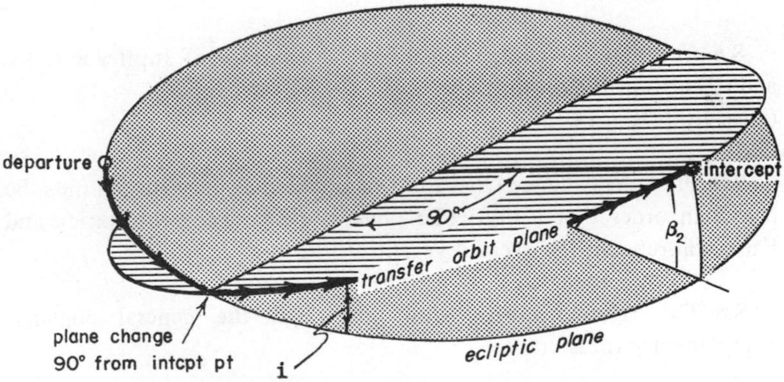

Figure 8.4-1 Optimum plane change point

Since the plane change is made 90° short of intercept, the required inclination is just equal to the ecliptic latitude, β_2, of the target planet at the time of intercept, t_2.

The Δv required to produce this midcourse plane change depends on the speed of the spacecraft at the time of the plane change and is

$$\Delta v = 2v \sin \frac{i}{2} \tag{8.4-1}$$

as derived in equation (3.3-1).

EXERCISES

8.1 Verify the synodic period of Venus in Table 8.3-2.

8.2 Calculate the velocity requirement to fly a solar probe into the surface of the Sun. Use a trajectory which causes the probe to fall directly into the Sun (a degenerate ellipse).

8.3 Calculate the radius of the Earth's sphere of influence with respect to the Sun (see equation (7.4-1)).

8.4 Discuss the advantages and disadvantages of the use of a Hohmann transfer for interplanetary travel.

8.5 Repeat the example problem at the end of section 8.3.5 for the planet Jupiter.

8.6 Calculate the escape speed from the surface of Jupiter in ft/sec and in Earth canonical units.
(Ans. 7.62 $DU_\oplus/TU_\oplus$)

8.7 Find the distance from the Sun at which a space station must be placed in order that a particular phase angle between the station and Earth will repeat itself every 4 years.

8.8 The figure shown below illustrates the general coplanar interplanetary transfer.

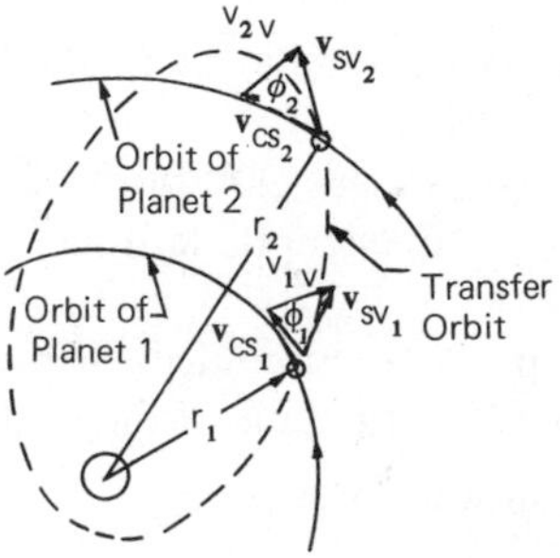

Show that

$$v_{iv} = \left\{ \mu_\odot \left[\frac{3}{r_i} - \frac{(1 - e_\odot^2)}{p_\odot} - 2\sqrt{\frac{p_\odot}{r_i^3}} \right] \right\}^{1/2} \text{AU/TU}$$

or

$$v_{iv} = \left\{ \mu_\odot \left[\frac{3}{r_i} + \frac{2\mathcal{E}_\odot}{\mu_\odot} - \frac{2h_\odot}{\mathbb{P}_i \mu_\odot} \right] \right\}^{1/2} \text{AU/TU}$$

where v_{iv} is the speed of the space vehicle relative to planet i.
r_i is the heliocentric orbit radius of planet i.
$e_\odot$ is the eccentricity of the heliocentric transfer orbit.
$p_\odot$ is the parameter of the heliocentric transfer orbit.
$\mathcal{E}_\odot$ is the specific mechanical energy of the heliocentric transfer orbit.
$h_\odot$ is the specific angular momentum of the heliocentric transfer orbit.
$\mathbb{P}_i$ is the period of planet i in *sidereal years.*

Note that $\mathbf{v}_{iv} = \mathbf{v}_\infty$ at planet i. Refer to material in Chapters 1 and 7 for treating the orbit inside the sphere of influence.

8.9 In preliminary planning for any space mission, it is necessary to see if we have the capability of actually producing the velocity required to accomplish this mission.

a. The space vehicle is in a circular orbit about the Earth of 1.1 $DU_\oplus$ radius. What is the speed necessary to place the vehicle on a parabolic escape path? What is the Δv required?
(Ans. $v_{esc} = 1.348\ DU_\oplus/TU_\oplus$, $\Delta v = 0.394\ DU_\oplus/TU_\oplus$)

b. Actually we want a hyperbolic excess speed of 5,000 ft/sec. What must our speed be as we leave the circular orbit? What is the Δv?
(Ans. $v = 39{,}900$ ft/sec, $\Delta v = 15{,}100$ ft/sec)

c. It can be shown that $\Delta v = c \ln M$ where c is the effective exhaust velocity of the engine and $M = \dfrac{m_o}{m_{bo}}$. If $c = 9{,}000$ ft/sec what is the ratio of the initial mass to burnout mass?
(Ans. $M = 5.35$)

8.10 A Venus probe departs from a $2DU_\oplus$ radius circular parking orbit with a burnout speed of 1.1 $DU_\oplus/TU_\oplus$. Find the hyperbolic excess speed in geocentric and heliocentric speed units. What is v_∞ in ft/sec?
(Ans. $v_\infty = 0.458\ DU_\oplus/TU_\oplus = 0.122\ AU/TU_\odot = 11{,}900$ ft/sec)

8.11 A Mariner space probe is to be sent from the Earth to Mars on a heliocentric transfer orbit with $p = 2/3\ DU_\odot$ and $e = 2/3$.

a. What will be the speed of the probe relative to the Earth after it has escaped the Earth's gravity?
(Ans. 0.731 $DU_\odot TU_\odot$)

b. What burnout speed is required near the surface of the Earth to inject the Mariner into its heliocentric orbit?
(Ans. 3.09 $DU_\oplus/TU_\oplus$)

8.12 A space probe is in an elliptical orbit with an apogee of 3 $DU_\oplus$ and a perigee of 1 $DU_\oplus$.

a. Determine the energy in the orbit before and after an impulsive Δv of .1 $DU_\oplus/TU_\oplus$ is applied at apogee. What is the $\Delta\mathcal{E}$?
(Ans. $\Delta\mathcal{E} = 0.046\ DU_\oplus^2/TU_\oplus^2$)

b. Find the $\Delta\mathcal{E}$ which results if the same Δv is applied at perigee.
(Ans. $\Delta\mathcal{E} = 0.128\ DU_\oplus^2/TU_\oplus^2$)

c. What is the Δv required to achieve escape speed from the original elliptical orbit at apogee?

d. What is the Δv required to escape from the above orbit at perigee?

e. Is it more efficient to apply a Δv at apogee or perigee?

8.13 (This problem is fairly long, so work carefully and follow any suggestions.) We wish to travel from the Earth to Mars. The mission will begin in a 1.5 $DU_\oplus$ circular orbit. The burnout speed after thrust application in the circular orbit is to be 1.5 $DU_\oplus/TU_\oplus$. The thrust is applied at the perigee of the escape orbit. The transfer orbit has an energy (with respect to the Sun) of -0.28 AU^2/TU^2.

a. Find the hyperbolic excess speed, v_∞.
(Ans. 0.956 $DU_\oplus/TU_\oplus$)

b. Find the hyperbolic excess speed in heliocentric units.
(Ans. 0.254 $AU/TU_\odot$)

c. Find the velocity of the satellite with respect to the Sun at its departure from the Earth.
(Ans. 1.2 $AU/TU_\odot$)

d. Find v_{sv_2} at arrival at the Mars orbit.
(Ans. 0.867 $AU/TU_\odot$)

e. Find the hyperbolic excess speed upon arrival at Mars. (Hint: Find ϕ_1 from the Law of Cosines, then find h, ϕ_2 and v_{mv} in that order.)
(Ans. v_{mv} = 0.373 $AU/TU_\odot$)

f. What will be the re-entry speed at the surface of Mars?
(Ans. 3.39 $DU_♂/TU_♂$)

8.14 a. What will be the time-of-flight to Mars on the heliocentric orbit of Problem 8.11 if the radius of Mars' orbit is 1.5 AU?
(Ans. 0.87 $TU_\odot$)

b. Compute the phase angle at departure, γ_1, for the above transfer.
(Ans. 0°)

8.15 To accomplish certain measurements of phenomena associated with sunspot activity, it is necessary to establish a heliocentric orbit with a perihelion of 0.5 AU. The departure from the Earth's orbit will be at apohelion.What must the burnout velocity)in $DU_\oplus/TU_\oplus$ and ft/sec) be at an altitude of 0.2 $DU_\oplus$ to accomplish this mission?
(Ans. 1.464 $DU_\oplus/TU_\oplus$ = 38,050 ft/sec)

8.16 A space probe can alter its flight path without adding propulsive energy by passing by a planet en route to its destination. Upon passing the planet it will be deflected some amount, called the turning angle. For a given planet, P, show that

$$\sin \delta = \frac{1}{1 + \frac{Dv^2}{\mu_p}}$$

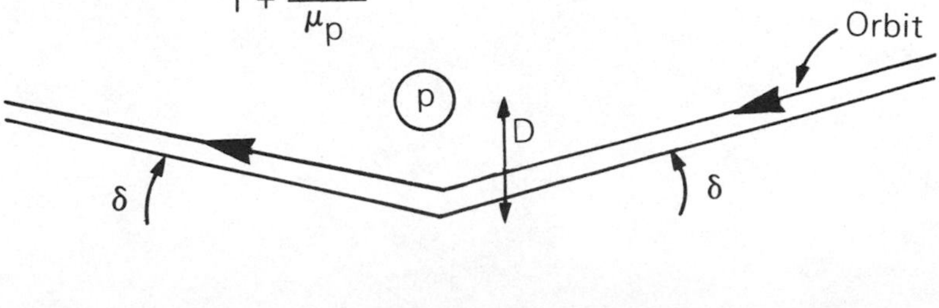

where D and δ are as defined in the drawing and v is the velocity relative to the planet and μ_p is the gravitational parameter of the planet.

* **8.17** With the use of Hohmann transfer analysis calculate an estimate of the total Δv required to depart from Earth and soft land a craft on Mars. What would be an estimate of the return Δv? Give the answer in km/sec.

* **8.18** It is desired to return to an inferior (inner) planet at the earliest possible time from a superior (outer) planet. The return is to be made on an ellipse with the same value of p and e as was used on the outbound journey. Derive an expression for the loiter time at the superior planet in terms of the phase angle, transfer time, and the angular rates of the planets. Assume circular, coplanar planet orbits.

LIST OF REFERENCES

1. Newman, James R. *The World of Mathematics*. Vol 2. New York, NY, Simon and Schuster, 1956.

2. *American Ephemeris and Nautical Almanac, 1969*. Washington, DC, US Government Printing Office, 1967.

3. *Explanatory Supplement to the American Ephemeris and Nautical Almanac*. London, Her Majesty's Stationery Office, 1961.

4. Fimple, W. R. "Optimum Midcourse Plane Change for Ballistic Interplanetary Trajectories." *AIAA Journal*, Vol 1, No 2, pp 430-434, 1963.

CHAPTER 9

PERTURBATIONS

"Life itself is a perturbation of the norm."
"An unperturbed existence leads to dullness of spirit and mind."
"To cause a planet or other celestial body to deviate from a theoretically regular orbital motion." —Webster

9.1 INTRODUCTION AND HISTORICAL BACKGROUND

A perturbation is a deviation from some normal or expected motion. Macroscopically, one tends to view the universe as a highly regular and predictable scheme of motion. Yet when it is analyzed from accurate observational data, it is found that there seem to be distinct, and at times unexplainable, irregularities of motion superimposed upon the average or mean movement of the celestial bodies. After working in great detail in this text with many aspects of the two-body problem, it may come as somewhat of a shock to realize that real-world trajectory problems cannot be solved accurately with the restricted two-body theory presented. The actual path will vary from the theoretical two-body path due to perturbations caused by other mass bodies (such as the Moon) and additional forces not considered in Keplerian motion (such as nonspherical Earth).

We are very familiar with perturbations in almost every area of life. Seldom does anything go exactly as planned, but is perturbed by various unpredictable circumstances. There is always some variation from the norm. An excellent example is an aircraft encountering a wind gust. The power and all directional controls are kept constant, yet the path may change abruptly from the theoretical. The wind gust was a perturbation. Fortunately, most of the perturbations we will need to consider in orbital flight are much more predictable and analytically tractable than the above example. Those which are unpredictable (wind gusts, meteoroid collisions, etc.) must be treated in a stochastic (probabilistic) manner. Some of the perturbations which could be considered are due to the presence of other attracting bodies, atmospheric drag and lift, the asphericity of the Earth or Moon, solar radiation effects, magnetic, and relativistic effects. Analytic formulations of some of these will be given in section 9.6.

It should not be supposed that perturbations are always small, for they can be as large as or larger than the primary attracting force. In fact, many interplanetary missions would miss their target entirely if the perturbing effect of other attracting bodies were not taken into account. Ignoring the effect of the oblateness of the Earth on an artificial satellite would cause us to completely fail in the prediction of its position over a long period of time. Without the use of perturbation methods of analysis it would be impossible to explain or predict the orbit of the Moon. Following are some specific examples of the past value of perturbation analysis. Newton had explained most of the variations in the Moon's orbit, except the motion of the perigee. In 1749 Clairant found that the second-order perturbation terms removed discrepancies between the observed and theoretical values, which had not been treated by Newton. Then, about a century later, the full explanation was found in an unpublished manuscript of Newton's. E. M. Brown's papers of 1897-1908 explain in great detail the perturbative effects of the oblateness of the Earth and Moon on the Moon's orbit and the effect of other planets. The presence of the planet Neptune was deduced analytically by Adams and by Leverrier from analysis of the perturbed motion of Uranus. The first accurate prediction of the return of Halley's Comet in 1759 by Clairant was made from calculation of the perturbations due to Jupiter and Saturn. He correctly predicted a possible error of 1 month due to mass uncertainty and other more distant planets. The shape of the Earth was deduced by an analysis of long-period perturbation terms in the eccentricity of the Earth's orbit. These are but a few of the examples of application of perturbation analysis. Note that these were studies of bodies over which we have no control. With the capability for orbital flight, knowledge of how to handle perturbations is a necessary skill for applying astrodynamics to real-world problems of getting from one place (or planet) to another. This chapter is a first step to reality from the simplified theoretical foundation laid earlier in this text.

There are two main categories of perturbation techniques. These are referred to as *special perturbations* and *general* (or *absolute*) *perturbations*. Special perturbations are techniques which deal with the direct numerical integration of the equations of motion including all necessary perturbing accelerations. This is the main emphasis of this chapter. General perturbation techniques involve an analytic integration of series expansions of the perturbing accelerations. This is a more difficult and lengthy technique than special perturbation techniques, but it leads to better understanding of the source of the perturbation. Most of the discoveries of the preceding paragraph are due to general perturbation studies.

The objective of this chapter is to present some of the more useful and well-known special perturbation techniques in such a way that the reader can determine when and how to apply them to a specific problem. In particular, the Cowell, Encke, and variation of elements techniques are discussed. Since numerical integration is necessary in many uses, a section discussing methods and errors is included. Analytical formulation of some of the more common perturbation accelerations is included to aid application to specific problems. A list of classical and current works on perturbations and integration techniques will be included at the end of the chapter for the reader who wishes to study perturbation analysis in greater depth. See section 1.2 for a table showing relative perturbation accelerations on a typical Earth satellite.

9.2 COWELL'S METHOD

This is the simplest and most straight forward of all the perturbation methods. It was developed by P. H. Cowell in the early 20th century and was used to determine the orbit of the eighth satellite of Jupiter. Cowell and Crommelin also used it to predict the return of Halley's Comet for the period 1759 to 1910. This method has been "rediscovered" many times in many forms. It is especially popular and useful now with faster and larger capacity computers becoming common.

The application of Cowell's method is simply to write the equations of motion of the object being studied, including all the perturbations, and then to integrate them step-by-step numerically. For the two-body problem with perturbations, the equation would be

$$\ddot{\mathbf{r}} + \frac{\mu}{r^3}\mathbf{r} = \mathbf{a}_p \tag{9.2-1}$$

For numerical integration this would be reduced to first-order differential equations.

$$\dot{\mathbf{r}} = \mathbf{v}$$
$$\dot{\mathbf{v}} = \mathbf{a}_p - \frac{\mu}{r^3}\mathbf{r} \tag{9.2-2}$$

where $\mathbf{r}$ and $\mathbf{v}$ are the radius and velocity of a satellite with respect to the larger central body. For numerical integration equation (9.2-2) would be further broken down into the vector components.

$$\dot{x} = v_x \qquad \dot{v}_x = a_{px} - \frac{\mu}{r^3}x$$

$$\dot{y} = v_y \qquad \dot{v}_y = a_{py} - \frac{\mu}{r^3}y \tag{9.2-3}$$

$$\dot{z} = v_z \qquad \dot{v}_z = a_{pz} - \frac{\mu}{r^3}z$$

where $r = (x^2 + y^2 + z^2)^{1/2}$

The perturbing acceleration, $\mathbf{a}_p$, could be due to the presence of other gravitational bodies such as the Moon, Sun and planets. Considering the Moon as the perturbing body, the equations would be

$$\dot{\mathbf{r}} = \mathbf{v}$$

$$\dot{\mathbf{v}} = -\frac{\mu_\oplus}{r^3}\,\mathbf{r} - \mu_m\left(\frac{\mathbf{r}_{ms}}{r_{ms}^3} - \frac{\mathbf{r}_{m\oplus}}{r_{m\oplus}^3}\right) \tag{9.2-4}$$

where r = radius from Earth to satellite
r_{ms} = radius from Moon to satellite
$r_{m\oplus}$ = radius from Moon to Earth
μ_m = gravitational parameter of the Moon.

Having the analytical formulation of the perturbation, the state (**r** and **v**) at any time can be found by applying one of the many available numerical integration schemes to equations (9.2-3).

The main advantage of Cowell's method is the simplicity of formulation and implementation. Any number of perturbations can be handled at the same time. Intuitively, one would suspect that no method so simple would also be free of shortcomings, and he would be correct. There are some disadvantages of the method. When motion is near a large attracting body, smaller integration steps must be taken which severely affect time and accumulative error due to roundoff. This method is approximately 10 times slower than Encke's method. It has been found that Cowell's method is not the best for lunar trajectories. Even in double precision computer arithmetic, roundoff error will soon take its toll in interplanetary flight calculations if step size is small.

Classically, Cowell's method has been applied in a Cartesian coordinate system as shown in equations (9.2-2). Some improvement can be made in trajectory problems by formulating the problem in polar or spherical coordinates. In this case the r will tend to vary slowly and the angle change often will be monotonic. This will often permit larger integration step sizes for the same truncation error. In spherical coordinates (r, θ, ϕ) the equations of motion are (for an equatorial coordinate system):

$$\ddot{r} - r\,(\dot{\theta}^2 \cos^2\phi + \dot{\phi}^2) = -\frac{\mu}{r^2}$$

$$r\ddot{\theta}\cos\phi + 2\dot{r}\dot{\theta}\cos\phi - 2r\dot{\theta}\dot{\phi}\sin\phi = 0 \tag{9.2-5}$$

$$r\ddot{\phi} + 2\dot{r}\dot{\phi} + r\dot{\theta}^2 \sin\phi \cos\phi = 0 \qquad (9.2\text{-}5)$$

As a reminder, one should always pick the coordinate system that best accommodates the problem formulation and the solution method.

9.3 ENCKE'S METHOD

Though the method of Encke is more complex than that of Cowell, it appeared over half a century earlier in 1857. Bond had actually suggested it in 1849–2 years before Encke's work became known.

In the Cowell method the sum of all accelerations was integrated together. In the Encke method, the difference between the primary acceleration and all perturbing accelerations is integrated. This implies

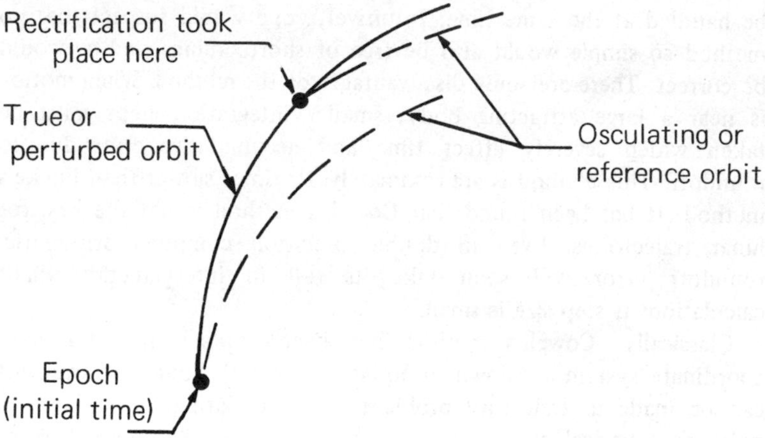

Figure 9.3-1 The osculating orbit with rectification

a reference orbit along which the object would move in the absence of all perturbing accelerations. Presumably, this reference orbit would be a conic section in an ideal Newtonian gravitational field. Thus all calculations and states (position and velocity) would be with respect to the reference trajectory. In most works the reference trajectory is called the *osculating* orbit. "Osculation" is the "scientific" term for kissing. The term connotes the sense of contact, and, in our case, contact between the reference (or osculating) orbit and the true perturbed orbit. The osculating orbit is the orbit that would result if all perturbing accelerations could be removed at a particular time (epoch). At that time, or epoch, the osculating and true orbits are in contact or coincide (see Figure 9.3-1).

Any particular osculating orbit is good until the true orbit deviates too far from it. Then a process called *rectification* must occur to continue the integration. This simply means that a new epoch and starting point will be chosen which coincide with the true orbital path. Then a new osculating orbit is calculated from the true radius and velocity vectors, neglecting the perturbations (see Figure 9.3-1).

Now we will consider the analytic formulation of Encke's method. The basic objective is to find an analytic expression for the difference between the true and reference orbits. Let **r** and $\boldsymbol{\rho}$ be the radius vectors to, respectively, the true (perturbed) and osculating (reference) orbits at a particular time τ ($\tau = t - t_0$). Then for the true orbit,

$$\ddot{\mathbf{r}} + \frac{\mu}{r^3}\mathbf{r} = \mathbf{a}_p \tag{9.3-1}$$

and for the osculating orbit,

$$\ddot{\boldsymbol{\rho}} + \frac{\mu}{\rho^3}\boldsymbol{\rho} = 0\,. \tag{9.3-2}$$

Note that at the epoch, $t_0 = 0$,

$$\mathbf{r}(t_0) = \boldsymbol{\rho}(t_0) \text{ and } \mathbf{v}(t_0) = \dot{\boldsymbol{\rho}}(t_0)\,.$$

Let the deviation from the reference orbit, $\delta\mathbf{r}$, be defined as (see Figure 9.3-2)

$$\delta\mathbf{r} = \mathbf{r} - \boldsymbol{\rho}$$

$$\text{and} \quad \delta\ddot{\mathbf{r}} = \ddot{\mathbf{r}} - \ddot{\boldsymbol{\rho}} \tag{9.3-3}$$

Substituting equations (9.3-1) and (9.3-2) into (9.3-3) we obtain

$$\ddot{\delta \mathbf{r}} = \mathbf{a}_p + \left(\frac{\mu}{\rho^3}\boldsymbol{\rho} - \frac{\mu}{r^3}\mathbf{r}\right)$$

$$= \mathbf{a}_p + \left[\frac{\mu}{\rho^3}(\mathbf{r} - \delta\mathbf{r}) - \frac{\mu}{r^3}\mathbf{r}\right]$$

$$\ddot{\delta \mathbf{r}} = \mathbf{a}_p + \frac{\mu}{\rho^3}\left[\left(1 - \frac{\rho^3}{r^3}\right)\mathbf{r} - \delta\mathbf{r}\right] . \qquad (9.3\text{-}4)$$

This is the desired differential equation for the deviation, $\delta\mathbf{r}$. For a given set of initial conditions, $\delta\mathbf{r}\ (t_o + \Delta t)$ can be calculated numerically. $\boldsymbol{\rho}$ is a known function of time, so $\mathbf{r}$ can be obtained from $\delta\mathbf{r}$ and $\boldsymbol{\rho}$. So, theoretically, we should be able to calculate the perturbed position and velocity of the object. However, one of the reasons for going to this method from Cowell's method was to obtain more accuracy, but the term $(1 - \frac{\rho^3}{r^3})$ is the difference of two very nearly equal quantities which requires many extra digits of computer accuracy on that one operation to maintain reasonable accuracy throughout. A standard method of treating the difference of nearly equal quantities is to define

$$2q = 1 - \frac{r^2}{\rho^2} \qquad (9.3\text{-}5)$$

thus $$\frac{\rho^3}{r^3} = (1 - 2q)^{-3/2} \quad . \qquad (9.3\text{-}6)$$

Equation (9.3-4) then becomes

$$\ddot{\delta \mathbf{r}} = \mathbf{a}_p + \frac{\mu}{\rho^3}\left\{\left[1 - (1 - 2q)^{-3/2}\right]\mathbf{r} - \delta\mathbf{r}\right\} \qquad (9.3\text{-}7)$$

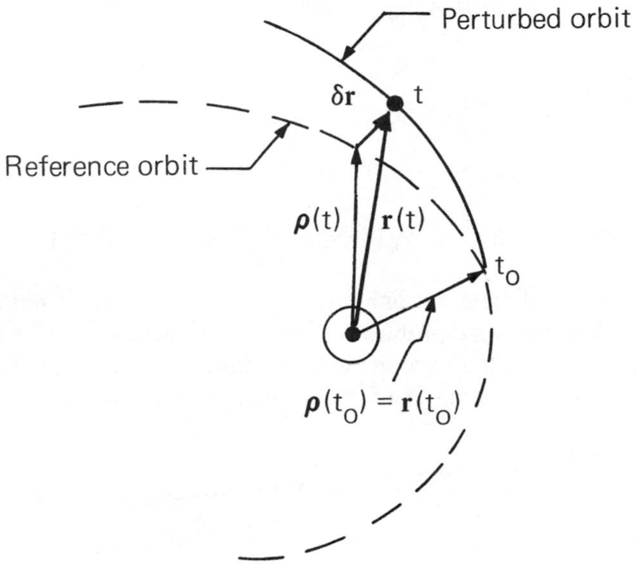

Figure 9.3-2 δ**r** deviation from reference orbit

Our problem is not yet solved, however, since q is a small quantity and the accuracy problem seems unsolved. Now some ways of calculating q must be found. In terms of its components

$$r^2 = x^2 + y^2 + z^2$$

$$= (\rho_x + \delta x)^2 + (\rho_y + \delta y)^2 + (\rho_z + \delta z)^2$$

where δx, δy, δz are the cartesian components of $\delta \mathbf{r}$.
Expanding, we get

$$r^2 = \rho_x^2 + \rho_y^2 + \rho_z^2 + \delta x^2 + \delta y^2 + \delta z^2 + 2\rho_x \delta x + 2\rho_y \delta y$$

$$+ 2\rho_z \delta z = \rho^2 + 2\,[\delta x(\rho_x + \tfrac{1}{2}\delta x) + \delta y(\rho_y + \tfrac{1}{2}\delta y)$$

$$+ \delta z(\rho_z + \tfrac{1}{2}\delta z)]$$

then $$\frac{r^2}{\rho^2} = 1 + \frac{2}{\rho^2}\,[\delta x(\rho_x + \tfrac{1}{2}\delta x) + \delta y(\rho_y + \tfrac{1}{2}\delta y)$$

$$+ \delta z(\rho_z + \tfrac{1}{2}\delta z)]$$

From equation (9.3-5) we find that

$$\frac{r^2}{\rho^2} = 1 - 2q$$

So

$$q = -\frac{1}{\rho^2}\left[\delta x(\rho_x + \tfrac{1}{2}\delta x) + \delta y(\rho_y + \tfrac{1}{2}\delta y) + \delta z(\rho_z + \tfrac{1}{2}\delta z)\right] \qquad (9.3\text{-}8)$$

Now, at any point where $\boldsymbol{\rho}$ and $\delta\mathbf{r}$ are known we can calculate q, but as was noted before the same problem of small differences still exists in equation (9.3-7). There are two methods of solution at this point. The first is to expand the term $(1 - 2q)^{-3/2}$ in a binomial series to get

$$1 - (1 - 2q)^{-3/2} = 3q - \frac{3 \cdot 5}{2!} q^2 + \frac{3 \cdot 5 \cdot 7}{3!} q^3 - \ldots \qquad (9.3\text{-}9)$$

Before the advent of high speed digital computers this was a very cumbersome task, so the second method was to define a function f as

$$f = \frac{1}{q}\left[1 - (1 - 2q)^{-3/2}\right] \qquad (9.3\text{-}10)$$

Then equation (9.3-7) becomes

$$\delta\ddot{\mathbf{r}} = \mathbf{a}_p + \frac{\mu}{\rho^3}(fq\mathbf{r} - \delta\mathbf{r}) \;. \qquad (9.3\text{-}11)$$

Tables of f vs q have been constructed (see *Planetary Coordinates*, 1960)[1] to avoid hand calculation. It should also be noted that when the deviation from the reference orbit is small (which should be the case most of the time) the δx^2, δy^2, δz^2 term can be neglected in equation (9.3-8) and

$$q = -\frac{\rho_x \delta x + \rho_y \delta y + \rho_z \delta z}{\rho^2} \qquad (9.3\text{-}12)$$

which is easily calculated. Then the tables could be used to find f. The method using equation (9.3-9) is recommended for use with the computer, although the use of equation (9.3-12) may be slightly faster. If equation (9.3-12) is used, care must be taken to see when the approximation is not valid.

The Encke formulation reduces the number of integration steps since the $\delta\mathbf{r}$ presumably changes more slowly than $\mathbf{r}$, so larger step sizes can be taken. The advantages of the method diminish if (1) $\mathbf{a}_p$ becomes much larger than $\frac{\mu}{\rho^3}\,(fq\mathbf{r} - \delta\mathbf{r})$ or (2) $\frac{\delta\mathbf{r}}{\rho}$ does not remain quite small. In the case of (1) it may be that the reference parameters (or orbit) need to be changed since the perturbations are becoming primary. In the case of (2) a new osculating orbit needs to be chosen using the values of $\mathbf{r}$ and $\mathbf{v}$ as described earlier. Rectification should be initiated when $\frac{\delta r}{\rho}$ is greater than or equal to some small constant (of the reader's selection depending on the accuracy and speed of the computing machinery used, but of the order of magnitude of 0.01).

Encke's method is generally much faster than Cowell's due to the ability to take larger integration step sizes when near a large attracting body. It will be about 10 times faster for interplanetary orbits, but only about 3 times faster for Earth satellites (see Baker, Vol 2, p 229).[2]

A computational algorithm for Encke's method can be outlined as follows;

a. Given the initial conditions $\mathbf{r}(t_o) = \boldsymbol{\rho}(t_o)$, $\mathbf{v}(t_o) = \dot{\boldsymbol{\rho}}(t_o)$ define the osculating orbit. $\delta\mathbf{r} = 0$, and $\delta r = 0$ at this point.

b. For an integration step, Δt, calculate $\delta\mathbf{r}(t_o + \Delta t)$, knowing $\boldsymbol{\rho}(t_o)$, $r(t_o)$; $q(t_o) = 0$.

c. Knowing $\delta\mathbf{r}(t_o + \Delta t)$ calculate

1. $\boldsymbol{\rho}(t_o + \Delta t)$
2. $q(t_o + \Delta t)$ from equation (9.3-8)
3. $f(t_o + \Delta t)$ from equation (9.3-10).

d. Integrate another Δt to get $\delta\mathbf{r}(t_o + k\Delta t)$.

e. If $\frac{\delta r}{\rho} >$ a specified constant, rectify and go to step a. Otherwise continue.

f. Calculate $\mathbf{r} = \boldsymbol{\rho} + \delta\mathbf{r}$, $\mathbf{v} = \dot{\boldsymbol{\rho}} + \delta\dot{\mathbf{r}}$.

g. Go to step c with Δt replaced by $k\Delta t$ where k is the step number.

This algorithm assumes that the perturbation accelerations are known in analytic form. In a lunar flight another check would be added to determine when the perturbation due to the Moon should become the primary acceleration. The Encke method has been found to be quite good for calculating lunar trajectories. See equation (9.2-3) for the addition of the Moon's gravitational acceleration. Similarly, from the Encke method

$$\delta\ddot{\mathbf{r}} = -\mu_m \left(\frac{\mathbf{r}_{ms}}{r_{ms}^3} - \frac{\mathbf{r}_{m\oplus}}{r_{m\oplus}^3}\right) + \frac{\mu_\oplus}{\rho^3}(fq\mathbf{r} - \delta\mathbf{r}) \qquad (9.3\text{-}13)$$

9.4 VARIATION OF PARAMETERS OR ELEMENTS

The method of the variation of parameters was first developed by Euler in 1748 and was the only successful method of perturbations used until the more recent development of the machine-oriented Cowell and Encke methods. The latter are concerned with a calculation of the coordinates whereas the variation of parameters calculates the orbital elements or any other consistent set of parameters which adequately describe the orbit. Though, at the outset, this method may seem more difficult to implement, it has distinct advantages in many problems where the perturbing forces are quite small.

One main difference between the Encke method and the variation of parameters method is that the Encke reference orbit is constant until rectification occurs. In variation of parameters the reference orbit is changing continuously. For instance, if at any two successive points along the perturbed trajectory one were to calculate the eccentricity of a reference trajectory (unperturbed) using the actual **r** and **v**, one would observe a small change in eccentricity due to the perturbing forces. Then $\dot{e} \approx \frac{\Delta e}{\Delta t}$ which is an approximation for the time rate of change of eccentricity. A similar check could be made on i, a, etc. Thus, the orbital parameters of the reference trajectory are changing with time. In the absence of perturbations, they would remain constant. From previous work we know that **r** and **v** can be calculated from the set of six orbital elements (parameters), so if the perturbed elements could be calculated as a function of time, then the perturbed state (**r** and **v**) would be known.

A two-body orbit can be described by *any* consistent set of six parameters (or constants) since it is described by three second order differential equations. The orbital elements are only one of many possible sets (see Baker, Vol 2, p 246, for a table of parameter sets).[2] The essence of the variation of parameters method is to find how the selected set of parameters vary with time due to the perturbations. This is done by finding analytic expressions for the rate-of-change of the parameters in terms of the perturbations. These expressions are then integrated numerically to find their values at some later time. Integrating the expressions *analytically* is the method of *general perturbations*. It is clear that the elements will vary slowly as compared to the position and velocity variations (e.g. the eccentricity may change only slightly in an entire orbit) so larger integration steps may be taken than in the cases where the total acceleration is being integrated such as in Cowell's method, or the perturbing acceleration is integrated as in Encke's method. In this section two sets of parameters will be treated. The first will be the classical orbital elements because of their familiarity and universality in the literature. The second will be the f and g expression variations which are of greater practical value and are easier to implement.

9.4.1 Variation of the Classical Orbital Elements. The standard orbital elements a, e, i, Ω, ω and T (or M) will be used, where

a = semi-major axis
e = eccentricity
i = orbit inclination
Ω = longitude of the ascending node
ω = argument of periapsis
T = time of periapsis passage
(M_0 = mean anomaly at epoch = $M - n(t - t_0)$).

The object is to find analytic expressions for $\frac{da}{dt}, \frac{de}{dt}, \frac{di}{dt}, \frac{d\Omega}{dt}, \frac{d\omega}{dt}$ and $\frac{dM_0}{dt}$. To do this some reference coordinate system is necessary. Eventually, reference to an "inertial" system may be desired, but any other system can be used in the derivation and the results transferred to the desired coordinate system rather easily. Therefore, it would be best for illustrative purposes to choose a system in which the derivations are the easiest to follow. This derivation partially follows one described in NASA CR-1005.[3] The coordinate system chosen has its principle axis, R (unit vector $\mathbf{R}$), along the instantaneous radius vector, $\mathbf{r}$. The axis S is rotated 90° from R in the direction of increasing true anomaly. The third axis, W, is perpendicular to both R and S. Note that this coordinate system is simply rotated ν from the PQW perifocal system described in Chapter 2.

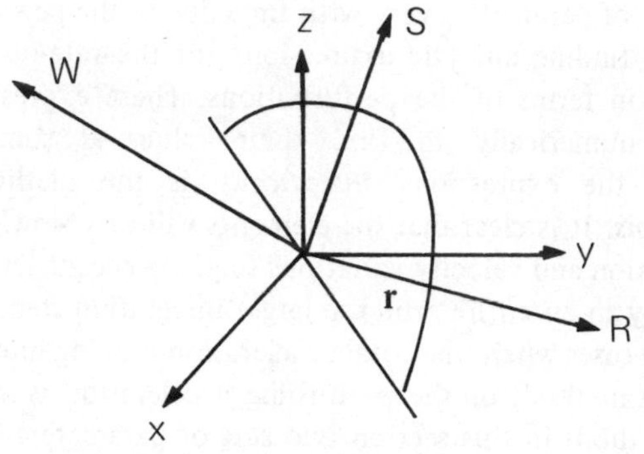

Figure 9.4-1 RSW coordinate system

In the RSW coordinate system the perturbing force is

$$\mathbf{F} = m\,(F_r\mathbf{R}+F_s\mathbf{S}+F_w\mathbf{W}) \text{ (in terms of specific forces)} \tag{9.4-1}$$

and $\mathbf{r} = r\mathbf{R}$

$$\mathbf{v} = \dot{r}\mathbf{R} + r\dot{\nu}\mathbf{S} \tag{9.4-2}$$

$$= \dot{\nu}\left(\frac{dr}{d\nu}\mathbf{R} + r\mathbf{S}\right) .$$

First we will consider the *derivation of* $\frac{da}{dt}$.

The time rate-of-change of energy per unit mass (a constant in the normal two-body problem) is a result of the perturbing force and can be expressed as

$$\frac{d\mathcal{E}}{dt} = \frac{\mathbf{F}\cdot\mathbf{V}}{m} = \dot{\nu}\left(\frac{dr}{d\nu}F_r + rF_s\right) \tag{9.4-3}$$

and

$$\mathcal{E} = -\frac{\mu}{2a} \quad \text{or} \quad a = -\frac{\mu}{2\mathcal{E}}. \tag{9.4-4}$$

Now

$$\frac{da}{dt} = \frac{da}{d\mathcal{E}}\frac{d\mathcal{E}}{dt} = \frac{\mu}{2\mathcal{E}^2}\frac{d\mathcal{E}}{dt}\ . \tag{9.4-5}$$

To express this in known terms, expressions for $\dot{\nu}$ and $\frac{dr}{d\nu}$ must be found. Differentiating the conic equation

$$\frac{dr}{d\nu} = \frac{re\sin\nu}{1 + e\cos\nu} \tag{9.4-6}$$

From the conservation of angular momentum

$$h = r^2\dot{\nu} = \sqrt{\mu p} = \sqrt{\mu a(1 - e^2)} = na^2\sqrt{1 - e^2} \tag{9.4-7}$$

where $n = \sqrt{\frac{\mu}{a^3}}$

Therefore $\dot{\nu} = \frac{na^2}{r^2}\sqrt{1 - e^2}$ (9.4-8)

Substituting in equation (9.4-5) from (9.4-3), (9.4-4), (9.4-6) and (9.4-8) we find

$$\boxed{\frac{da}{dt} = \frac{2e\sin\nu}{n\sqrt{1-e^2}}F_r + \frac{2a\sqrt{1-e^2}}{nr}F_S} \tag{9.4-9}$$

Consider now the *derivation of* $\frac{de}{dt}$.

In deriving the remaining variations, the time rate-of-change of angular momentum is needed. This rate is expressed as the moment of all perturbing forces acting on the system, so

$$\frac{d\mathbf{h}}{dt} = \frac{1}{m}(\mathbf{r}\times\mathbf{F}) = rF_S\mathbf{W} - rF_W\mathbf{S} \tag{9.4-10}$$

(This could also have been developed from

$$\frac{d\mathbf{h}}{dt} = \frac{d}{dt}(\mathbf{r}\times\mathbf{v}) = \left(\frac{d\mathbf{r}}{dt}\times\mathbf{v}\right)^{0} + \mathbf{r}\times\frac{d\mathbf{v}}{dt}$$

where $\frac{d\mathbf{v}}{dt} = \frac{\mathbf{F}}{m} = \mathbf{a}_p$.)

Since $\mathbf{h} = h\mathbf{W}$, the vector time derivative can be expressed as the change of length in $\mathbf{W}$ and its transverse component along the plane of rotation of $\mathbf{h}$, so

$$\frac{d\mathbf{h}}{dt} = \dot{h}\mathbf{W} + h\frac{d\alpha}{dt}\mathbf{S} \tag{9.4-11}$$

where α is the angle of rotation. Note that the change of $\mathbf{h}$ must be in the S-W plane since the perturbing force is applied to $\mathbf{r}$ and the momentum change is the cross product of $\mathbf{r}$ and $\frac{\mathbf{F}}{m}$. Comparing components of equations (9.4-11) and (9.4-10),

$$\frac{dh}{dt} = rF_S \tag{9.4-12}$$

From the expression $p = a(1 - e^2)$

$$e = \left(1 - \frac{p}{a}\right)^{1/2} = \left(1 - \frac{h^2}{\mu a}\right)^{1/2}. \tag{9.4-13}$$

So $\frac{de}{dt} = -\frac{h}{2\mu ae}\left(2\frac{dh}{dt} - \frac{h}{a}\frac{da}{dt}\right)$

$$= - \frac{\sqrt{1 - e^2}}{2na^2 e} \left(2 \frac{dh}{dt} - na\sqrt{1 - e^2} \frac{da}{dt}\right) . \qquad (9.4\text{-}14)$$

Substituting for $\frac{dh}{dt}$ and $\frac{da}{dt}$

$$\boxed{\frac{de}{dt} = \frac{\sqrt{1 - e^2} \sin \nu}{na} F_r + \frac{\sqrt{1 - e^2}}{na^2 e} \left[\frac{a^2(1 - e^2)}{r} - r\right] F_s.}$$

(9.4-15)

The *derivation of* $\frac{di}{dt}$ could be found geometrically from the relationship of $\mathbf{h}, \Omega$ and i, but it is more direct to do it analytically (see NASA CR-1005 for a geometric development).[3] From the chapter on orbit determination recall that

$$\cos i = \frac{\mathbf{h} \cdot \mathbf{K}}{h} \qquad (9.4\text{-}16)$$

Differentiating both sides of equation (9.4-16) we obtain

$$- \sin i \frac{di}{dt} = \frac{h(\frac{d\mathbf{h}}{dt} \cdot \mathbf{K}) - (\mathbf{h} \cdot \mathbf{K}) \frac{dh}{dt}}{h^2}$$

(9.4-17)

$$= \frac{h(rF_s\mathbf{W} - rF_w\mathbf{S}) \cdot \mathbf{K} - h \cos i \, rF_s}{h^2}$$

But $\mathbf{W} \cdot \mathbf{K} = \cos i$ and $\mathbf{S} \cdot \mathbf{K} = \sin i \cos u$ where u is the argument of latitude (angle from the ascending node to $\mathbf{r}$), so

$$- \sin i \frac{di}{dt} = - \frac{rF_w \sin i \cos u}{na^2 \sqrt{1-e^2}} \qquad (9.4\text{-}18)$$

and $$\boxed{\frac{di}{dt} = \frac{rF_W \cos u}{na^2\sqrt{1-e^2}}} \; . \tag{9.4-18}$$

As was expected, changes in i result only from a perturbation along **W**.

In the *derivation of* $\frac{d\Omega}{dt}$ we note from the orbit determination chapter that

$$\cos \Omega = \frac{\mathbf{I} \cdot (\mathbf{K} \times \mathbf{h})}{|\mathbf{K} \times \mathbf{h}|} \; . \tag{9.4-19}$$

Differentiating both sides,

$$- \sin \Omega \frac{d\Omega}{dt}$$

$$= \frac{\mathbf{I} \cdot (\mathbf{K} \times \frac{d\mathbf{h}}{dt}) \; |\mathbf{K} \times \mathbf{h}| - \mathbf{I} \cdot (\mathbf{K} \times \mathbf{h}) \frac{d}{dt} |\mathbf{K} \times \mathbf{h}|}{|\mathbf{K} \times \mathbf{h}|^2}$$

$$= \Big\{ \mathbf{I} \cdot [\mathbf{K} \times (rF_S \mathbf{W} - rF_W \mathbf{S})] \, h \sin i$$

$$- h \cos \Omega \sin i \left(\frac{dh}{dt} \sin i + h \cos i \frac{di}{dt}\right) \Big\} \frac{1}{h^2 \sin^2 i} \; .$$

But $\mathbf{I} \cdot \mathbf{K} \times \mathbf{W} = \cos \Omega \sin i$ and $\mathbf{I} \cdot \mathbf{K} \times \mathbf{S} = \mathbf{I} \times \mathbf{K} \cdot \mathbf{S} = -\mathbf{J} \cdot \mathbf{S} = \sin \Omega \sin u - \cos \Omega \cos u \cos i$ (see Chapter 2 for the coordinate transformation for the PQW system and replace ω with u using spherical trigonometery).

Substituting for $\frac{di}{dt}$ and $\frac{dh}{dt}$ from equations (9.4-18) and (9.4-12) and cancelling terms,

$$\frac{d\Omega}{dt} = \frac{rF_W \sin u}{h \sin i} \tag{9.4-20}$$

$$\boxed{\frac{d\Omega}{dt} = \frac{rF_W \sin u}{na^2\sqrt{1-e^2}\,\sin i}} . \tag{9.4-20}$$

In the following *derivation of* $\frac{d\omega}{dt}$ an important difference from the previous derivations becomes apparent. Up to now "constants" such as a, e, i have been differentiated and the variation has obviously been due to the perturbations. Now, however, the state vectors of position and velocity will appear in the expression. Since we are considering only the time rate-of-change due to the perturbative force, not the changes due to the 2-body reference motion, **r** will *not change to first order as a result of the perturbations* ($\frac{d\mathbf{r}}{dt} = 0$) but the derivative is only with respect to the perturbation forces, so

$$\frac{d\mathbf{v}}{dt} = \frac{\mathbf{F}_p}{m} = \mathbf{a}_p.$$

Without this explanatory note some of the following would be rather mysterious. This procedure is necessary to show the perturbation from the two-body reference orbit.

Consider the expression from orbit determination for finding u ($u = \omega + \nu$)

$$\frac{(\mathbf{K}\times\mathbf{h})\cdot\mathbf{r}}{|\mathbf{K}\times\mathbf{h}|} = r\cos(\omega+\nu) \tag{9.4-21}$$

Differentiating,

$$\frac{|\mathbf{K}\times\mathbf{h}|\,(\mathbf{K}\times\frac{d\mathbf{h}}{dt}\cdot\mathbf{r}) - (\mathbf{K}\times\mathbf{h}\cdot\mathbf{r})\,\frac{d}{dt}|\mathbf{K}\times\mathbf{h}|}{|\mathbf{K}\times\mathbf{h}|^2} = -r\sin(\omega+\nu)\left(\frac{d\omega}{dt} + \frac{d\nu}{dt}\right)$$

$$\frac{d\omega}{dt} = \frac{-|\mathbf{K}\times\mathbf{h}|\,(\mathbf{K}\times\frac{d\mathbf{h}}{dt}\cdot\mathbf{r}) + (\mathbf{K}\times\mathbf{h}\cdot\mathbf{r})\,\frac{d}{dt}|\mathbf{K}\times\mathbf{h}| - \frac{d\nu}{dt}|\mathbf{K}\times\mathbf{h}|^2\, r\sin(\omega+\nu)}{|\mathbf{K}\times\mathbf{h}|^2\, r\sin(\omega+\nu)}$$

$$\frac{d\omega}{dt} = \left\{\frac{1}{h^2 \sin^2 i \; r \sin u}\right\} \left\{- h \sin i \; [\mathbf{K}\times(rF_S\mathbf{W} - rF_W\mathbf{S}) \cdot \mathbf{r}]\right.$$

$$\left. +(\mathbf{K}\times\mathbf{h}\cdot\mathbf{r}) [\frac{dh}{dt}\sin i + h \cos i \frac{di}{dt}] - \frac{d\nu}{dt} h^2 r \sin^2 i \sin u \right\} \tag{9.4-22}$$

where $\mathbf{K}\times\mathbf{W} \cdot \mathbf{r} = r \sin i \cos u$
$\mathbf{K}\times\mathbf{S} \cdot \mathbf{r} = r\mathbf{S}\times\mathbf{R} \cdot \mathbf{K} = - r\mathbf{W} \cdot \mathbf{K} = - r \cos i$
$\mathbf{K}\times\mathbf{h} \cdot \mathbf{r} = rh \sin i \cos u$

and $\frac{dh}{dt}$ and $\frac{di}{dt}$ are given by equations (9.4-12) and (9.4-18). Now an expression for $\frac{d\nu}{dt}$ must be found. The true anomaly is affected by the perturbation due to movement of periapsis and also the node. From the conic equation,

$$r(1 + e \cos \nu) = \frac{h^2}{\mu} \tag{9.4-23}$$

Differentiating the terms which are affected by the perturbations,

$$r(\frac{de}{dt} \cos \nu - e \sin \nu \frac{d\nu}{dt}) = \frac{2h}{\mu}\frac{dh}{dt} \tag{9-4.24}$$

$$re \sin \nu \frac{d\nu}{dt} = r \cos \nu \frac{de}{dt} - \frac{2h}{\mu}\frac{dh}{dt} \tag{9.4-25}$$

At this point equation (9.4-25) could be solved for $\frac{d\nu}{dt}$ and put in equation (9.4-22) and a correct expression would result. However, experience has shown that this is algebraically rather complex and the resulting expression is not as simple as will result from another formulation of $\frac{d\nu}{dt}$ Differentiate the identity

$$\mu re \sin \nu = h \; \mathbf{r}\cdot\mathbf{v} \tag{9.4-26}$$

$$\mu r \frac{de}{dt} \sin \nu + \mu re \cos \nu \frac{d\nu}{dt} = \frac{dh}{dt} \mathbf{r} \cdot \mathbf{v} + hr \cdot \dot{\mathbf{v}} \tag{9.4-27}$$

Now multiply equation (9.4-25) by $\mu \sin \nu$ and equation (9.4-27) by $\cos \nu$ and add to get

$$\frac{d\nu}{dt} = \frac{1}{reh} \left[p \cos \nu\, \mathbf{r} \cdot \dot{\mathbf{v}} - (p+r) \sin \nu \frac{dh}{dt}\right] \tag{9.4-28}$$

Now equation (9.4-22) becomes

$$\frac{d\omega}{dt} = \left[\frac{1}{h^2 r \sin^2 i \sin u}\right] \Big\{ - h \sin i \;[r^2 F_s \sin i \cos u + r^2 F_w \cos i]$$

$$+ rh \sin i \cos u\, [rF_s \sin i + rF_w \cos i \cos u] \tag{9.4-29}$$

$$- \frac{1}{reh} (p \cos \nu\, rF_r - (p+r) \sin \nu\, rF_s)\, h^2 r \sin^2 i \sin u \Big\}$$

Writing $\left(\frac{d\omega}{dt}\right)_r$, $\left(\frac{d\omega}{dt}\right)_s$ and $\left(\frac{d\omega}{dt}\right)_w$ as the variations due to the three components of perturbation forces, after algebraic reduction,

$$\left(\frac{d\omega}{dt}\right)_r = -\frac{\sqrt{1-e^2}\cos \nu}{nae} F_r \tag{9.4-30}$$

$$\left(\frac{d\omega}{dt}\right)_s = \frac{p}{eh} \left[\sin \nu \left(1 + \frac{1}{1+e\cos \nu}\right)\right] F_s \tag{9.4-31}$$

$$\left(\frac{d\omega}{dt}\right)_w = -\frac{r \cot i \sin u}{na^2\sqrt{1-e^2}} F_w \tag{9.4-32}$$

$$\frac{d\omega}{dt} = \left(\frac{d\omega}{dt}\right)_r + \left(\frac{d\omega}{dt}\right)_s + \left(\frac{d\omega}{dt}\right)_w \tag{9.4-33}$$

As an aside, note that it can be shown that the change in ω due to the in-plane perturbations is directly related to the change in the true anomaly due to the perturbation,

$$\left(\frac{d\omega}{dt}\right)_{r,s} = -\frac{d\nu}{dt} \tag{9.4-34}$$

The *derivation of* $\frac{dM_o}{dt}$ follows from the differentiation of the mean anomaly at epoch,

$$M_o = E - e \sin E - n\,(t - t_o) \tag{9.4-35}$$

$$\frac{dM_o}{dt} = \frac{dE}{dt} - \frac{de}{dt} \sin E - e \cos E \frac{dE}{dt} - \frac{dn}{dt}\,(t - t_o)$$

$\frac{de}{dt}$ is known, and $\frac{dn}{dt} = -\frac{3\mu}{2na^4} \frac{da}{dt}$, $\frac{dE}{dt}$, sin E and

Cos E can be obtained from the expressions

$$\cos \nu = \frac{\cos E - e}{1 - e \cos E} \tag{9.4-36}$$

$$\sin \nu = \frac{\sqrt{1 - e^2}\,\sin E}{1 - e \cos E} \tag{9.4-37}$$

The result is (for ellipitcal orbits only)

$$\frac{dMo}{dt} = -\frac{1}{na}\left(\frac{2r}{a} - \frac{1 - e^2}{e} \cos \nu\right) F_r$$
$$- \frac{(1 - e^2)}{nae}\left(1 + \frac{r}{a(1 - e^2)}\right) \sin \nu\, F_s - t \frac{dn}{dt} \tag{9.4-38}$$

To use the variation of parameters formulation the procedure is as follows:

a. At $t = t_0$, calculate the six orbital elements.

b. Compute the perturbation force and transform it at $t = t_0$ to the RSW system.

c. Compute the six rates-of-change of the elements (right side of equations).

d. Numerically integrate the equations one step.

e. The integration has been for the perturbation of the elements, so the changes in elements must be added to the old values at each step.

f. From the new values of the orbital elements, calculate a position and velocity.

g. Go to step b and repeat until the final time is reached.

There are some limitations on this set of parameters. Note that some of them break down when $e = 1$ or $e = 0$. Special formulations using other parameter sets have been developed to handle near-circular and parabolic orbits. Note also that the $\frac{dM_o}{dt}$ was derived for elliptic orbits only.

9.4.2 Variation of Parameters in the Universal Variable Formulation. Since there are many restrictions on the use of the perturbation equations developed in the previous sections, equations relating to the universal variables are developed here. There are no restrictions on the results of this development.

The parameters to be varied in this case are the six components of $\mathbf{r}_0$ and $\mathbf{v}_0$ (these certainly describe the orbit as well as the orbital elements). Thus, we would like to find the time derivative of $\mathbf{r}_0$ and $\mathbf{v}_0$ due to the perturbation forces. In Chapter 4 we described the position at any time as a function of $\mathbf{r}_0$, $\mathbf{v}_0$, f, $\dot{f}$, g and $\dot{g}$, where the f and g expressions were in terms of the universal variable, x. We can express $\mathbf{r}_0$ and $\mathbf{v}_0$ as

$$\begin{aligned} \mathbf{r}_0 &= F\mathbf{r} + G\mathbf{v} \\ \mathbf{v}_0 &= \dot{F}\mathbf{r} + \dot{G}\mathbf{v} \end{aligned} \qquad (9.4\text{-}39)$$

F, G, $\dot{F}$, $\dot{G}$ are the same as f, g, $\dot{f}$, $\dot{g}$ except t is replaced by -t and x is replaced by - x and r and r_0 are interchanged. Thus

$$F = 1 - \frac{x^2}{r} C(z)$$

$$G = -\left[(t - t_0) - \frac{x^3}{\sqrt{\mu}} S(z)\right] \qquad (9.4\text{-}40)$$

$$\dot{F} = \frac{\sqrt{\mu}x}{rr_0}(1 - zS(z))$$

$$\dot{G} = 1 - \frac{x^2}{r_0} C(z) \tag{9.4-40}$$

Note that

$$z = \frac{x^2}{a} = \alpha x^2 \tag{9.4-41}$$

$$\text{where } \alpha = \frac{1}{a} = \frac{2}{r} - \frac{v^2}{\mu} \tag{9.4-42}$$

α is used to permit treatment of parabolic orbits. In fact, in the calculation of the derivatives of equation (9.4-39), $\frac{d(\alpha \mathbf{r}_0)}{dt}$ and $\frac{d(\alpha \mathbf{v}_0)}{dt}$ will be computed. In the differentiation we consider $\mathbf{r}$ to be constant and the acceleration $\dot{\mathbf{v}}$ to result only from the perturbing forces and do not consider changes due to the 2-body reference motion. In this development we see from equation (9.4-42) that

$$\frac{d\alpha}{dt} = -\frac{2}{\mu}\mathbf{v}\cdot\dot{\mathbf{v}} \tag{9.4-43}$$

Differentiating equation (9.4-39),

$$\frac{d(\alpha \mathbf{r}_0)}{dt} = \frac{d(\alpha F)}{dt}\mathbf{r} + \frac{d(\alpha G)}{dt}\mathbf{v} + \alpha G\dot{\mathbf{v}}$$

$$\frac{d(\alpha \mathbf{v}_0)}{dt} = \frac{d(\alpha \dot{F})}{dt}\mathbf{r} + \frac{d(\alpha \dot{G})}{dt}\mathbf{v} + \alpha \dot{G}\dot{\mathbf{v}} \tag{9.4-44}$$

The derivatives in equation (9.4-44) can be found from equations (9.4-40). After algebraic reduction,

$$\frac{d(\alpha F)}{dt} = \frac{1}{\mu}\left(\frac{xr_0\dot{F}}{\sqrt{\mu}} - 2\right)\mathbf{v}\cdot\dot{\mathbf{v}} - \frac{r_0\dot{F}}{\sqrt{\mu}}\left(\alpha\frac{dx}{dt}\right) \tag{9.4-45}$$

$$\frac{d(\alpha G)}{dt} = \frac{-1}{\mu}\left[\frac{xr(1-F)}{\sqrt{\mu}} - (3t - 3t_0 + G)\right]\mathbf{v}\cdot\dot{\mathbf{v}} + \frac{r(1-F)}{\sqrt{\mu}}\left(\alpha\frac{dx}{dt}\right)$$

$$\frac{d(\alpha \dot{F})}{dt} = \frac{1}{rr_0\sqrt{\mu}}\left[\alpha\sqrt{\mu}(G + t - t_0) + \alpha x r(1-F) - 2x\right]\mathbf{v}\cdot\dot{\mathbf{v}} \tag{9.4-45}$$

$$+\frac{\sqrt{\mu}}{rr_0}[1 - \alpha r(1-F)]\,\alpha\frac{dx}{dt} - \frac{\dot{F}}{r_0}\left(\alpha\frac{dr_0}{dt}\right)$$

$$\frac{d(\alpha \dot{G})}{dt} = \frac{1}{\mu}\left(\frac{xr\dot{F}}{\sqrt{\mu}} - 2\right)\mathbf{v}\cdot\dot{\mathbf{v}} - \frac{r\dot{F}}{\sqrt{\mu}}\left(\alpha\frac{dx}{dt}\right) + \frac{r(1-F)}{r_0^2}\left(\alpha\frac{dr_0}{dt}\right)$$

It is still necessary to find $\frac{dx}{dt}$ and $\frac{dr_0}{dt}$ Interchanging the roles of $\mathbf{r}_0$, $\mathbf{v}_0$ and $\mathbf{r}$, $\mathbf{v}$ as before, equation (4.4-12), which is Kepler's equation, becomes

$$\sqrt{\mu}(t - t_0) = (1 - r\alpha)x^3 S(z) + rx - \frac{\mathbf{r}\cdot\mathbf{v}}{\sqrt{\mu}}x^2 C(z). \tag{9.4-46}$$

Also equation (4.4-13) becomes

$$r_0 = x^2\, C(z) - \frac{\mathbf{r}\cdot\mathbf{v}}{\sqrt{\mu}}\, x(1 - zS(z)) + r(1 - zC(z)). \tag{9.4-47}$$

Differentiating equation (9.4-47),

$$\alpha\frac{dr_0}{dt} = \frac{1}{\mu}\,[2r(1-F) + \alpha x\sqrt{\mu}(t - t_0) + \alpha\mathbf{r}\cdot\mathbf{v}(G + t - t_0)$$

$$- \alpha x^2 r]\,\mathbf{v}\cdot\mathbf{v} - \frac{\alpha r r_0 \dot{F}}{\mu}\,\mathbf{r}\cdot\dot{\mathbf{v}} + \left(x - \alpha\sqrt{\mu}(t - t_0) - \frac{\mathbf{r}\cdot\mathbf{v}}{\sqrt{\mu}}\right)\left(\alpha\frac{dx}{dt}\right) \tag{9.4-48}$$

Differentiating equation (9.4-46),

$$\alpha \frac{dx}{dt} = \frac{1}{\mu r_0} [x(r + r_0) - 2\sqrt{\mu}(t - t_0)$$

$$- \sqrt{\mu}(1 + r\alpha)(G + t - t_0)]\,\mathbf{v}\cdot\dot{\mathbf{v}} + \frac{\alpha r(1 - F)}{r_0\sqrt{\mu}}\,\mathbf{r}\cdot\dot{\mathbf{v}} \tag{9.4-49}$$

Note that $\dot{\mathbf{v}}$ includes only the perturbative accelerations. Now the perturbed position and velocity can be calculated by numerical integration of equations (9.4-44). The corresponding x is determined from Kepler's equation (4.4-12), and then position and velocity are obtained from equations (4.4-18) and (4.4-19). Note that equation (9.4-43) also must be integrated to obtain α .

9.4.3 Example to Introduce General Perturbations. The variation of parameters method of the previous section is classified as a special perturbation technique only because the final integration is accomplished numerically. If these equations could be integrated analytically, the technique would be that of general perturbations. To permit an analytic treatment the perturbation forces can be expressed in a power series and integrated analytically. Though this sounds simple enough, in practice the integration may be quite difficult. Also, a series solution method of integration is frequently used. In special perturbations the result is an answer to one specific problem or set of initial conditions. In general perturbations the result will cover many cases and will give a great deal more information on the perturbed orbit. This is especially true for long duration calculations. To give the reader a beginning understanding of general perturbations, a very simple example will be treated here which uses the variation of parameters method and a series expansion of the perturbing force.

Consider the equation

$$\dot{x} - \alpha = \frac{\nu}{(1 + \mu x)^2} \tag{9.4-50}$$

where α is a constant and μ and ν are small numbers. Consider the term on the right to be a perturbation. In the absence of the perturbation, the solution would be

$$x = \alpha t + c \tag{9.4-51}$$

where c will be the parameter. The constant of integration, c, is analogous to the orbital elements in orbital mechanics. Since μ and ν are small, the solution will vary only slightly from equation (9.4-51), so the variation of c to satisfy equation (9.4-50) should be small. From equation (9.4-51), allowing c to vary with time,

$$\dot{x} = \alpha + \dot{c}$$

Substituting $\dot{x}$ and x into equation (9.4-50) and solving for $\dot{c}$ we find

$$\dot{c} = \frac{\nu}{[1 + \mu(\alpha t + c)]^2} \tag{9.4-52}$$

The right-hand side of equation (9.4-52) is the perturbation and can be expanded as

$$\dot{c} = \nu[1 - 2\mu(\alpha t + c) + 3\mu^2(\alpha t + c)^2 - \ldots\ldots.] \tag{9.4-53}$$

The lowest order approximation would be $\dot{c} = \nu$, but more practically we would consider

$$\dot{c} = \nu - 2\mu\nu(\alpha t + c)$$

$$\dot{c} + 2\mu\nu c - \nu = -2\mu\nu\alpha t \tag{9.4-54}$$

which can be easily solved for c.

$$c = \nu e^{-2\mu\nu t} - \alpha t + \frac{\alpha + \nu}{2\mu\nu} - \frac{\alpha}{2\mu\nu} e^{-2\mu\nu t} \quad \text{for } c(0) = \nu + \frac{1}{2\mu}$$

Then x will be be given by equation (9.4-51) correct to first order in μ. Though this example has no physical significance it clearly demonstrates how the perturbation would be treated analytically using a series expansion, which is a common method of expression of the perturbation.

For the reader who wishes to pursue the general perturbation analysis Brouwer and Clemence,[4] Moulton[5] and many papers discuss it in detail. Many other approaches could be treated in this chapter, such as the disturbing function, Lagrange's planetary variables or derivations from Hamiltonian mechanics, but to pursue the topic further here would be beyond the scope of this text.

9.5 COMMENTS ON INTEGRATION SCHEMES AND ERRORS

The topic of numerical integration in the context of special perturbations is both relevant and necessary. No matter how elegant the analytic foundations of a particular special perturbation method, the results after numerical integration can be meaningless or, at best, highly questionable, if the integration scheme is not accurate to a sufficient degree. Many results fall into this questionable category because the investigator has not understood the limitations of, or has not carefully selected, his integration method. Unfortunately, such understanding requires a combination of knowledge of numerical analysis and experience. The purpose of this section is not the analytic development of various integration schemes but is to summarize several methods and help the reader to benefit from others' experience so he will be able to choose the kind of integration method which will best suit his needs.

9.5.1 Criteria for Choosing a Numerical Integration Method. The selection of an adequate integration scheme for a particular problem is too often limited to what is readily available for use rather than what is best. In theory, efficiency, not expediency, should be the criteria. The problem to be solved should first be analyzed. The following questions may help to narrow the selection possibilities:

a. Is a wide range in the independent variables required (thus a large number of integration steps)?

b. Is the problem extremely sensitive to small errors?

c. Are the dependent variables changing rapidly (i.e., will a constant step-size be satisfactory)?

Factors to be considered for any given method are speed, accuracy, storage and complexity. Desirable qualities may be in opposition to each other such that they all cannot be realized to the fullest extent. Some of the desirable qualities of a good integration scheme are:

a. allow as large a step-size as possible,

b. variable step-size provision which is simple and fast,

c. economical in computing time,

d. the method should be stable such that errors do not exhibit exponential growth.

e. it should be as insensitive as possible to roundoff errors, and
f. it should have a maximum and minimum control on truncation error.

Obviously, all of these cannot be perfectly satisfied in any one method, but they are qualities to consider when evaluating an integration method.

9.5.2 Integration Errors. Before evaluating any particular method of numerical integration it is important to have some understanding of the kind of errors involved.

In any numerical integration method there are two primary kinds of errors that will always be present to some degree. They are roundoff and truncation errors. Roundoff errors result from the fact that a computer can carry only a finite number of digits of any number. For instance, suppose that the machine can carry six digits and we wish to add 4.476,276 + 3.388,567 = 7.864,843. In the machine the numbers would be rounded off to six significant digits,

$$4.476{,}28 + 3.388{,}57 = 7.86485$$

The true answer rounded off would have a 4 in the sixth digit, but the rounding error has caused it to be a 5. It is clear that the occurrence of this roundoff many times in the integration process can result in significant errors. Brouwer and Clemence[4] (p 158) give a formula for the *probable* error after n steps as $.1124n^{3/2}$ (in units of the last decimal) or $\log(.1124n^{3/2})$ in number of decimal places. Baker (v 2, p 270) gives the example of error in 200 integration steps as

$$\log\,[.1129\,(200)^{3/2}] \approx 2.5 \tag{9.5-1}$$

If six places of accuracy are required then $6 + 2.5 \approx 9$ places must be carried in the calculations. The lesson here is that the fewer integration steps taken the smaller the accumulated roundoff error. The best inhibitor to the significant accumulation of roundoff error is the use of double precision arithmetic.

Truncation error is a result of an inexact solution of the differential equation. In fact, the numerical integration is actually an exact solution of some *difference* equation which imperfectly represents the actual differential equation. Truncation error results from not using all of the series expression employed in the integration method. The larger the step-size, the larger the truncation error, so the ideal here is to have small step-sizes. Note that this is in opposition to the cure of roundoff errors. Thus, errors are unavoidable in numerical integration and the object is to use a method that minimizes the sum of roundoff and truncation error. Roundoff error is primarily a function of the machine (except for some resulting from poor programming technique) and truncation error is a function of the integration method.

9.6 NUMERICAL INTEGRATION METHODS

Numerical integration methods can be divided into single-step and multi-step categories. The multi-step methods usually require a single-step method to start them at the beginning and after each step-size change. Some representative single-step methods are Runge-Kutta, Gill, Euler-Cauchy and Bowie. Some of the multi-step methods are Milne, Adams-Moulton, Gauss-Jackson (sum-squared), Obrechkoff and Adams-Bashforth. The multi-step methods are often referred to as predictor-corrector methods. A few of the more common methods will be discussed here. In general, we will consider the integration of systems of first order differential equations though there are methods available to integrate some special second order systems. Of course, any second order system can be written as a system of first order equations.

We will treat the first order equation

$$\frac{dx}{dt} = f(t, x) \tag{9.6-1}$$

with initial conditions $x(t_0) = x_0$ where x and f could be vectors.

9.6.1 Runge-Kutta Method. There is actually a family of Runge-Kutta methods of varying orders. The technique approximates a Taylor series extrapolation of a function by several evaluations of the first derivatives at points within the interval of extrapolation. The order of a particular member of the family is the order of the highest power of the step-size, h, in the equivalent Taylor series expansion. The formulas for the standard fourth order method are:

$$x_{n+1} = x_n + \frac{1}{6}(k_1 + 2k_2 + 2k_3 + k_4) \tag{9.6-2}$$

where $k_1 = hf(t_n, x_n)$ (9.6-3)

$$k_2 = hf(t_n + \frac{h}{2}, x_n + \frac{k_1}{2})$$

$$k_3 = hf(t_n + \frac{h}{2}, x_n + \frac{k_2}{2}) \qquad (9.6\text{-}3)$$

$$k_4 = hf(t_n + h, x_n + k_3)$$

and n is the increment number.

Another member of the Runge-Kutta family is the Runge-Kutta-Gill formula. It was developed for high-speed computers to use a minimum number of storage registers and to help control the growth of roundoff errors. This formula is

$$x_{n+1} = x_n + \frac{1}{6}k_1 + \frac{1}{3}(1 - \sqrt{1/2})\, k_2 + \frac{1}{3}(1 + \sqrt{1/2})\, k_3 + \frac{1}{6}k_4 \qquad (9.6\text{-}4)$$

where $k_1 = hf(t_n, x_n)$

$$k_2 = hf(t_n + \frac{h}{2}, x_n + \frac{k_1}{2})$$

$$k_3 = hf(t_n + \frac{h}{2}, x_n + k_1[\sqrt{1/2} - 1/2] + k_2[1 - \sqrt{1/2}]) \qquad (9.6\text{-}5)$$

$$k_4 = hf(t_n + h, x_n - \frac{k_2}{\sqrt{2}} + k_3[1 + \sqrt{1/2}])$$

The Runge-Kutta methods are stable and they do not require a starting procedure. They are relatively simple, easy to implement, have a relatively small truncation error, and step-size is easily changed. One of the disadvantages is that there is no simple way to determine the truncation error, so it is difficult to determine the proper step-size. One obvious failure is that use is made only of the last calculated step. This idea leads to multi-step or predictor-corrector methods.

9.6.2 Adams-Bashforth Multi-step Method. The multi-step methods take advantage of the history of the function being integrated. In general, they are faster than the single-step methods, though at a cost of greater complexity and a requirement for a starting procedure at the beginning and after each step-size change. Some multi-step methods calculate a predicted value for x_{n+1} and then substitute x_{n+1} in the differential equation to get $\dot{x}_{n+1}$, which is in turn used to calculate a corrected value of x_{n+1}. These are naturally called "predictor-corrector" methods.

The Adams (sometimes called Adams-Bashforth) formula is

$$x_{n+1} = x_n + h \sum_{k=0}^{N} \alpha_k \nabla^k f_n \tag{9.6-6}$$

where N = the number of terms desired

h = step-size

n = step number

α = 1, 1/2, 5/12, 3/8, 251/720, 95/288

$k = 0, \ldots 5$

$\nabla^k f_n$ = backwards difference operator

$= \nabla^{k-1} f_n - \nabla^{k-1} f_{n-1}$ and $\nabla^0 f_n = f_n$

$f_n = f(t_n, X_n)$

The first few terms are

$$x_{n+1} = x_n + h\left(1 + \tfrac{1}{2}\nabla + \frac{5}{12}\nabla^2 + \frac{3}{8}\nabla^3 + \frac{251}{720}\nabla^4 + \frac{95}{288}\nabla^5\right) f_n \tag{9.6-7}$$

A method is available for halving or doubling the step-size without a complete restart. Also the local truncation error can be calculated (see references for detailed formulas).[3]

9.6.3 Adams–Moulton Formulas. The Adams-Bashforth method is only a multi-step predictor scheme. In 1926 Moulton added a corrctor formula to the method. The fourth order formulas retaining third differences are given here. The predictor is

$$x_{n+1}^{(P)} = x_n + \frac{h}{24}(55f_n - 59f_{n-1} + 37f_{n-2} - 9f_{n-3})$$
$$+ \frac{251}{720} h^5 \frac{d^5x(\xi)}{dt^5} \qquad (9.6\text{-}8)$$

and the corrector is

$$x_{n+1}^{(C)} = x_n + \frac{h}{24}(9f_{n+1} + 19f_n - 5f_{n-1} + f_{n-2})$$
$$\frac{19}{720} h^5 \frac{d^5x(\xi)}{dt^5} \qquad (9.6\text{-}9)$$

where the last term in each of the equations is the truncation error term. In equation (9.6-9), f_{n+1} is found from the predicted value

$$f_{n+1} = f(t_{n+1}, x_{n+1}^{(P)}) \qquad (9.6\text{-}10)$$

In using the formulas, the error terms are not generally known, so the value used is the expression without those terms (for a discussion of the error terms see, for instance, Ralston's text on Numerical Analysis).[6] An estimate of the truncation error for the step from t_n to t_{n+1} is

$$\frac{19}{270}\left|x_{n+1}^{(C)} - x_{n+1}^{(P)}\right| \qquad (9.6\text{-}11)$$

To use any predictor-corrector scheme the predicted value is computed first. Then the derivative corresponding to the predicted value is found and used to find the corrected value using the corrector formula. It is possible to iterate on the corrector formula until there is no significant change in x_{n+1} on successive iterations. The step-size can be changed if the truncation error is larger than desired. This method, like other multi-step methods, requires a single step method to start it to obtain f_n, f_{n-1}, etc. The Runge-Kutta method is suggested for a starter. The Adams-Moulton formulation is one of the more commonly used integration methods.

9.6.4 The Gauss-Jackson or Sum Squared (Σ^2) Method. The Gauss-Jackson method is one of the best, and most used, numerical integration methods for trajectory problems of the Cowell and Encke type. It is designed for the integration of systems of second order equations and is faster than integrating two first order equations. Its predictor alone is generally more accurate than the predictor and corrector of other methods, though it also includes a corrector. It exhibits especially good control of accumulated roundoff error effect. It is, though, more complex and difficult for the beginner to implement than the other methods. The general formula for solving the equation $\ddot{x} = f(t, x, \dot{x})$ is

$$x_n = h^2\left(\sum{}^2 \ddot{x}_n + \frac{1}{12}\ddot{x}_n - \frac{1}{240}\delta^2\ddot{x}_n + \frac{31}{60480}\delta^4\ddot{x}_n - \frac{289}{36288}\delta^6\ddot{x}_n + \dots\right) \tag{9.6-12}$$

where $\ddot{x}_n = f(t_n, x_n, \dot{x}_n) = f(t_n) = f_n$

and the central differences are defined as

$$\delta^0 f(t_n) = f(t_n)$$

$$\delta^1 f(t_n) = f(t_n + \frac{h}{2}) - f(t_n - \frac{h}{2})$$

$$\delta^2 f(t_n) = f(t_n + h) + f(t_n - h) - 2f(t_n)$$

$$= f_{n+1} + f_{n-1} - 2f_n$$

$$\delta^k f(t_n) = \delta^{k-1} f(t_n + \frac{h}{2}) - \delta^{k-1} f(t_n - \frac{h}{2})$$

The definition of $\Sigma^2 \ddot{x}_n$ can be found in several references (Baker, V. 2 and NASA CR-1005).[2, 3] The procedure for calculating it will not be reiterated here since it becomes fairly involved and the reader can find it easily in the above-mentioned references.

9.6.5 Numerical Integration Summary. Experience has shown that the Gauss-Jackson method is clearly superior for orbital problems using the Cowell and Encke techniques. For normal integration of first order equations such as occur in the variation of parameters technique, the Adams-Moulton or Adams (Adams-Bashforth) methods are preferred. The Runge-Kutta method is suggested for starting the multi-step methods. Roundoff error is a function of the number of integration steps and is most effectively controlled by using double-precision arithmetic. Local truncation error can and should be calculated for the multi-step methods and should be used as a criterion for changing step-size.

Table 9.6-1 gives a tabular comparison of a number of integration methods, including some which were not discussed in this chapter.

9.7 ANALYTIC FORMULATION OF PERTURBATIVE ACCELERATIONS

A few of the perturbation accelerations commonly used in practice will be presented in this section. Some of them become quite complex, but only the simpler forms will be treated. They will be stated with a minimal amount of discussion, but in sufficient detail such that they can be used in the methods of this chapter without extensive reference to more detailed works.

9.7-1 The Nonspherical Earth. The simplified gravitational potential of the Earth, μ/r, is due to a spherically symmetric mass body and results in conic orbits. However, the Earth is not a spherically symmetric body but is bulged at the equator, flattened at the poles and is generally asymmetric. If the potential function, ϕ, is known, the accelerations can be found from

$$\mathbf{a} = \nabla\phi = \frac{\delta\phi}{\delta x}\mathbf{I} + \frac{\delta\phi}{\delta y}\mathbf{J} + \frac{\delta\phi}{\delta z}\mathbf{K} \tag{9.7-1}$$

One such potential function, according to Vinti, is

$$\phi = \frac{\mu}{r}\left[1 - \sum_{n=2}^{\infty} J_n \left(\frac{r_e}{r}\right)^n P_n \sin L\right] \tag{9.7-2}$$

COMPARISON OF INTEGRATION METHODS
(From NASA SP-33, Vol 1, Part 1)[7]

Method of Numerical Integration	Truncation Error	Ease of Changing Step-Size	Speed	Stability	Round-off Error Accumulation
Single Step Methods					
Runge-Kutta	h^5	*	Slow	Stable	Satisfactory
Runge-Kutta-Gill	h^5	*	Slow	Stable	Satisfactory
Bowie	h^3	Trivial (step-size varied by error control)	Fast	Stable	Satisfactory
Fourth Order Multi-Step Predictor-Corrector					
Milne	h^5	Excellent	Very fast	Unstable	Poor
Adams-Moulton	h^5	Excellent	Very fast	Unconditionally stable	Satisfactory
Higher Order Multi-Step					
Adams Backward Difference	Arbitrary	Good	Very fast	Moderately stable	Satisfactory
Gauss-Jackson**	Arbitrary	Awkward and expensive	Fast	Stable	Excellent
Obrechkoff	h^7	Excellent	***	Stable	Satisfactory
Special Second Order Equations $[\ddot{z} = f(t,z)]$					
Special Runge-Kutta	h^5	*	Slow	Stable	Satisfactory
Milne-Stormer	h^6	Excellent	Very fast	Moderately stable	Poor

*R-K (single-step) trivial to change steps, very difficult to determine proper size.
**Gauss-Jackson is for second order equations.
***Speed of Obrechkoff depends on complexity of the higher order derivatives required; it could be very fast.

Table 9.6-1

where μ = gravitational parameter

J_n = coefficients to be determined by experimental observation

r_e = equatorial radius of the earth

P_n = Legendre polynomials

L = geocentric latitude

$\sin L = \frac{z}{r}$

The first 7 terms of the expression are

$$\phi = \frac{\mu}{r}\left[1 - \frac{J_2}{2}\left(\frac{r_e}{r}\right)^2 (3\sin^2 L - 1)\right.$$

$$- \frac{J_3}{2}\left(\frac{r_e}{r}\right)^3 (5\sin^3 L - 3\sin L)$$

$$- \frac{J_4}{8}\left(\frac{r_e}{r}\right)^4 (35\sin^4 L - 30\sin^2 L + 3)$$

$$- \frac{J_5}{8}\left(\frac{r_e}{r}\right)^5 (63\sin^5 L - 70\sin^3 L + 15\sin L)$$

$$\left. - \frac{J_6}{16}\left(\frac{r_e}{r}\right)^6 (231\sin^6 L - 315\sin^4 L + 105\sin^2 L - 5)\right] \tag{9.7-3}$$

Taking the partial derivative of ϕ,

$$\ddot{x} = \frac{\delta\phi}{\delta x} = -\frac{\mu x}{r^3}\left[1 - J_2\frac{3}{2}\left(\frac{r_e}{r}\right)^2\left(5\frac{z^2}{r^2} - 1\right)\right.$$

$$+ J_3\frac{5}{2}\left(\frac{r_e}{r}\right)^3\left(3\frac{z}{r} - 7\frac{z^3}{r^3}\right) - J_4\frac{5}{8}\left(\frac{r_e}{r}\right)^4\left(3 - 42\frac{z^2}{r^2} + 63\frac{z^4}{r^4}\right)$$

$$- J_5\frac{3}{8}\left(\frac{r_e}{r}\right)^5\left(35\frac{z}{r} - 210\frac{z^3}{r^3} + 231\frac{z^5}{r^5}\right)$$

$$+ J_6\frac{1}{16}\left(\frac{r_e}{r}\right)^6\left(35 - 945\frac{z^2}{r^2}\right.$$

$$\left.\left. + 3465\frac{z^4}{r^4} - 3003\frac{z^6}{r^6}\right) + \ldots\right] \tag{9.7-4}$$

$$\ddot{y} = \frac{\delta\phi}{\delta y} = \frac{y}{x}\ddot{x} \tag{9.7-5}$$

$$\ddot{z} = \frac{\delta\phi}{\delta z} = -\frac{\mu z}{r^3}\Big[1 + J_2 \frac{3}{2}\left(\frac{r_e}{r}\right)^3\left(3 - 5\frac{z^2}{r^2}\right)$$

$$+ J_3 \frac{3}{2}\left(\frac{r_e}{r}\right)^2\left(10\frac{z}{r} - \frac{35}{3}\frac{z^3}{r^3} - \frac{r}{z}\right)$$

$$- J_4 \frac{5}{8}\left(\frac{r_e}{r}\right)^4\left(15 - 70\frac{z^2}{r^2} + 63\frac{z^4}{r^4}\right)$$

$$- J_5 \frac{1}{8}\left(\frac{r_e}{r}\right)^5\left(315\frac{z}{r} - 945\frac{z^3}{r^3} + 693\frac{z^5}{r^5} - 15\frac{r}{z}\right)$$

$$+ J_6 \frac{1}{16}\left(\frac{r_e}{r}\right)^6\Big(315 - 2205\frac{z^2}{r^2} \tag{9.7-6}$$

$$+ 4851\frac{z^4}{r^4} - 3003\frac{z^6}{r^6}\Big) + \ldots\Big]$$

Note that sin L is replaced by z/r. The first term in $\ddot{x}$, $\ddot{y}$, $\ddot{z}$ is the 2-body acceleration and the remaining terms are the perturbation accelerations due to the Earth's nonsphericity. There have been various determinations of the J coefficients, which are slightly in variance, but a representative set of values will be given here (see Baker, Vol 1, p 175).[8]

$J_2 = (1082.64 \pm 0.03) \times 10^{-6}$

$J_3 = (-2.5 \pm 0.1) \times 10^{-6}$

$J_4 = (-1.6 \pm 0.5) \times 10^{-6}$

$J_5 = (-0.15 \pm 0.1) \times 10^{-6}$

$J_6 = (0.57 \pm 0.1) \times 10^{-6}$

$J_7 = (-0.44 \pm 0.1) \times 10^{-6}$

It is obvious that the confidence factor diminishes beyond J_4. These equations include only the zonal harmonics—that is those harmonics which are dependent only on that mass distribution which is symmetric about the north-south axis of the Earth (they are not longitude dependent).

The even numbered harmonics are symmetric about the equatorial plane and the odd numbered harmonics are antisymmetric. There are also tesseral harmonics (dependent on *both* latitude and longitude) and sectorial harmonics (dependent on longitude *only*). A general expression to account for all three classes of harmonics in the potential function is given by Baker (Vol 2, p 147),[2] but will not be presented here.

There are a few cautions to be pointed out in the use of the above formulation. The equations are formulated in the geocentric, equatorial coordinate system. Therefore, care must be taken to know the Earth's current relationship to this inertial frame due to rotation. The zonal harmonics are not longitude dependent, so the direction of the x axis need not point to a fixed geographic point. However, if tesseral and sectorial harmonics were considered, then the frame would need to be fixed to the rotation of the Earth. To use the acceleration equations in the perturbation methods it is necessary to transform the perturbation portion (not the $\frac{-\mu}{r^3}$ term) to the RSW system. To do this, simply use the IJK to PQW transformation of Chapter 2—substituting u (argument of latitude) for ω. Expressions for the Moon's triaxial ellipsoid potential can be found in Baker, Vol 1.[8]

9.7.2 Atmospheric Drag. The formulation of atmospheric drag equations is plagued with uncertainties of atmospheric fluctuations, frontal areas of orbiting object (if not constant), the drag coefficient, and other parameters. A fairly simple formulation will be given here (see NASA SP-33).[7] Drag, by definition, will be opposite to the velocity of the vehicle relative to the atmosphere. Thus, the perturbative acceleration is

$$\ddot{\mathbf{r}} = -\frac{1}{2} C_D \frac{A}{m} \rho v_a \dot{\mathbf{r}}_a \qquad (9.7\text{-}7)$$

where C_D = the dimensionless drag coefficient associated with A

A = the cross-sectional area of the vehicle perpendicular to the direction of motion

m = vehicle mass

$$\frac{C_D A}{m g} = \frac{1}{\text{ballistic coefficient}}$$

ρ = atmospheric density at the vehicle's altitude.

$$v_a = |\dot{\mathbf{r}}_a| = \text{speed of vehicle relative to the rotating atmosphere}$$

and $$\dot{\mathbf{r}}_a = \begin{bmatrix} \dot{x} + \dot{\theta} y \\ \dot{y} - \dot{\theta} x \\ \dot{z} \end{bmatrix}$$

$$\dot{r} = \begin{bmatrix} \dot{x} \\ \dot{y} \\ \dot{z} \end{bmatrix}, \text{ the inertial velocity}$$

$\dot{\theta}$ = rate of rotation of the earth

Once again the x, y, z refer to the geocentric, equatorial coordinate system.

This formulation could be greatly complicated by the addition of an expression for theoretical density, altitude above an oblate Earth, etc., but the reader can find this in other documents. Equation (9.7-7) can give the reader a basic understanding of drag effects. Equations for lifting forces will not be presented here.

9.7.3 Radiation Pressure. Radiation pressure, though small, can have considerable effect on large area/mass ratio satellites (such as Echo). In the geocentric, equatorial coordinate system the perturbative accelerations are

$$\ddot{x} = f \cos A_{\odot}$$

$$\ddot{y} = f \cos i_{\oplus} \sin A_{\odot}$$

$$\ddot{z} = f \sin i_{\oplus} \sin A_{\odot} \qquad (9.7\text{-}8)$$

where $f = -4.5 \times 10^{-5} \left(\frac{A}{m}\right)$ cm/sec^2

A = cross section of vehicle exposed to sun (cm).

m = mass of vehicle (grams).

$A_{\odot}$ = mean right ascension of the sun during computation.

$i_{\oplus}$ = inclination of equator to ecliptic = 23.4349°

9.7.4 Thrust. Thrust can be handled quite directly by resolving the thrust vector into x, y, z directions. Thus the perturbing acceleration is

$$\ddot{x} = \frac{T_x}{m}, \quad \ddot{y} = \frac{T_y}{m}, \quad \ddot{z} = \frac{T_z}{m} \tag{9.7-9}$$

Low thrust problems can be treated using a perturbation approach like Encke or Variation of Parameters, but high thrust should be treated using the Cowell technique since the thrust is no longer a small perturbation, but a major force.

EXERCISES

9.1 Verify the development of equation (9.4-15), the variation of eccentricity.

9.2 In the variation of parameters using the universal variable it is asserted that one can develop f and g expressions to find $\mathbf{r}_0$ and $\mathbf{v}_0$ in terms of $\mathbf{r}$ and $\mathbf{v}$ by t and x by $(-t)$ and $(-x)$. Prove this assertion.

9.3 Show how the potential function for the nonspherical Earth would be used in conjunction with Cowell's method. Write out the specific equations that would be used in a form suitable for programming.

9.4 Develop the equations of motion in spherical coordinates. See equation (9.2-5).

9.5 Describe the process of rectification as used in the Encke method and variation of parameters method.

9.6 Develop a computer flow diagram for the Encke method including rectification of the reference orbit.

* **9.7** Program the Cowell method including the perturbation due to the Moon.

* **9.8** Program the Encke method for the above problem and compare results for a) a near Earth satellite of 100 nm altitude and b) a flight toward the Moon with an apogee of approximately 150,000 n mi.

* **9.9** Verify at least one of the equations in equations (9.4-45).

LIST OF REFERENCES

1. *Planetary Coordinates for the Years 1960-1980*, Her Majesty's Stationery Office, London.

2. Baker, Robert M. L., Jr, *Astrodynamics: Applications and Advanced Topics*. New York, Academic Press, 1967.

3. Allione, M. S., Blackford, A. L., Mendez, J. C. and Whittouck, M. M. *Guidance, Flight Mechanics and Trajectory Optimization,* Vol VI, "The N-Body Problem and Special Perturbation Techniques." NASA CR-1005, National Aeronautics and Space Administration, Washington, DC, Feb 1968.

4. Brouwer, D. and Clemence, G. M. *Methods of Celestial Mechanics.* New York, Academic Press, 1961.

5. Moulton, F. R. *An Introduction to Celestial Mechanics*. New York, The Macmillan Co, 1914.

6. Ralston, A. and Wilf. *Mathematical Methods for Digital Computers.* New York, John Wiley and Sons, Inc, 1960.

7. Townsend, G. E., Jr. *Orbital Flight Handbook*, Vol 1, Part 1. NASA SP-33. National Aeronautics and Space Administration, Washington, DC, 1963.

8. Baker, Robert M. L., Jr and Makemson, Maud W. *An Introduction to Astrodynamics*, 2nd ed. New York, Academic Press, 1967.

9. Battin, R. H. *Astronautical Guidance*. New York, McGraw-Hill, 1964.

10. Herget, Paul. *The Computation of Orbits*. Published privately by the Author, Ann Arbor, Michigan, 1948.

11. Plumber, M. C. *Dynamical Astronomy*. New York, Cambridge University Press, 1918 (also available in paperback from Dover).

12. Smart, W. M. *Celestial Mechanics*. New York, Longmans, 1953.

13. Hildebrand, F. B. *Introduction to Numerical Analysis*. New York, McGraw-Hill, 1956.

14. Dubyago, A. D. *The Determination of Orbits*. New York, The Macmillan Co, 1961.

15. Escobal, P. R. *Methods of Orbit Determination*. New York, John Wiley and Sons, Inc, 1965.

16. Escobal, P. R. *Methods of Astrodynamics*. New York, John Wiley and Sons, Inc, 1968.

APPENDIX A

ASTRODYNAMIC CONSTANTS*

	Canonical Units	English Units	Metric Units
Geocentric			
Mean Equatorial Radius, $r_\oplus$	1 $DU_\oplus$	2.092567257×10^7 ft 3963.195563 miles 3443.922786 NM	6378.145 km
Time Unit	1 $TU_\oplus$	13.44686457 min	806.8118744 sec
Speed Unit	$1 \frac{DU_\oplus}{TU_\oplus}$	$25936.24764 \frac{ft}{sec}$	$7.90536828 \frac{km}{sec}$
Gravitational Parameter, $\mu_\oplus$	$1 \frac{DU_\oplus^3}{TU_\oplus^2}$	$1.407646882 \times 10^{16} \frac{ft^3}{sec^2}$	$3.986012 \times 10^5 \frac{km^3}{sec^2}$
Angular Rotation, $\omega_\oplus$	$0.0588336565 \frac{rad}{TU_\oplus}$	$.2506844773 \frac{deg}{min}$	$7.292115856 \times 10^{-5} \frac{rad}{sec}$
Heliocentric			
Mean Distance, Earth to Sun	1 AU	4.9081250×10^{11} ft	1.4959965×10^8 km
Time Unit	1 $TU_\odot$	58.132821 days	5.0226757×10^6 sec
Speed Unit	$1 \frac{AU}{TU_\odot}$	$9.7719329 \times 10^4 \frac{ft}{sec}$	$29.784852 \frac{km}{sec}$
Gravitational Parameter, $\mu_\odot$	$1 \frac{AU^3}{TU_\odot^2}$	$4.6868016 \times 10^{21} \frac{ft^3}{sec^2}$	$1.3271544 \times 10^{11} \frac{km^3}{sec^2}$

*Derived from those values of μ, $r_\oplus$, and AU in use at the NORAD/USAF Space Defense Center, 1975. Each of the two sets is consistent within itself.

APPENDIX B

MISCELLANEOUS CONSTANTS AND CONVERSIONS

second =	$1.239446309 \times 10^{-3}$	$TU_\oplus$
degree/sec =	14.08152366	$radian/TU_\oplus$
$radian/TU_\oplus$ =	0.0710150424	degree/sec
foot =	$4.77881892 \times 10^{-8}$	$DU_\oplus$
(statute) mile =	$2.52321639 \times 10^{-4}$	$DU_\oplus$
nautical mile =	$2.903665564 \times 10^{-4}$	$DU_\oplus$
foot/sec =	$3.85560785 \times 10^{-5}$	$DU_\oplus/TU_\oplus$
km/sec =	$.1264963205$	$DU_\oplus/TU_\oplus$
nautical mile/sec =	0.2342709	$DU_\oplus/TU_\oplus$

$$\frac{2\pi}{\sqrt{\mu_\oplus}} = \begin{cases} 5.295817457 \times 10^{-8} & sec/ft^{3/2} \\ 9.952004586 \times 10^{-3} & sec/km^{3/2} \end{cases}$$

The above values are consistent with the $r_\oplus$ and $\mu_\oplus$ of Appendix A.

π =	3.14159 26535 9	(radians)
degree =	0.01745 32925 199	radians
radian =	57.29577 95131	degrees

The following values have been established by international agreement and are exact as shown, i.e., there is no roundoff or truncation.

foot =	0.3048	(exact) meters
(statute) mile =	1609.344	(exact) meters
nautical mile =	1852	(exact) meters

APPENDIX C

VECTOR REVIEW

Vector analysis is a branch of mathematics which saves time and space in the derivation of relationships involving vector quantities. In three-dimensional space each vector equation represents three scalar equations. Thus, vector analysis is a powerful tool which makes many derivations easier and much shorter than otherwise would be possible.

In this section the fundamental vector operations will be discussed. A good understanding of these operations will be of significant help to the student in the use of this text.

C.1 DEFINITIONS

Vector. A vector is a quantity having both magnitude and direction. Examples are displacement, velocity, force and acceleration. The symbol used in the text to represent a vector quantity will be a bold-faced letter or a letter with a bar over the top in some illustrations.

Scalar. A scalar is a quantity having only magnitude. Examples are mass, length of a vector, speed, temperature, time or any real signed number. The symbol used in the text to represent a scalar quantity will be any signed number or a letter not bold-faced or with no bar over it. (Example: 3 or A). An exception to this symbology is sometimes desirable in which case the symbol for the magnitude of **A** will be $|\mathbf{A}|$.

Unit Vector. A unit vector is a vector having unit (1) magnitude. If **A** is a vector, not a null or zero vector, then

$$\frac{\mathbf{A}}{|\mathbf{A}|} = \mathbf{1}_A$$

where $\mathbf{1}_A$ is the symbol used to denote the unit vector in the direction of **A**. Certain sets of vectors are reserved for unit vectors in rectangular coordinate systems such as **(I, J, K)** and **(P, Q, W)**.

Equality of Vectors. Two vectors **A** and **B** are said to be equal if and only if they are parallel, have the same sense of direction and the same magnitude, regardless of the position of their origins.

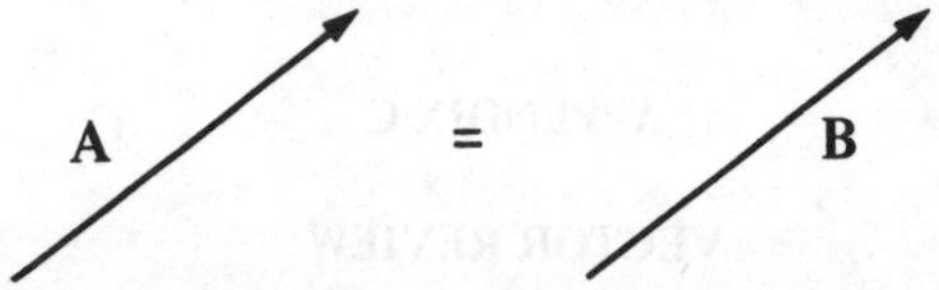

Coplanar Vectors. Vectors that are parallel to the same straight line are said to be collinear. Two collinear vectors differ only by a scalar factor. Vectors that are parallel to the same plane are said to be coplanar (not necessarily parallel vectors).

If **A**, **B** and **C** are coplanar vectors, and **A** and **B** are not collinear, it is possible to express **C** in terms of **A** and **B**.

The vector **C** may be resolved into components $\mathbf{C}_1$ and $\mathbf{C}_2$ parallel to **A** and **B** respectively so that

$$\mathbf{C} = \mathbf{C}_1 + \mathbf{C}_2$$

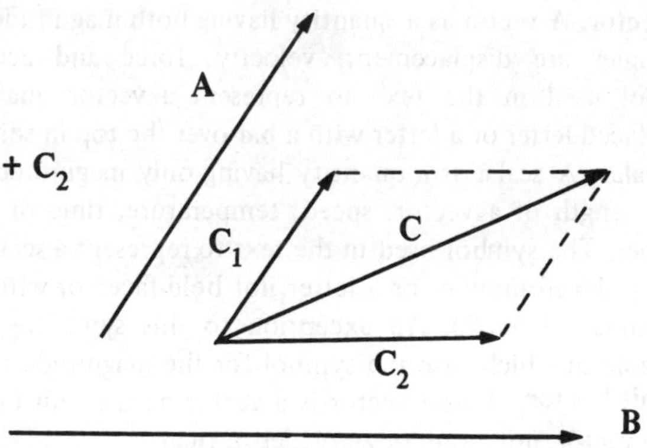

Since $\mathbf{C}_1$ and **A** are collinear vectors, they differ only by a scalar factor a, such that

$$\mathbf{C}_1 = a\mathbf{A},$$

Similarly, since $\mathbf{C}_2$ and **B** are collinear we may write

$$\mathbf{C}_2 = b\mathbf{B},$$

Then $\mathbf{C} = \mathbf{C}_1 + \mathbf{C}_2 = a\mathbf{A} + b\mathbf{B}$. (C.1-1)

C.2 VECTOR OPERATIONS

Addition of Vectors. The sum or result of two vectors **A** and **B** is a vector **C** obtained by constructing a triangle with **A** and **B** forming two sides of the triangle, **B** adjoined to **A**. The resultant **C** starts from the origin of **A** and ends at the terminus of **B**.

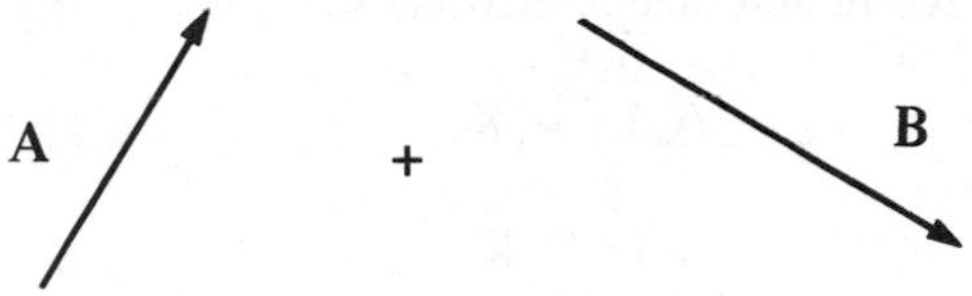

$\mathbf{C} = \mathbf{A} + \mathbf{B}$

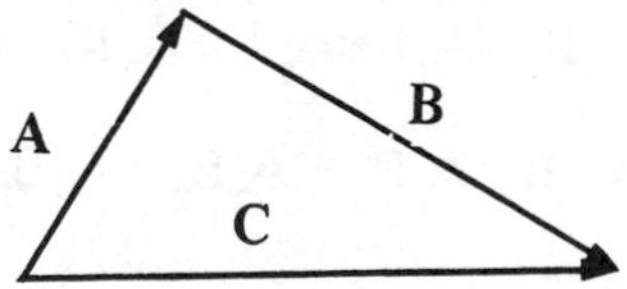

An obvious extension of this idea is the difference of two vectors. Suppose we wish to determine the resultant of $\mathbf{A} - \mathbf{B}$ which is the same as $\mathbf{A} + (-\mathbf{B})$.

Scalar or Dot Product. The "dot" product of two vectors **A** and **B**, denoted by $\mathbf{A} \cdot \mathbf{B}$, is a scalar quantity defined by

$$\mathbf{A} \cdot \mathbf{B} = |\mathbf{A}||\mathbf{B}|\cos\theta \tag{C.2-1}$$

where θ is the angle between the two vectors when drawn from a common origin.

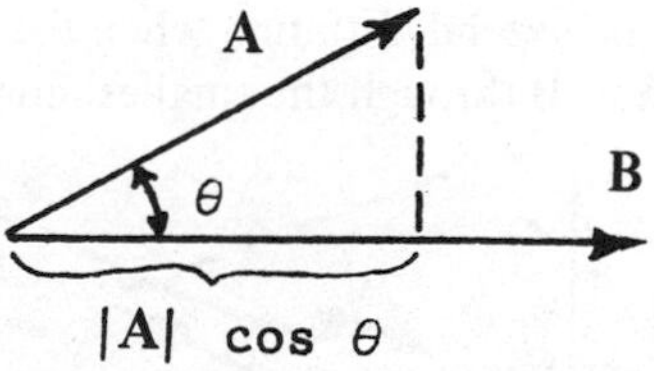

From the definition of the dot product the following laws are derived:

(a) $\mathbf{A}\cdot\mathbf{B} = \mathbf{B}\cdot\mathbf{A}$ (Commutative law)

(b) $\mathbf{A}\cdot(\mathbf{B}+\mathbf{C}) = \mathbf{A}\cdot\mathbf{B}+\mathbf{A}\cdot\mathbf{C}$ (Distributive law)

(c) $m(\mathbf{A}\cdot\mathbf{B}) = (m\mathbf{A})\cdot\mathbf{B} = \mathbf{A}\cdot(m\mathbf{B}) = (\mathbf{A}\cdot\mathbf{B})m$ (Associative law)

(d) $\mathbf{I}\cdot\mathbf{I} = \mathbf{J}\cdot\mathbf{J} = \mathbf{K}\cdot\mathbf{K} = 1$

$\mathbf{I}\cdot\mathbf{J} = \mathbf{J}\cdot\mathbf{K} = \mathbf{K}\cdot\mathbf{I} = 0$

where **I, J, K** are unit orthogonal vectors.

(e) If $\mathbf{A} = A_1\mathbf{I}+A_2\mathbf{J}+A_3\mathbf{K}$

$\mathbf{B} = B_1\mathbf{I}+B_2\mathbf{J}+B_3\mathbf{K}$

then $\mathbf{A}\cdot\mathbf{B} = (A_1\mathbf{I}+A_2\mathbf{J}+A_3\mathbf{K})\cdot(B_1\mathbf{I}+B_2\mathbf{J}+B_3\mathbf{K})$

$$\mathbf{A}\cdot\mathbf{B} = A_1B_1 + A_2B_2 + A_3B_3 \qquad \text{(C.2-2)}$$

(f) $\mathbf{A}\cdot\mathbf{A} = A^2$ (C.2-3)

(g) Differentiating both sides of the equation above yields

$$\mathbf{A}\cdot\dot{\mathbf{A}} = A\dot{A} \qquad \text{(C.2-4)}$$

Vector or Cross Product. The cross product of two vectors **A** and **B**, denoted by **A** × **B**, is a *vector* **C** such that the magnitude of **C** is the product of the magnitudes of **A** and **B** and the sine of the angle between them. The direction of the vector **C** is perpendicular to both **A** and **B** such that **A, B** and **C** form a right handed system. That is, if we say "**A** crossed with **B**" we mean that the resulting vector **C** will be in the direction of the extended thumb when the fingers of the right hand are closed from **A** to **B** through the smallest angle possible.

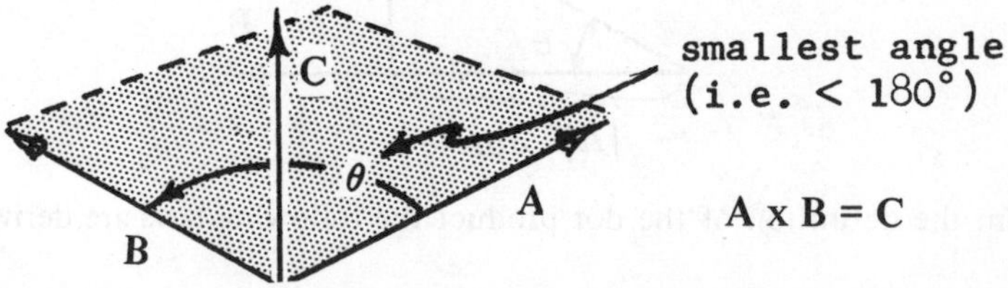

$$|\mathbf{C}| = |\mathbf{A}|\,|\mathbf{B}| \sin\theta \qquad \text{(C.2-5)}$$

If **A** and **B** are parallel, then $\mathbf{A} \times \mathbf{B} = \mathbf{0}$. In particular

$$\mathbf{A} \times \mathbf{A} = \mathbf{0} \qquad \text{(C.2-6)}$$

From the above figure it is easy to see that using the right hand rule:

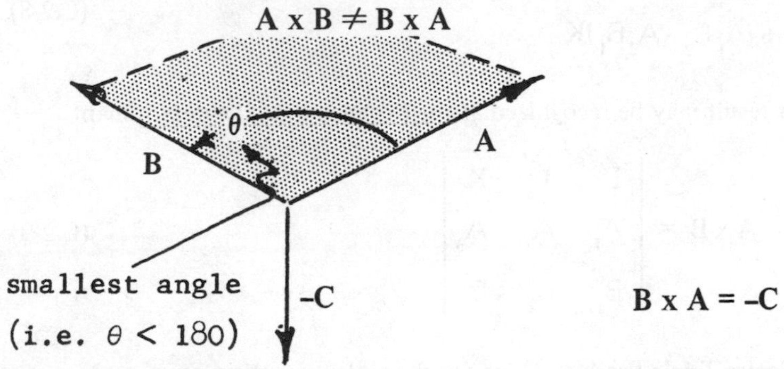

In fact $\mathbf{A} \times \mathbf{B} = -(\mathbf{B} \times \mathbf{A})$ (C.2-7)

since in either case the magnitudes are identical, the only difference is the sense of the resulting vector **C**.

Other laws valid for the cross product are:

(a) $\mathbf{A} \times (\mathbf{B} + \mathbf{C}) = \mathbf{A} \times \mathbf{B} + \mathbf{A} \times \mathbf{C}$ (Distributive law)

(b) $m(\mathbf{A} \times \mathbf{B}) = (m\,\mathbf{A}) \times \mathbf{B} = \mathbf{A} \times (m\mathbf{B})$ (Associative law)

$= (\mathbf{A} \times \mathbf{B})m$

(c) $\mathbf{I} \times \mathbf{I} = \mathbf{J} \times \mathbf{J} = \mathbf{K} \times \mathbf{K} = 0$

$\mathbf{I} \times \mathbf{J} = \mathbf{K}$

$\mathbf{J} \times \mathbf{K} = \mathbf{I}$

$\mathbf{K} \times \mathbf{I} = \mathbf{J}$

where **I, J, K** are unit orthogonal vectors.

(d) If $\mathbf{A} = A_1\mathbf{I} + A_2\mathbf{J} + A_3\mathbf{K}$

$\mathbf{B} = B_1\mathbf{I} + B_2\mathbf{J} + B_3\mathbf{K}$

then

$$\mathbf{A} \times \mathbf{B} = (A_1\mathbf{I} + A_2\mathbf{J} + A_3\mathbf{K}) \times (B_1\mathbf{I} + B_2\mathbf{J} + B_3\mathbf{K})$$

or
$$\mathbf{A} \times \mathbf{B} = (A_2B_3 - A_3B_2)\mathbf{I} + (A_3B_1 - A_1B_3)\mathbf{J} + (A_1B_2 - A_2B_1)\mathbf{K} \qquad \text{(C.2-8)}$$

This result may be recognized as the expansion of the determinant

$$\mathbf{A} \times \mathbf{B} = \begin{vmatrix} \mathbf{I} & \mathbf{J} & \mathbf{K} \\ A_1 & A_2 & A_3 \\ B_1 & B_2 & B_3 \end{vmatrix} \qquad \text{(C.2-9)}$$

Scalar Triple Product. Consider the scalar quantity

$$\mathbf{A} \cdot (\mathbf{B} \times \mathbf{C})$$

If we let θ be the angle between **B** and **C** and α the angle between the resultant of $(\mathbf{B} \times \mathbf{C})$ and the vector **A** then, from the definitions of dot and cross products

$$\mathbf{A} \cdot (\mathbf{B} \times \mathbf{C}) = A\,(BC \sin\theta) \cos\alpha$$

thus if

$$\mathbf{A} = A_1\mathbf{I} + A_2\mathbf{J} + A_3\mathbf{K}$$
$$\mathbf{B} = B_1\mathbf{I} + B_2\mathbf{J} + B_3\mathbf{K}$$
$$\mathbf{C} = C_1\mathbf{I} + C_2\mathbf{J} + C_3\mathbf{K}$$

then

$$\mathbf{A} \cdot (\mathbf{B} \times \mathbf{C}) = (A_1\mathbf{I} + A_2\mathbf{J} + A_3\mathbf{K}) \cdot \begin{vmatrix} \mathbf{I} & \mathbf{J} & \mathbf{K} \\ B_1 & B_2 & B_3 \\ C_1 & C_2 & C_3 \end{vmatrix}$$

$$= A_1 (B_2C_3 - B_3C_2)$$
$$+ A_2 (B_3C_1 - B_1C_3)$$
$$+ A_3 (B_1C_2 - B_2C_1)$$

But this is simply the expansion of the determinant

$$\mathbf{A} \cdot (\mathbf{B} \times \mathbf{C}) = \begin{vmatrix} A_1 & A_2 & A_3 \\ B_1 & B_2 & B_3 \\ C_1 & C_2 & C_3 \end{vmatrix} \qquad \text{(C.2-10)}$$

You may easily verify that

$$\begin{vmatrix} A_1 & A_2 & A_3 \\ B_1 & B_2 & B_3 \\ C_1 & C_2 & C_3 \end{vmatrix} = \begin{vmatrix} C_1 & C_2 & C_3 \\ A_1 & A_2 & A_3 \\ B_1 & B_2 & B_3 \end{vmatrix} = \begin{vmatrix} B_1 & B_2 & B_3 \\ C_1 & C_2 & C_3 \\ A_1 & A_2 & A_3 \end{vmatrix}$$

or $$\mathbf{A} \cdot (\mathbf{B} \times \mathbf{C}) = \mathbf{C} \cdot (\mathbf{A} \times \mathbf{B}) = \mathbf{B} \cdot (\mathbf{C} \times \mathbf{A}) \qquad \text{(C.2-11)}$$

Vector Triple Product. The vector triple produce, denoted by $\mathbf{A} \times (\mathbf{B} \times \mathbf{C})$, is a *vector* perpendicular to both $(\mathbf{B} \times \mathbf{C})$ and $\mathbf{A}$ such that it lies in the plane of **B** and **C**. The properties associated with this quantity are:

(a) $$\mathbf{A} \times (\mathbf{B} \times \mathbf{C}) = (\mathbf{A} \cdot \mathbf{C})\mathbf{B} - (\mathbf{A} \cdot \mathbf{B})\mathbf{C} \qquad \text{(C.2-12)}$$

(b) $$(\mathbf{A} \times \mathbf{B}) \times \mathbf{C} = (\mathbf{A} \cdot \mathbf{C})\mathbf{B} - (\mathbf{B} \cdot \mathbf{C})\mathbf{A} \qquad \text{(C.2-13)}$$

(c) $$\mathbf{A} \times (\mathbf{B} \times \mathbf{C}) \neq (\mathbf{A} \times \mathbf{B}) \times \mathbf{C} \qquad \text{(C.2-14)}$$

C.3 VELOCITY

Velocity is the rate of change of position and is a directed quantity. Consider the case of a point P moving along the space curve shown below. The *position* of P is denoted by

$$\overline{\mathbf{OP}} = \mathbf{r} = x(t)\mathbf{I} + y(t)\mathbf{J} + z(t)\mathbf{K}$$

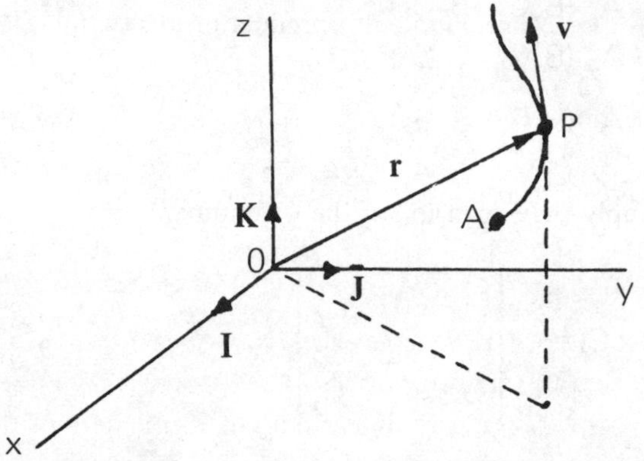

Then the velocity relative to the (**I**, **J**, **K**) coordinate system is:

$$\mathbf{v} = \frac{d\mathbf{r}}{dt} = \frac{dx}{dt}\mathbf{I} + \frac{dy}{dt}\mathbf{J} + \frac{dz}{dt}\mathbf{K}$$

and the components of the particle velocity are

$$v_x = \frac{dx}{dt}, \quad v_y = \frac{dy}{dt}, \quad v_z = \frac{dz}{dt}$$

If we now let s be the arc length measured from some fixed point A on the curve to the point P, then the coordinates of P are functions of s:

$$\mathbf{r} = x(s)\mathbf{I} + y(s)\mathbf{J} + z(s)\mathbf{K}$$

and $$\frac{d\mathbf{r}}{ds} = \frac{dx}{ds}\mathbf{I} + \frac{dy}{ds}\mathbf{J} + \frac{dz}{ds}\mathbf{K}$$

If we now take the dot product of $\frac{d\mathbf{r}}{ds}$ with itself,

$$\frac{d\mathbf{r}}{ds} \cdot \frac{d\mathbf{r}}{ds} = \left(\frac{dx}{ds}\right)^2 + \left(\frac{dy}{ds}\right)^2 + \left(\frac{dz}{ds}\right)^2$$

But $dx^2 + dy^2 + dz^2 = ds^2$, thus

$$\frac{d\mathbf{r}}{ds} \cdot \frac{d\mathbf{r}}{ds} = \frac{ds^2}{ds^2} = 1$$

which, from the definition of a dot product means that

$$\frac{d\mathbf{r}}{ds} = \text{unit vector} \triangleq \mathbf{T}$$

tangent to the curve at point P. Thus we may write

$$\mathbf{v} = \frac{d\mathbf{r}}{dt} = \frac{d\mathbf{r}}{ds}\,\frac{ds}{dt} = \frac{ds}{dt}\mathbf{T}$$

which more simply stated means that the velocity is the rate of change of arc length traveled and is always tangent to the path of the particle.

This result may be applied to a unit vector (which has constant magnitude). The instantaneous velocity of a unit vector is the rate of change of arc length traveled and is tangent to the path. In this special case then, the velocity associated with a unit vector is always perpendicular to the unit vector and has a magnitude of ω, the instantaneous angular velocity of the unit vector.

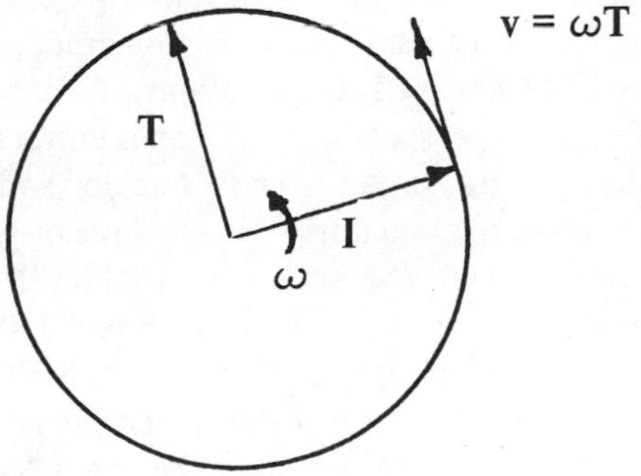

APPENDIX D

SUGGESTED PROJECTS

This appendix presents a series of computer projects which have proven to be very valuable in understanding much of the material covered in this text. They are especially valuable in developing a practical feel for the application of many of the methods. It is suggested that universal variables be used wherever possible.

In describing the projects the term "procedure" has the same meaning as "subroutine" in many computer languages.

D.1 PROJECT SITE/TRACK

The geographic location of a radar tracking site on the surface of the earth is known. Its geodetic latitude, longitude, and altitude above mean sea level are specified. Observations of a satellite are made by a radar at this site at a specified date and universal time. The radar determines ρ, $\dot{\rho}$, El, $\dot{\text{El}}$, Az, $\dot{\text{Az}}$ from its tracking and doppler capability. Compute the geocentric-equatorial components of position and velocity of the radar site (RS and VS), and of the satellite (R and V).

Make two separate procedures for SITE and TRACK such that they can be used in a larger program. The input to SITE should be latitude, altitude and local sidereal time of the launch site and the output should be position and velocity of the site. The input to TRACK should be range, range rate, elevation, elevation rate, azimuth, azimuth rate, latitude of the launch site, radius vector of the launch site and local sidereal time of the launch site. The output will be the radius and velocity vectors of the satellite.

Use standard unit conversions given in the text. An additional procedure will be needed to compute local sidereal time (LST) from the longitude (LONG, positive to the east) of the launch site, the day (DAY where 1 January is day zero) and the universal time (UT). From the

American Ephemeris and Nautical Almanac the Greenwich Sidereal Time (GST) on 1 January 1970 at 0000:00 is 1.74933340. GST, LST and LONG are in radians. An algorithm to do this is (in ALGOL)

```
PROCEDURE LSTIME (LONG, DAY, UT, LST);
REAL LONG, UT, LST; INTEGER DAY;
  BEGIN
  REAL D, PI, GST;
  PI ← 3.14159265359; GST ← 1.74933340;
  D ← DAY + (UT DIV 100) / 24 +
     (ENTIER (UT) MOD 100) / 1440 +
     (UT - ENTIER (UT)) / 864;
  LST ← LONG + GST + 1.0027379093⊗2⊗PI⊗D;
  LST ← LST MOD (2⊗PI);
END OF LSTIME;
```

If the above program is used, a VALUE declaration must be used in the main program for LST. Input UT as a decimal (i.e., 2210.575 for data set 3).

Following are five sample sets of data for site location, observation time and radar tracking data.

Data Set	1	2	3	4	5
Lat (deg)	39.007 N	37.8 S	29.8 N	0 N	45.7 N
Long (deg)	104.883 W	175.9 W	78.5 W	80.040075 E	72.9 E
Elev (ft)	7180	0	15	0	3610
UT	0317:02	1905:15	2210:57.5	0000:00	1024:30
DAY (1970)	244 (2 Sep)	280 (8 Oct)	360 (27 Dec)	0 (1 Jan)	328 (15 Nov)
ρ (km)	504.68	300	1510	6378.165	897.5
$\dot{\rho}$ (km/sec)	2.08	-5	4.5	0	-.57
El (deg)	30.7	45	135	90	76.7
$\dot{\text{El}}$ (deg/sec)	.07	-.3	.53	-.1	.48
Az (deg)	105.6	315	0	120	201.7
$\dot{\text{Az}}$ (deg/sec)	.05	-.2	.5	0	-.75

Answers for data set number 1 in x, y, z components are:

RS = (.20457216, -.75100391, .62624920)
VS = (.04418440, .01203575, 0)
R = (.27907599, -.77518019, .63745829)
V = (.26347198, -.14923608, .05195238)

D.2 PROJECT PREDICT

Project PREDICT is probably the easiest of all the projects in this appendix. It requires finding various orbital parameters from given vector position and velocity.

For a number of unidentified space objects the three components of vector position and velocity are generated from radar observations (see Project SITE/TRACK). From this information you are required to find:

a. The type of trajectory (circular, rectilinear, elliptical, parabolic, hyperbolic).

b. Position and velocity vectors at impact or closest approach in the geocentric-equatorial coordinate system.

c. Time for object to go from its observed position to point of impact or closest approach.

d. The total change in true anomaly from the observed position to impact or point of closest approach. Check to see if trajectory is going away from the earth.

Input data for Project PREDICT is stated in terms of the geocentric-equatorial inertial components in canonical units:

Object Number	RI	RJ	RK	VI	VJ	VK
1	-.1	1	0	-1.2	-.01	0
2	0	0	2	0	-.49	-.1
3	.49	.48	.9	0	0	1.01
4	0	4	0	-.5	-.5	0
5	0	0	2	.8	0	.6
6	-2.414	-2.414	0	.707	.293	0
7	0	2.1	.001	-.703	-.703	.001
8	0	0	530	-.00001	-.05	-1
9	-65.62	22.9	0	.01745	.000305	0

The answers for object number 1 are:

R = (.41359317, .91046180, 0)

V = (-1.12957919, .41722472, 0)

Object impacts at 3 hours 21 minutes 18.998 seconds
True anomaly change is 329.858654 degrees
Trajectory is an ellipse.

D.3 PROJECT KEPLER

Given the position and velocity vectors of a satellite at a particular instant of time, you are to determine the position and velocity vectors after an interval of time, Δt. To solve this problem use the universal variable approach and refer closely to section 4.4 of the text for guidance. Use a Newtonian iteration to solve for the value of x corresponding to the given Δt. You will need to write a special procedure to calculate the functions S(z) and C(z). To determine when convergence has occurred, use

$$\left| \frac{\Delta t - \Delta t_{calc}}{\Delta t} \right| \leq 10^{-7}$$

where Δt is the given time interval and it is compared to the calculated Δt. Write the program as a procedure or subprogram.

If your iteration process is efficient, you should achieve convergence in 5 or 6 iterations. To aid you in this goal the following suggestions should be considered.

a. Put upper and lower bounds on z determined by the computer being used. When these bounds are violated you must reduce the stepsize in the Newton iteration such that you go in the correct direction but stay within bounds.

b. If Δt is greater than the period (if it is an ellipse) you should reduce the flight time an integral number of periods.

c. Also for an ellipse, x should never be greater than $2\pi\sqrt{a}$.

d. Note that the sign of Δt should always be the same as the sign of x.

On debugging your program it will be helpful for you to print the values of x, Δt (calculated), and dt/dx for each iteration. If you do not get convergence in 50 iterations, write a message to that effect and go to the next data set. The following data is given in earth canonical units.

Data Set	1	2	3	4	5	6
RI	0	0	.3	.5	.025917	-.5
RJ	1	0	1	.7	-.150689	0
RK	0	-.5	0	.8	1.138878	0

Data Set	1	2	3	4	5	6
VI	0	0	3	0	.000361	0
VJ	0	2	0	.1	.001074	1.999
VK	1	0	0	.9	.002177	0
DT	π	10^6	5	-20	1.5	10^3

Answers for the second data set are: parabolic trajectory.

R2 = (0, 181.70655823, 16508.1362596)
V2 = (0, 0.00006057, 0.01100642)

Note: You can make a check of your results by comparing energy or angular momentum at both points. Care should also be taken as to your initial choice of x as described in the text.

D.4 PROJECT GAUSS

A satellite in a conic orbit leaves position $\mathbf{r}_1$ and arrives at position $\mathbf{r}_2$ at time Δt later. It can travel the "short way" ($\Delta\nu < \pi$) or the "long way" ($\Delta\nu \geqslant \pi$). Given $\mathbf{r}_1$, $\mathbf{r}_2$, Δt, and sign ($\pi - \Delta\nu$), find $\mathbf{v}_1$ and $\mathbf{v}_2$.

Write your program as a procedure or subroutine. Use the universal variable method to solve the problem. You will need to write a procedure to calculate the transcendental functions S(z) and C(z) and their derivatives. In this problem you will need to use a Newton iteration to iterate on z. For convergence criteria use

$$|t - t_c| \leqslant 10^{-6} \text{ for } t > 10^6$$

$$\left|\frac{t - t_c}{t}\right| \leqslant 10^{-6} \text{ for } 1 \leqslant t \leqslant 10^6$$

$$|t - t_c| \leqslant 10^{-6} \text{ for } t < 1$$

where t_c is the time-of-flight which results from the trial value of z and t is the given time-of-flight. If there is no convergence in 50 iterations, write a message to that effect and go to the next data set. Follow carefully the suggestions given in Chapter 5. For instance note that y can never be negative. For ellipses limit z to $(2\pi)^2$. Be careful near the point where the value of z corresponds to the t = 0 point in Figure 5.3-1. One of the data sets tests this situation.

For trouble shooting purposes you should print out the values of z, y, x, t_c, dt/dz and z_{new} for each iteration. Also print $\Delta\nu$ and a for each data set.

The plane of the transfer orbit is not uniquely defined if $\Delta\nu = \pi$. Check for this case, print out a message and go to the next data set. If the orbit is rectilinear (parallel position vectors) the problem can still be solved with some special provisions in your program. This is not specifically required in this problem, so a check should be made for that condition, a message printed and you should go to the next data set.

The following data is in canonical units. DM = +1 specifies a short way trajectory and DM = -1 specifies the long way.

Data Set	1	2	3	4	5	6
R1(I)	.5	.3	.5	-.2	1	-.4
R1(J)	.6	.7	.6	.6	0	.6
R1(K)	.7	.4	.7	.3	0	-1.201
R2(I)	0	.6	0	.4	0	.2
R2(J)	-1	-1.4	1	1.2	1	-.3
R2(K)	0	.8	0	.6	0	.6
DT	20	5	1.2	50	.0001	5
DM	-1	+1	-1	+1	+1	+1

Answers for data set number 1 are:

V1 = (0.12298144, 1.19216212, -.17217401)
V2 = (.66986992, .48048471, .93781789)
$\Delta\nu$ = 4.10335237 radians
a = -.66993197

D.5 PROJECT INTERCEPT

Write a computer program that takes as its input radar tracking data on a target satellite, location of the tracking site, time of radar observation, and the location of an interceptor launch site. The output is to be the impulsive velocity change required for both intercept and intercept-plus-rendezvous for various combinations of launch time and interceptor time-of-flight. Neglect the atmosphere and assume impulsive velocity change from the launch site and at the target. See section 5.7 for further discussion of this project.

The accomplishment of this project will involve the use of the procedures SITE, TRACK, KEPLER and GAUSS. The launch time of the interceptor will be specified in terms of "reaction time" which is the time elapsed between tracking of the target and launch of the interceptor.

Specific requirements for the project are as follows:

a. Remove all intermediate write statements from all procedures except the no convergence messages from GAUSS and KEPLER.

b. To make your program flexible, read in on separate data cards the desired starting value of reaction time, the desired increment size for reaction time, the number of reaction times desired, the starting value of time-of-flight increment size, and the number of time-of-flight increments desired. Times should be read in as minutes.

c. Compute the delta-Vs for both directions (long and short way) and save the smaller values in doubly subscripted arrays so that they may be printed all at one time. Make any delta-V that corresponds to a long way around transfer orbit negative so you can identify it when printed out. Print out the velocities in km/sec even though calculations are in canonical units.

d. If the interceptor strikes the earth enroute to the target, assign delta-V some large value such as 10^{10}. If the interceptor strikes the earth enroute to the target for both directions of motion this will cause asterisks to be printed in your output (in most computers).

e. When you point out your arrays, list reaction times down the left side and time-of-flight across the top. Print out only 3 places after the decimal for the delta-Vs.

f. Use the accompanying flow diagram as a guide for your program.

g. Make some provision to print out all of your input data, all variables of interest in the calculations (for the first delta-V calculation only) and the local sidereal time of the launch site for each reaction time. This will aid you in verifying your solution.

Use data as follows:

Location of launch site (Johnston Island):

Latitude: 16.45° N	Longitude: 169.32° W
Elevation: 5 feet	

Radar tracking site data (Shemya):

Latitude: 52.45° N	Longitude: 174.05° E
Elevation: 52 feet	Day: 104 (15 Apr 70)
Universal time: 0600:00	

$\rho = 186.613$ km $\quad\quad \dot{\rho} = 4.012$ km/sec
El = 56.95 degrees $\quad\quad \dot{El} = -1.92$ deg/sec
Az = 152.44 degrees $\quad\quad \dot{Az} = 1.09$ deg/sec

	Data Set No 1	**Data Set No 2**
First Reaction time	10 minutes	20 minutes
Increment size	5 minutes	1 minutes
No. of Reaction Times	40	40
First time-of-flight	5 minutes	5 minutes
Increment Size	5 minutes	1 minute
No. of time-of-flights	14	14

Note that data set number 2 is a "blow-up" of a selected region of data set number 1.

When the problem is complete you will have four arrays of Figure 5.7-4. You could draw contour lines through equal delta-Vs. The shaded area is equivalent to the asterisks in your printout.

Your first few lines of array output for the intercept only case for data set number 1 should be as follows (in km/sec):

0	5	10	15	20	25	30	35
10	9.338	8.262	7.981	7.879	7.844	7.844	7.868
15	****	****	****	****	****	****	****
20	****	****	****	****	****	****	****
25	****	****	****	****	****	****	****
30	****	****	****	****	****	****	****
35	****	****	****	****	****	****	8.038

GROUND TO SATELLITE INTERCEPT AND RENDEZVOUS

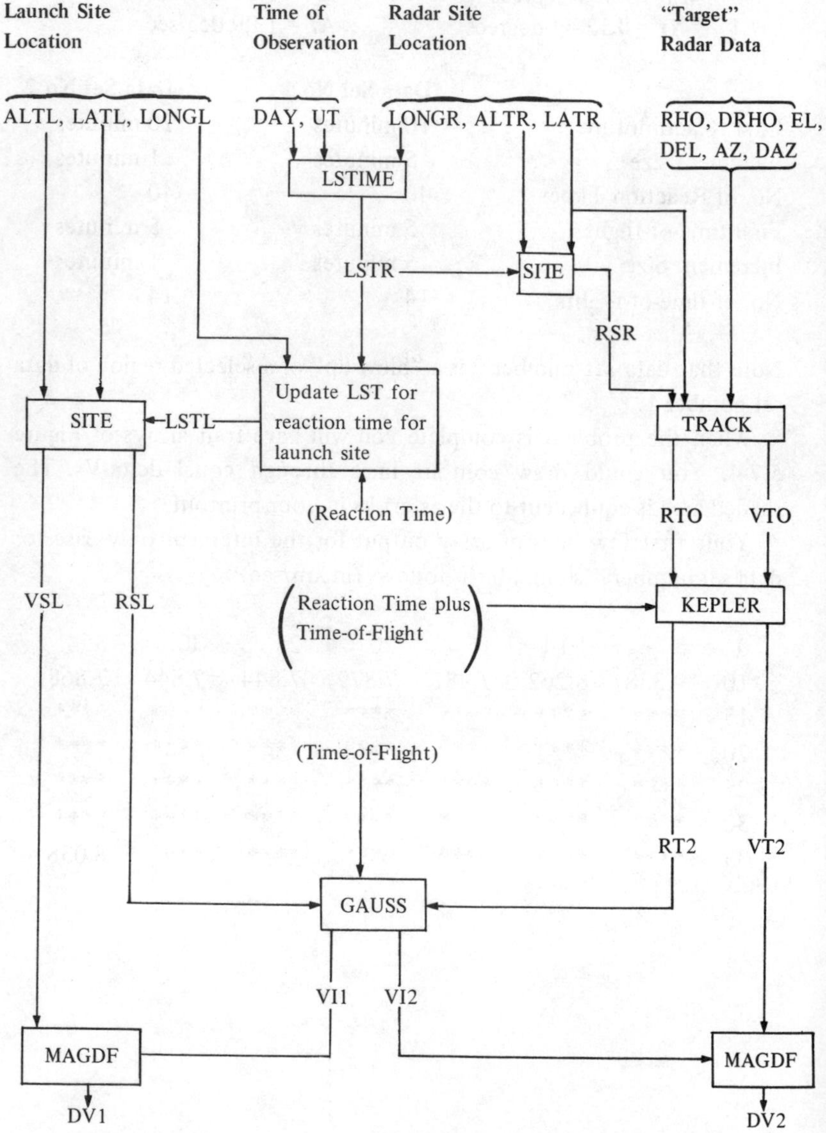

L Suffix stands for Launch Site
R Suffix stands for Radar Site

INDEX

INDEX

A CATALOGUE OF SELECTED DOVER BOOKS IN ALL FIELDS OF INTEREST

A CATALOGUE OF SELECTED DOVER BOOKS IN ALL FIELDS OF INTEREST

CELESTIAL OBJECTS FOR COMMON TELESCOPES, T. W. Webb. The most used book in amateur astronomy: inestimable aid for locating and identifying nearly 4,000 celestial objects. Edited, updated by Margaret W. Mayall. 77 illustrations. Total of 645pp. 5⅜ x 8½.
20917-2, 20918-0 Pa., Two-vol. set $9.00

HISTORICAL STUDIES IN THE LANGUAGE OF CHEMISTRY, M. P. Crosland. The important part language has played in the development of chemistry from the symbolism of alchemy to the adoption of systematic nomenclature in 1892. ". . . wholeheartedly recommended,"—Science. 15 illustrations. 416pp. of text. 5⅝ x 8¼. 63702-6 Pa. $6.00

BURNHAM'S CELESTIAL HANDBOOK, Robert Burnham, Jr. Thorough, readable guide to the stars beyond our solar system. Exhaustive treatment, fully illustrated. Breakdown is alphabetical by constellation: Andromeda to Cetus in Vol. 1; Chamaeleon to Orion in Vol. 2; and Pavo to Vulpecula in Vol. 3. Hundreds of illustrations. Total of about 2000pp. 6⅛ x 9¼.
23567-X, 23568-8, 23673-0 Pa., Three-vol. set $27.85

THEORY OF WING SECTIONS: INCLUDING A SUMMARY OF AIRFOIL DATA, Ira H. Abbott and A. E. von Doenhoff. Concise compilation of subatomic aerodynamic characteristics of modern NASA wing sections, plus description of theory. 350pp. of tables. 693pp. 5⅜ x 8½.
60586-8 Pa. $8.50

DE RE METALLICA, Georgius Agricola. Translated by Herbert C. Hoover and Lou H. Hoover. The famous Hoover translation of greatest treatise on technological chemistry, engineering, geology, mining of early modern times (1556). All 289 original woodcuts. 638pp. 6¾ x 11.
60006-8 Clothbd. $17.95

THE ORIGIN OF CONTINENTS AND OCEANS, Alfred Wegener. One of the most influential, most controversial books in science, the classic statement for continental drift. Full 1966 translation of Wegener's final (1929) version. 64 illustrations. 246pp. 5⅜ x 8½. 61708-4 Pa. $4.50

THE PRINCIPLES OF PSYCHOLOGY, William James. Famous long course complete, unabridged. Stream of thought, time perception, memory, experimental methods; great work decades ahead of its time. Still valid, useful; read in many classes. 94 figures. Total of 1391pp. 5⅜ x 8½.
20381-6, 20382-4 Pa., Two-vol. set $13.00

THE PHILOSOPHY OF HISTORY, Georg W. Hegel. Great classic of Western thought develops concept that history is not chance but a rational process, the evolution of freedom. 457pp. 5⅜ x 8½. 20112-0 Pa. $4.50

LANGUAGE, TRUTH AND LOGIC, Alfred J. Ayer. Famous, clear introduction to Vienna, Cambridge schools of Logical Positivism. Role of philosophy, elimination of metaphysics, nature of analysis, etc. 160pp. 5⅜ x 8½. (Available in U.S. only) 20010-8 Pa. $2.00

A PREFACE TO LOGIC, Morris R. Cohen. Great City College teacher in renowned, easily followed exposition of formal logic, probability, values, logic and world order and similar topics; no previous background needed. 209pp. 5⅜ x 8½. 23517-3 Pa. $3.50

REASON AND NATURE, Morris R. Cohen. Brilliant analysis of reason and its multitudinous ramifications by charismatic teacher. Interdisciplinary, synthesizing work widely praised when it first appeared in 1931. Second (1953) edition. Indexes. 496pp. 5⅜ x 8½. 23633-1 Pa. $6.50

AN ESSAY CONCERNING HUMAN UNDERSTANDING, John Locke. The only complete edition of enormously important classic, with authoritative editorial material by A. C. Fraser. Total of 1176pp. 5⅜ x 8½. 20530-4, 20531-2 Pa., Two-vol. set $16.00

HANDBOOK OF MATHEMATICAL FUNCTIONS WITH FORMULAS, GRAPHS, AND MATHEMATICAL TABLES, edited by Milton Abramowitz and Irene A. Stegun. Vast compendium: 29 sets of tables, some to as high as 20 places. 1,046pp. 8 x 10½. 61272-4 Pa. $14.95

MATHEMATICS FOR THE PHYSICAL SCIENCES, Herbert S. Wilf. Highly acclaimed work offers clear presentations of vector spaces and matrices, orthogonal functions, roots of polynomial equations, conformal mapping, calculus of variations, etc. Knowledge of theory of functions of real and complex variables is assumed. Exercises and solutions. Index. 284pp. 5⅝ x 8¼. 63635-6 Pa. $5.00

THE PRINCIPLE OF RELATIVITY, Albert Einstein et al. Eleven most important original papers on special and general theories. Seven by Einstein, two by Lorentz, one each by Minkowski and Weyl. All translated, unabridged. 216pp. 5⅜ x 8½. 60081-5 Pa. $3.50

THERMODYNAMICS, Enrico Fermi. A classic of modern science. Clear, organized treatment of systems, first and second laws, entropy, thermodynamic potentials, gaseous reactions, dilute solutions, entropy constant. No math beyond calculus required. Problems. 160pp. 5⅜ x 8½. 60361-X Pa. $3.00

ELEMENTARY MECHANICS OF FLUIDS, Hunter Rouse. Classic undergraduate text widely considered to be far better than many later books. Ranges from fluid velocity and acceleration to role of compressibility in fluid motion. Numerous examples, questions, problems. 224 illustrations. 376pp. 5⅝ x 8¼. 63699-2 Pa. $5.00

THE COMPLETE BOOK OF DOLL MAKING AND COLLECTING, Catherine Christopher. Instructions, patterns for dozens of dolls, from rag doll on up to elaborate, historically accurate figures. Mould faces, sew clothing, make doll houses, etc. Also collecting information. Many illustrations. 288pp. 6 x 9. 22066-4 Pa. $4.50

THE DAGUERREOTYPE IN AMERICA, Beaumont Newhall. Wonderful portraits, 1850's townscapes, landscapes; full text plus 104 photographs. The basic book. Enlarged 1976 edition. 272pp. 8¼ x 11¼. 23322-7 Pa. $7.95

CRAFTSMAN HOMES, Gustav Stickley. 296 architectural drawings, floor plans, and photographs illustrate 40 different kinds of "Mission-style" homes from *The Craftsman* (1901-16), voice of American style of simplicity and organic harmony. Thorough coverage of Craftsman idea in text and picture, now collector's item. 224pp. 8⅛ x 11. 23791-5 Pa. $6.00

PEWTER-WORKING: INSTRUCTIONS AND PROJECTS, Burl N. Osborn. & Gordon O. Wilber. Introduction to pewter-working for amateur craftsman. History and characteristics of pewter; tools, materials, step-by-step instructions. Photos, line drawings, diagrams. Total of 160pp. 7⅞ x 10¾. 23786-9 Pa. $3.50

THE GREAT CHICAGO FIRE, edited by David Lowe. 10 dramatic, eyewitness accounts of the 1871 disaster, including one of the aftermath and rebuilding, plus 70 contemporary photographs and illustrations of the ruins—courthouse, Palmer House, Great Central Depot, etc. Introduction by David Lowe. 87pp. 8¼ x 11. 23771-0 Pa. $4.00

SILHOUETTES: A PICTORIAL ARCHIVE OF VARIED ILLUSTRATIONS, edited by Carol Belanger Grafton. Over 600 silhouettes from the 18th to 20th centuries include profiles and full figures of men and women, children, birds and animals, groups and scenes, nature, ships, an alphabet. Dozens of uses for commercial artists and craftspeople. 144pp. 8⅜ x 11¼. 23781-8 Pa. $4.50

ANIMALS: 1,419 COPYRIGHT-FREE ILLUSTRATIONS OF MAMMALS, BIRDS, FISH, INSECTS, ETC., edited by Jim Harter. Clear wood engravings present, in extremely lifelike poses, over 1,000 species of animals. One of the most extensive copyright-free pictorial sourcebooks of its kind. Captions. Index. 284pp. 9 x 12. 23766-4 Pa. $8.95

INDIAN DESIGNS FROM ANCIENT ECUADOR, Frederick W. Shaffer. 282 original designs by pre-Columbian Indians of Ecuador (500-1500 A.D.). Designs include people, mammals, birds, reptiles, fish, plants, heads, geometric designs. Use as is or alter for advertising, textiles, leathercraft, etc. Introduction. 95pp. 8¾ x 11¼. 23764-8 Pa. $3.50

SZIGETI ON THE VIOLIN, Joseph Szigeti. Genial, loosely structured tour by premier violinist, featuring a pleasant mixture of reminiscenes, insights into great music and musicians, innumerable tips for practicing violinists. 385 musical passages. 256pp. 5⅝ x 8¼. 23763-X Pa. $4.00

TONE POEMS, SERIES II: TILL EULENSPIEGELS LUSTIGE STREICHE, ALSO SPRACH ZARATHUSTRA, AND EIN HELDENLEBEN, Richard Strauss. Three important orchestral works, including very popular *Till Eulenspiegel's Marry Pranks,* reproduced in full score from original editions. Study score. 315pp. 9⅜ x 12¼. (Available in U.S. only)
23755-9 Pa. $8.95

TONE POEMS, SERIES I: DON JUAN, TOD UND VERKLARUNG AND DON QUIXOTE, Richard Strauss. Three of the most often performed and recorded works in entire orchestral repertoire, reproduced in full score from original editions. Study score. 286pp. 9⅜ x 12¼. (Available in U.S. only)
23754-0 Pa. $7.50

11 LATE STRING QUARTETS, Franz Joseph Haydn. The form which Haydn defined and "brought to perfection." *(Grove's).* 11 string quartets in complete score, his last and his best. The first in a projected series of the complete Haydn string quartets. Reliable modern Eulenberg edition, otherwise difficult to obtain. 320pp. 8⅜ x 11¼. (Available in U.S. only)
23753-2 Pa. $7.50

FOURTH, FIFTH AND SIXTH SYMPHONIES IN FULL SCORE, Peter Ilyitch Tchaikovsky. Complete orchestral scores of Symphony No. 4 in F Minor, Op. 36; Symphony No. 5 in E Minor, Op. 64; Symphony No. 6 in B Minor, "Pathetique," Op. 74. Bretikopf & Hartel eds. Study score. 480pp. 9⅜ x 12¼.
23861-X Pa. $10.95

THE MARRIAGE OF FIGARO: COMPLETE SCORE, Wolfgang A. Mozart. Finest comic opera ever written. Full score, not to be confused with piano renderings. Peters edition. Study score. 448pp. 9⅜ x 12¼. (Available in U.S. only)
23751-6 Pa. $11.95

"IMAGE" ON THE ART AND EVOLUTION OF THE FILM, edited by Marshall Deutelbaum. Pioneering book brings together for first time 38 groundbreaking articles on early silent films from *Image* and 263 illustrations newly shot from rare prints in the collection of the International Museum of Photography. A landmark work. Index. 256pp. 8¼ x 11.
23777-X Pa. $8.95

AROUND-THE-WORLD COOKY BOOK, Lois Lintner Sumption and Marguerite Lintner Ashbrook. 373 cooky and frosting recipes from 28 countries (America, Austria, China, Russia, Italy, etc.) include Viennese kisses, rice wafers, London strips, lady fingers, hony, sugar spice, maple cookies, etc. Clear instructions. All tested. 38 drawings. 182pp. 5⅜ x 8.
23802-4 Pa. $2.50

THE ART NOUVEAU STYLE, edited by Roberta Waddell. 579 rare photographs, not available elsewhere, of works in jewelry, metalwork, glass, ceramics, textiles, architecture and furniture by 175 artists—Mucha, Seguy, Lalique, Tiffany, Gaudin, Hohlwein, Saarinen, and many others. 288pp. 8⅜ x 11¼.
23515-7 Pa. $6.95

THE CURVES OF LIFE, Theodore A. Cook. Examination of shells, leaves, horns, human body, art, etc., in "*the* classic reference on how the golden ratio applies to spirals and helices in nature . . . ."—Martin Gardner. 426 illustrations. Total of 512pp. 5⅜ x 8½. 23701-X Pa. $5.95

AN ILLUSTRATED FLORA OF THE NORTHERN UNITED STATES AND CANADA, Nathaniel L. Britton, Addison Brown. Encyclopedic work covers 4666 species, ferns on up. Everything. Full botanical information, illustration for each. This earlier edition is preferred by many to more recent revisions. 1913 edition. Over 4000 illustrations, total of 2087pp. 6⅛ x 9¼. 22642-5, 22643-3, 22644-1 Pa., Three-vol. set $25.50

MANUAL OF THE GRASSES OF THE UNITED STATES, A. S. Hitchcock, U.S. Dept. of Agriculture. The basic study of American grasses, both indigenous and escapes, cultivated and wild. Over 1400 species. Full descriptions, information. Over 1100 maps, illustrations. Total of 1051pp. 5⅜ x 8½. 22717-0, 22718-9 Pa., Two-vol. set $15.00

THE CACTACEAE,, Nathaniel L. Britton, John N. Rose. Exhaustive, definitive. Every cactus in the world. Full botanical descriptions. Thorough statement of nomenclatures, habitat, detailed finding keys. The one book needed by every cactus enthusiast. Over 1275 illustrations. Total of 1080pp. 8 x 10¼. 21191-6, 21192-4 Clothbd., Two-vol. set $35.00

AMERICAN MEDICINAL PLANTS, Charles F. Millspaugh. Full descriptions, 180 plants covered: history; physical description; methods of preparation with all chemical constituents extracted; all claimed curative or adverse effects. 180 full-page plates. Classification table. 804pp. 6½ x 9¼. 23034-1 Pa. $12.95

A MODERN HERBAL, Margaret Grieve. Much the fullest, most exact, most useful compilation of herbal material. Gigantic alphabetical encyclopedia, from aconite to zedoary, gives botanical information, medical properties, folklore, economic uses, and much else. Indispensable to serious reader. 161 illustrations. 888pp. 6½ x 9¼. (Available in U.S. only) 22798-7, 22799-5 Pa., Two-vol. set $13.00

THE HERBAL or GENERAL HISTORY OF PLANTS, John Gerard. The 1633 edition revised and enlarged by Thomas Johnson. Containing almost 2850 plant descriptions and 2705 superb illustrations, Gerard's *Herbal* is a monumental work, the book all modern English herbals are derived from, the one herbal every serious enthusiast should have in its entirety. Original editions are worth perhaps $750. 1678pp. 8½ x 12¼. 23147-X Clothbd. $50.00

MANUAL OF THE TREES OF NORTH AMERICA, Charles S. Sargent. The basic survey of every native tree and tree-like shrub, 717 species in all. Extremely full descriptions, information on habitat, growth, locales, economics, etc. Necessary to every serious tree lover. Over 100 finding keys. 783 illustrations. Total of 986pp. 5⅜ x 8½. 20277-1, 20278-X Pa., Two-vol. set $11.00

AMERICAN BIRD ENGRAVINGS, Alexander Wilson et al. All 76 plates. from Wilson's *American Ornithology* (1808-14), most important ornithological work before Audubon, plus 27 plates from the supplement (1825-33) by Charles Bonaparte. Over 250 birds portrayed. 8 plates also reproduced in full color. 111pp. 9⅜ x 12½. 23195-X Pa. $6.00

CRUICKSHANK'S PHOTOGRAPHS OF BIRDS OF AMERICA, Allan D. Cruickshank. Great ornithologist, photographer presents 177 closeups, groupings, panoramas, flightings, etc., of about 150 different birds. Expanded *Wings in the Wilderness*. Introduction by Helen G. Cruickshank. 191pp. 8¼ x 11. 23497-5 Pa. $6.00

AMERICAN WILDLIFE AND PLANTS, A. C. Martin, et al. Describes food habits of more than 1000 species of mammals, birds, fish. Special treatment of important food plants. Over 300 illustrations. 500pp. 5⅜ x 8½. 20793-5 Pa. $4.95

THE PEOPLE CALLED SHAKERS, Edward D. Andrews. Lifetime of research, definitive study of Shakers: origins, beliefs, practices, dances, social organization, furniture and crafts, impact on 19th-century USA, present heritage. Indispensable to student of American history, collector. 33 illustrations. 351pp. 5⅜ x 8½. 21081-2 Pa. $4.50

OLD NEW YORK IN EARLY PHOTOGRAPHS, Mary Black. New York City as it was in 1853-1901, through 196 wonderful photographs from N.-Y. Historical Society. Great Blizzard, Lincoln's funeral procession, great buildings. 228pp. 9 x 12. 22907-6 Pa. $8.95

MR. LINCOLN'S CAMERA MAN: MATHEW BRADY, Roy Meredith. Over 300 Brady photos reproduced directly from original negatives, photos. Jackson, Webster, Grant, Lee, Carnegie, Barnum; Lincoln; Battle Smoke, Death of Rebel Sniper, Atlanta Just After Capture. Lively commentary. 368pp. 8⅜ x 11¼. 23021-X Pa. $8.95

TRAVELS OF WILLIAM BARTRAM, William Bartram. From 1773-8, Bartram explored Northern Florida, Georgia, Carolinas, and reported on wild life, plants, Indians, early settlers. Basic account for period, entertaining reading. Edited by Mark Van Doren. 13 illustrations. 141pp. 5⅜ x 8½. 20013-2 Pa. $5.00

THE GENTLEMAN AND CABINET MAKER'S DIRECTOR, Thomas Chippendale. Full reprint, 1762 style book, most influential of all time; chairs, tables, sofas, mirrors, cabinets, etc. 200 plates, plus 24 photographs of surviving pieces. 249pp. 9⅞ x 12¾. 21601-2 Pa. $7.95

AMERICAN CARRIAGES, SLEIGHS, SULKIES AND CARTS, edited by Don H. Berkebile. 168 Victorian illustrations from catalogues, trade journals, fully captioned. Useful for artists. Author is Assoc. Curator, Div. of Transportation of Smithsonian Institution. 168pp. 8½ x 9½. 23328-6 Pa. $5.00

PRINCIPLES OF ORCHESTRATION, Nikolay Rimsky-Korsakov. Great classical orchestrator provides fundamentals of tonal resonance, progression of parts, voice and orchestra, tutti effects, much else in major document. 330pp. of musical excerpts. 489pp. 6½ x 9¼. 21266-1 Pa. **$7.50**

TRISTAN UND ISOLDE, Richard Wagner. Full orchestral score with complete instrumentation. Do not confuse with piano reduction. Commentary by Felix Mottl, great Wagnerian conductor and scholar. Study score. 655pp. 8⅛ x 11. 22915-7 Pa. $13.95

REQUIEM IN FULL SCORE, Giuseppe Verdi. Immensely popular with choral groups and music lovers. Republication of edition published by C. F. Peters, Leipzig, n. d. German frontmaker in English translation. Glossary. Text in Latin. Study score. 204pp. 9⅜ x 12¼.
23682-X Pa. $6.00

COMPLETE CHAMBER MUSIC FOR STRINGS, Felix Mendelssohn. All of Mendelssohn's chamber music: Octet, 2 Quintets, 6 Quartets, and Four Pieces for String Quartet. (Nothing with piano is included). Complete works edition (1874-7). Study score. 283 pp. 9⅜ x 12¼.
23679-X Pa. **$7.50**

POPULAR SONGS OF NINETEENTH-CENTURY AMERICA, edited by Richard Jackson. 64 most important songs: "Old Oaken Bucket," "Arkansas Traveler," "Yellow Rose of Texas," etc. Authentic original sheet music, full introduction and commentaries. 290pp. 9 x 12. 23270-0 Pa. **$7.95**

COLLECTED PIANO WORKS, Scott Joplin. Edited by Vera Brodsky Lawrence. Practically all of Joplin's piano works—rags, two-steps, marches, waltzes, etc., 51 works in all. Extensive introduction by Rudi Blesh. Total of 345pp. 9 x 12. 23106-2 Pa. $14.95

BASIC PRINCIPLES OF CLASSICAL BALLET, Agrippina Vaganova. Great Russian theoretician, teacher explains methods for teaching classical ballet; incorporates best from French, Italian, Russian schools. 118 illustrations. 175pp. 5⅜ x 8½. 22036-2 Pa. $2.50

CHINESE CHARACTERS, L. Wieger. Rich analysis of 2300 characters according to traditional systems into primitives. Historical-semantic analysis to phonetics (Classical Mandarin) and radicals. 820pp. 6⅛ x 9¼.
21321-8 Pa. $10.00

EGYPTIAN LANGUAGE: EASY LESSONS IN EGYPTIAN HIEROGLYPHICS, E. A. Wallis Budge. Foremost Egyptologist offers Egyptian grammar, explanation of hieroglyphics, many reading texts, dictionary of symbols. 246pp. 5 x 7½. (Available in U.S. only)
21394-3 Clothbd. $7.50

AN ETYMOLOGICAL DICTIONARY OF MODERN ENGLISH, Ernest Weekley. Richest, fullest work, by foremost British lexicographer. Detailed word histories. Inexhaustible. Do not confuse this with *Concise Etymological Dictionary*, which is abridged. Total of 856pp. 6½ x 9¼.
21873-2, 21874-0 Pa., Two-vol. set $12.00

CATALOGUE OF DOVER BOOKS

GEOMETRY, RELATIVITY AND THE FOURTH DIMENSION, Rudolf Rucker. Exposition of fourth dimension, means of visualization, concepts of relativity as Flatland characters continue adventures. Popular, easily followed yet accurate, profound. 141 illustrations. 133pp. 5⅜ x 8½.
23400-2 Pa. $2.75

THE ORIGIN OF LIFE, A. I. Oparin. Modern classic in biochemistry, the first rigorous examination of possible evolution of life from nitrocarbon compounds. Non-technical, easily followed. Total of 295pp. 5⅜ x 8½.
60213-3 Pa. $4.00

PLANETS, STARS AND GALAXIES, A. E. Fanning. Comprehensive introductory survey: the sun, solar system, stars, galaxies, universe, cosmology; quasars, radio stars, etc. 24pp. of photographs. 189pp. 5⅜ x 8½. (Available in U.S. only)
21680-2 Pa. $3.75

THE THIRTEEN BOOKS OF EUCLID'S ELEMENTS, translated with introduction and commentary by Sir Thomas L. Heath. Definitive edition. Textual and linguistic notes, mathematical analysis, 2500 years of critical commentary. Do not confuse with abridged school editions. Total of 1414pp. 5⅜ x 8½. 60088-2, 60089-0, 60090-4 Pa., Three-vol. set $18.50

Prices subject to change without notice.

Available at your book dealer or write for free catalogue to Dept. GI, Dover Publications, Inc., 31 East Second Street, Mineola, N.Y. 11501. Dover publishes more than 175 books each year on science, elementary and advanced mathematics, biology, music, art, literary history, social sciences and other areas.